QUANTITATIVE CHEMICAL ANALYSIS

QUANTITATIVE CHEMICAL ANALYSIS

Na Li
Peking University, China

John J. Hefferren
Peking University, China & University of Kansas, USA

Ke'an Li
Peking University, China

Published by

World Scientific Publishing Co. Pte. Ltd.
5 Toh Tuck Link, Singapore 596224
USA office: 27 Warren Street, Suite 401-402, Hackensack, NJ 07601
UK office: 57 Shelton Street, Covent Garden, London WC2H 9HE

Library of Congress Cataloging-in-Publication Data
Li, Na, 1965–
Quantitative chemical analysis / by Na Li (Peking University, China), John J Hefferren (Peking University, China & University of Kansas, USA) & Ke'an Li (Peking University, China).
pages cm
Includes bibliographical references and index.
Summary: "Covers both fundamental and practical aspects of chemical analysis. A textbook for Freshmen or sophomores"-- Provided by publisher.
ISBN 978-9814452281 (alk. paper)
1. Chemistry, Analytic--Quantitative. I. Hefferren, John J. II. Li, Ke'an. III. Title.
QD101.L63 2013
543'.1--dc23

2013004455

British Library Cataloguing-in-Publication Data
A catalogue record for this book is available from the British Library.

Copyright © 2013 by Na Li, John J. Hefferren, Ke'an Li

The Work was originally published by Peking University Press in 2009.
This edition is published by World Scientific Publishing Co Pte Ltd by arrangement with Peking University Press, Beijing, China.
All rights reserved. No reproduction and distribution without permission.

Printed in Singapore by B & Jo Enterprise Pte Ltd

PREFACE

This text brings together the individuals and the desire to develop a text for undergraduate students who have English as a second language. Our initial focus was undergraduate students with chemistry major in the College of Chemistry and Molecular Engineering, Peking University, Beijing, China, but now we hope and expect that other undergraduate students may be able to learn more easily with this text as they cope with the English language and the essentials of analytical chemistry.

Lecture Series Leading to Text:

The one semester course in analytical chemistry in English for undergraduate students was initiated by Professor Li Ke'an in February 2005 with Dr. Li Na as the presenter of one 2 hour lecture each week for 15 weeks. The size of the lecture room limited the number of students to 50.

Each year the students prepare presentations of their science project reports. The audience of their peers grades the oral presentations of those students who volunteered and were selected to give oral presentations. Competition for being included among the oral presenters has been impressive. Student discussion, grading of the presenters and the presentations bring forth a profound bonding. Each of us in the lecture room feels the shoes worn by another.

As we introduced different examples and illustrations to the lecture series, these quickly became ideas for the coming analytical chemistry text. In Chapter 1, the Human Genome Project was used to show the power and success possible when analytical chemists join forces to bring the minds and resources of the academic community to focus on a goal. The project is indeed a road map of problem solving using new and different technologies plus automation to resolve analytical road blocks to meet the time constraints of the Genome Project thus opening research opportunities for decades. A global environmental need brought together another group of scientists in the concluding Chapter 10 to address the ever present need to monitor drinking water contamination throughout the world. We selected arsenic as one example of the world-wide need for simple, sensitive, cost effective analytical methods to monitor drinking water. Again, we tried to show the rationale dictating the path taken by scientists to deal with the specifics of water treatment analysis. Fundamental to analytical chemistry are the problem solving tools that scientific minds select for each set of circumstances. This analytical chemistry text lists many of the scientific minds of analytical chemistry from Berthelot and Jungfleisch in 1872 to those of 2009 who have provided the best tool for the situation.

Peking University Lecture Team:

This text has been assembled and is based upon more than a quarter of century of teaching analytical chemistry at Peking University. Many examples and illustrative problems in this text have been taken from previous textbooks in Chinese written by the Peking University Team Teaching Program. This text was written by Dr. Li Na and edited by Professor John Hefferren with the guidance of Professor Li Ke'an. Some of the Chinese teaching style has been used in the text to maintain ties to the Chinese traditions; however, references and terminology of established analytical chemistry books published in the world have also been used. It is hoped that this blend of perspective is helpful and interesting to all readers.

Authors:

Li Na is associate professor, Analytical Chemistry Division, College of Chemistry and Molecular Engineering, Peking University, Beijing, China. John J. Hefferren is guest professor of Peking University and adjunct professor, Department of Pharmaceutical Chemistry, School of Pharmacy, University of Kansas, Lawrence, Kansas, USA. Li Ke'an is professor and vice provost, Peking University, Beijing, China.

Acknowledgements:

Graduate students Zou Mingjian, Xu Xiao, Wang Xinyin, Chen Qiang and Chen Yang helped in many ways to the completion of the text. The text was supported in part by Peking University and prepared by the editorial staff.

We gratefully acknowledge the influence of those that we consider our chemistry colleagues who have written the texts of the world. It is our profound hope that each of you will join us in continuing to nurture the rich tradition of analytical chemistry.

Li Na, Ph. D.
John J. Hefferren, Ph. D.
Li Ke'an, professor

CONTENTS

CHAPTER 1 INTRODUCTION OF ANALYTICAL CHEMISTRY 1

1.1 What is Analytical Chemistry 2
1.2 Steps in the Development of an Analytical Method 5
1.3 Classification of Quantitative Analytical Methods 7
- 1.3.1 Chemical Analysis 7
- 1.3.2 Instrumental Analysis 8

1.4 Principles of Volumetric Titration 8
- 1.4.1 Basic Terms 9
- 1.4.2 Requirements of Titration Reactions 9
- 1.4.3 Classification of Titration Processes 10
- 1.4.4 Primary Standards and Standard Solutions 10
- 1.4.5 Basic Apparatus in Chemical Analyses 11

1.5 Calculations in Volumetric Titration 15
- 1.5.1 Preparation of Standard Solutions 15
- 1.5.2 Titration Results 18

CHAPTER 2 DATA ANALYSIS 22

2.1 Error and Classification 23
- 2.1.1 Accuracy and Precision 23
- 2.1.2 Errors and Deviation 24
- 2.1.3 Systematic and Random Errors 25

2.2 Distribution of Random Errors 26
- 2.2.1 Frequency Distribution 27
- 2.2.2 Normal Distribution 28
- 2.2.3 Predicting the Probability of Random Errors—Area under Gaussian Curve 30

2.3 Statistical Data Treatment 31
- 2.3.1 Estimation of Population Mean (μ) and Population Standard Deviation (σ) 31
- 2.3.2 Confidence Interval for Population Mean 34
- 2.3.3 Statistical Aids to Hypothesis Testing 37
- 2.3.4 Detection of Gross Errors 42

2.4 Propagation of Error 43

2.4.1 Systematic Errors 43

2.4.2 Random Errors (Standard Deviation) 43

2.4.3 Maximum Errors (E_R) 44

2.4.4 Distribution of Errors 44

2.5 Significant Figure Convention 45

2.5.1 Significant Figures 45

2.5.2 Numerical Rounding in Calculations 47

CHAPTER 3 ACID-BASE EQUILIBRIUM 50

3.1 Equilibrium Constants and Effect of Electrolytes 51

3.2 Acid-base Reactions and Equilibria 53

3.2.1 Acid and Base—Brønsted Concept 53

3.2.2 Dissociation of Acid or Base and Acid-base Equilibria 55

3.2.3 Magnitude of Dissociating Species at a Given pH: x-values 57

3.3 Solving Equilibrium Calculations Using pH Calculations as an Example 61

3.3.1 General Approaches (Systematic Approaches) 61

3.3.2 pH Calculations 64

3.4 Buffer Solutions 71

3.4.1 pH Calculations of Buffer Solutions 71

3.4.2 Buffer Capacity 72

3.4.3 Preparation of Buffers 74

CHAPTER 4 ACID-BASE TITRATION 78

4.1 Acid/Base Indicators 79

4.1.1 Principle 79

4.1.2 Examples 80

4.1.3 Titration Errors 82

4.1.4 Factors Influencing Performance 82

4.2 Titration Curves and Selection of Indicators 83

4.2.1 Strong Acids (Bases) 83

4.2.2 Monoprotic Acids (Bases) 86

4.2.3 Strong and Weak Acids (Bases) 91

4.2.4 Polyfunctional Weak Acids (Bases) 92

4.2.5 Mixture of Weak Acids (Bases) 95

4.3 Titration Error Calculations 95

4.3.1 Strong Acids (Bases) 95

4.3.2 Monoprotic Weak Acids (Bases) 96
4.3.3 Polyfunctional Acids (Bases) 97
4.4 Preparation of Standard Solutions 98
4.4.1 Standard Acid Solutions 98
4.4.2 Standard Base Solutions 99
4.4.3 The Carbonate Error 100
4.5 Examples of Acid-base Titrations 101
4.5.1 Determination of Total Alkalinity 101
4.5.2 Determination of Nitrogen 102
4.5.3 Determination of Boric Acid 103
4.6 Acid-base Titrations in Non-aqueous Solvents 104
4.6.1 Non-aqueous Solvents 104
4.6.2 Examples of Non-aqueous Titrations 105

CHAPTER 5 COMPLEXATION REACTION AND COMPLEXOMETRIC TITRATION 108

5.1 Complexes and Formation Constants 109
5.1.1 Formation Constants 109
5.1.2 Concentration of ML_n in Complexation Equilibria 111
5.1.3 Ethylenediaminetetraacetic Acid (EDTA) and Metal-EDTA Complexes 113
5.1.4 Side Reaction Coefficients and Conditional Formation Constants in Complexation Reactions 115
5.2 Metallochromic Indicators 122
5.2.1 How a Metallochromic Indicator Works 122
5.2.2 Color Transition Point pM ($(pM)_t$) for Metallochromic Indicators 123
5.2.3 Frequently Used Metallochromic Indicators 125
5.3 Titration Curves and Titration Errors 126
5.3.1 Titration Curves 126
5.3.2 Titration Errors 128
5.3.3 pH Control in Complexometric Titrations 129
5.4 Selective Titrations of Metal Ions in the Presence of Multiple Metal Ions 130
5.4.1 Selective Titration by Regulating pH 131
5.4.2 Selective Titration Using Masking Reagents 133
5.5 Applications of Complexometric Titrations 137
5.5.1 Buffer Selection in Complexometric Titrations 137
5.5.2 Titration Methods and Applications 138
5.5.3 Preparation of Standard Solutions 142

CHAPTER 6 REDOX EQUILIBRIUM AND TITRATION 146

6.1 Standard Electrode Potentials, Formal Potentials and Redox Equilibria 147

6.1.1 Standard Electrode Potentials 147

6.1.2 The Nernst Equation and Formal Potentials 149

6.1.3 Factors Affecting the Formal Potential 150

6.1.4 The Equilibrium Constant of Redox Reaction 154

6.2 Factors Affecting the Reaction Rate 155

6.2.1 Concentrations 156

6.2.2 Temperature 157

6.2.3 Catalysts and Reaction Rate 157

6.2.4 Induced Reaction 157

6.3 Redox Titrations 158

6.3.1 Constructing Redox Titration Curves 158

6.3.2 Indicators 162

6.3.3 Auxiliary Oxidizing and Reducing Agents 164

6.4 Examples of Redox Titrations 165

6.4.1 Potassium Permanganate ($KMnO_4$) 165

6.4.2 Potassium Dichromate ($K_2Cr_2O_7$) 168

6.4.3 Iodine: Iodimetry and Iodometry 169

6.4.4 Potassium Bromate ($KBrO_3$) 173

6.4.5 Ceric Sulfate ($Ce(SO_4)_2$) 174

CHAPTER 7 PRECIPITATION EQUILIBRIUM, TITRATION, AND GRAVIMETRY 177

7.1 Precipitation Equilibria and Solubility 178

7.1.1 Solubility of Precipitates in Pure Water 178

7.1.2 Ionic Strength and the Solubility of Precipitates 178

7.1.3 Common Ion and the Solubility of Precipitates 179

7.1.4 pH and the Solubility of Precipitates 179

7.1.5 Complexing Agents and the Solubility of Precipitates 182

7.2 Precipitation Titrations 184

7.2.1 Titration Curves 184

7.2.2 Examples of Methods Classified by Endpoint Indication 186

7.2.3 Preparation of Standard Solutions 189

7.3 Precipitation Gravimetry 190

7.3.1 Classification of Gravimetric Methods of Analysis 190

7.3.2 General Procedure and Requirements for Precipitation 190

7.3.3 Precipitate Formation 192
7.3.4 Obtaining High Purity Precipitates 193
7.3.5 Experimental Considerations 197
7.3.6 Examples of Organic Precipitating Reagents 200

CHAPTER 8 SPECTROPHOTOMETRY 206

8.1 Principle of Spectrochemical Analysis 207
8.1.1 Properties of Electromagnetic Radiation 207
8.1.2 Interaction of Electromagnetic Radiation with Matter 208
8.1.3 Beer's Law, the Quantitative Principle of Light Absorption 213
8.1.4 Limitations to Beer's Law 216
8.2 Principles of Instrumentation 217
8.2.1 Instrumentation 217
8.2.2 Instrumental Errors in Absorption Measurement 226
8.3 Applications of Spectrophotometry 226
8.3.1 Single Component Analyses 226
8.3.2 Multicomponent Analyses 228
8.3.3 Spectrophotometric Titrations 230
8.3.4 Studies of Complex Formation in Solutions 231
8.3.5 Measurements of Dissociation Constants of Organic Acids/Bases 233

CHAPTER 9 INTRODUCTION TO ANALYTICAL SEPARATION 238

9.1 General Considerations of Separation Efficiency 239
9.2 Separation by Precipitation 241
9.2.1 Inorganic Precipitants 241
9.2.2 Organic Precipitants 242
9.2.3 Coprecipitation of Species in Trace Amounts for Separation 243
9.2.4 Improving the Selectivity of Precipitation Separation 244
9.3 Separation by Extraction 245
9.3.1 Principles for Liquid-liquid Extraction 245
9.3.2 Percent Extraction 247
9.3.3 Extraction of Inorganic Species 249
9.3.4 Other Extraction Methods 254
9.4 Separation by Ion Exchange 257
9.4.1 Ion Exchange Resins 257
9.4.2 Cross-linkage and Exchange Capacity 259

9.4.3 Ion Exchange Equilibria 260
9.4.4 Applications of Ion Exchange Separation 261
9.5 Separation by Chromatography 263
9.5.1 Classification 263
9.5.2 Chromatogram 264
9.5.3 Column Chromatography 265
9.5.4 Planar Chromatography 266

CHAPTER 10 SOLVING A REAL ANALYTICAL PROBLEM 271

10.1 Definition of the Analytical Problem 272
10.2 Literature Review 273
10.3 Choosing a Method 275
10.4 Developing and Evaluating the Method 276
10.4.1 Selectivity 276
10.4.2 Accuracy 277
10.4.3 Sensitivity and Linear Dynamic Range 279
10.5 Conclusion 280

APPENDICES 281
Appendix A References 281
Appendix B Indicators 283
Appendix C Activity Coefficients(γ) for Ions at 25℃ 285
Appendix D Constants for Acid-base, Complexometric, Redox, and Precipitation Titrimetry 286
Appendix E Molecular Masses 299

ANSWERS 302

INDEX 305

PERIODIC TABLE OF THE ELEMENTS 309

CHAPTER 1

INTRODUCTION OF ANALYTICAL CHEMISTRY

1.1 What is Analytical Chemistry

1.2 Steps in the Development of an Analytical Method

1.3 Classification of Quantitative Analytical Methods

1.3.1 Chemical Analysis

1.3.2 Instrumental Analysis

1.4 Principles of Volumetric Titration

1.4.1 Basic Terms

1.4.2 Requirements of Titration Reactions

1.4.3 Classification of Titration Processes

1.4.4 Primary Standards and Standard Solutions

1.4.5 Basic Apparatus in Chemical Analyses

1.5 Calculations in Volumetric Titration

1.5.1 Preparation of Standard Solutions

1.5.2 Titration Results

This chapter deals with the basis of analytical chemistry including the definition of analytical chemistry, steps of developing an analytical method, classification of the analytical methods, and introduction of volumetric analysis.

1.1 WHAT IS ANALYTICAL CHEMISTRY

Analytical chemistry is a measurement science, responsible for characterizing the composition of natural and artificial materials, both qualitatively (what is present) and quantitatively (how much is present). With the advancement of science, the current definition of analytical chemistry would be "analytical chemistry is the science of inventing and applying the concepts, principles, and strategies for measuring the characteristics of chemical systems and species" (Murray R W. Anal Chem, 1991, 63: 271A). Analytical chemistry widely applies the knowledge of all the other chemical disciplines, physics, biology, information theory and many other technical fields to the development of analytical methods for all the sciences and human activities with information on the character and amount of chemical species and their distribution in space and time. In return, analytical chemistry is of fundamental importance to academia and industry of almost all disciplines, naming a few: biological sciences, engineering, medicine, public health, and the environment, as well as homeland and food safety.

The human genome project (HGP) is one of the best examples to demonstrate how analytical chemistry came to the rescue to solve the human genome mysteries. The human genome project started in 1990 and was scheduled to finish in 2005. There are approximately 20000～25000 genes in human DNA which is composed of 3 billion base pairs (http://www. ornl. gov/sci/tech resources/Human_Genome/project/).

DNA, as shown in Figure 1. 1, is a polymer with the monomer units called nucleotides. Each nucleotide consists of a pentacarbon sugar (deoxyribose), a nitrogen containing base attached to the sugar, and a phosphate group. There are four different types of nucleotides found in DNA, differing only in the nitrogenous base, adenine (A), thymine (T), guanine (G), and cytosine (C). DNA is a normally double stranded macromolecule. Within the DNA double helix, A forms two hydrogen bonds with T on the opposite strand, and G forms three hydrogen bonds with C on the opposite strand. A DNA sequence or genetic sequence is a succession of letters (A,T,G, and C) representing the primary structure of a real or hypothetical DNA molecule or strand, with the capacity to carry information as described by the central dogma of molecular biology.

At the early stage, human DNA was sequenced by the Sanger method which used gel electrophoresis with radioactive labeled dideoxynucleotides (ddNTPs) to produce fragments terminated at each of the four bases in separate lanes. This method was time consuming, labor intensive, and expensive. It is hard to imagine that HGP can be accomplished in time using this traditional method.

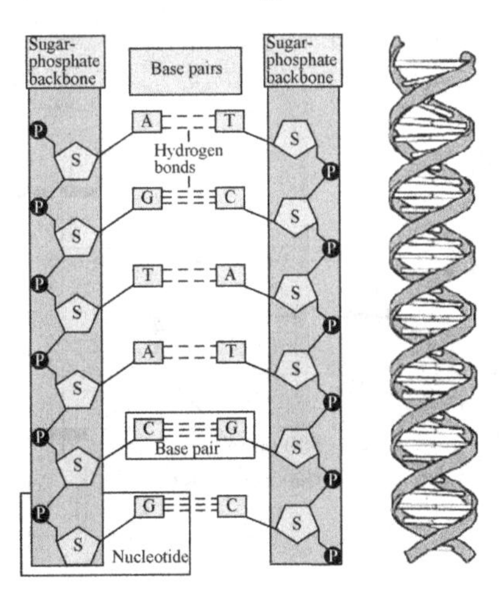

Figure 1.1 Illustration diagram for DNA structure.

Fortunately, analytical chemists worked together to make this happen. Professor Lloyd Smith at the University of Wisconsin-Madison developed the strategy of four-color DNA sequencing which made it possible to run the entire sequencing reaction in one lane rather than keeping the A, G, C, T bases in separate channels (Figure 1.2). Laser was induced to ensure a high sensitivity of this technique. Professor Barry Karger of Northwestern University developed a modified linear polyacrylamide matrix (LPA) DNA separation in a capillary. This polymer provided high-resolution separation performance that was easy to reload the capillaries by simply blowing out the polymer solution. Many analytical chemists worked on developing capillary arrays for high throughput assays. In 1992, Professor Rich Mathies and his colleagues at the University of California-Berkeley developed a 25-capillary array system with scanning detection (Figure 1.3). The confocal microscope set-up together with laser induced fluorescence detection provided very high spatial resolution and sensitivity, arming this system with higher throughput. The capillary electrophoresis combined with laser induced fluorescence detection by labeling four bases with different fluorescent dyes, and made it possible to run the sequencing in one lane or channel (Elizabeth Zubritsky. How analytical chemists saved the human genome project... or at least gave it a helping hand. Anal Chem, 2002, 74 (1): 22 A-26 A).

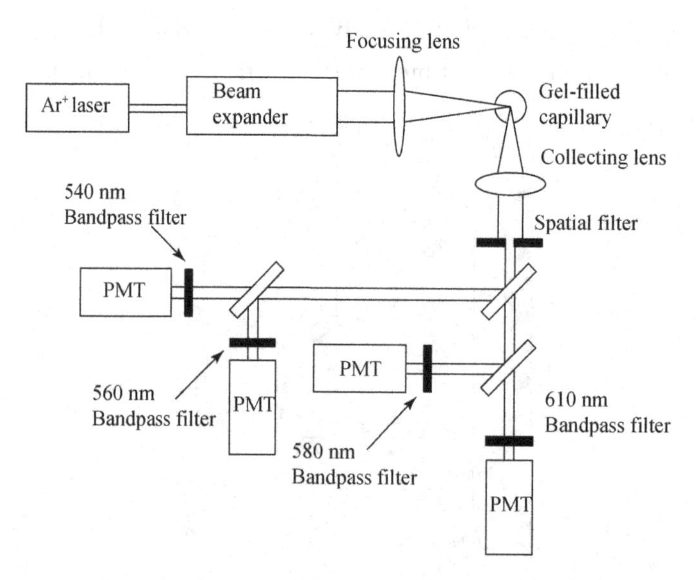

Figure 1. 2 Optical system employed for multiple wavelength fluorescence detection in capillary electrophoresis of DNA. From Luckey J A, Drossman H, Kostichka A J, et al. High-speed DNA sequencing by capillary electrophoresis. Nucleic Acids Res, 1990,18(15): 4417-4421.

The human genome project was completed in 2003, two years ahead of the projected deadline, which is best demonstration of how analytical chemists contributed by improving the throughput, speed, sensitivity, reproducibility of the analytical method for DNA sequencing, and an appropriate way (in this author's eyes) to commemorate the 50th anniversary of the discovery of DNA-helix structure in 1953 by James Watson and Francis Crick who shared the 1962 Nobel Prize.

Another good example would be the food-borne pathogen detection. The detection of pathogenic bacteria is the key step in prevention and identification of problems related to public health and safety. The most frequently used methods are based on culture and colony counting methods. While this method gives reliable results, it is very time consuming, and usually needs a few days to one week to give a result. Biosensor technology can provide equally reliable results in a much shorter time period. For detection of *Legionella pneumophila*, the colony count method needs 5～14 days to give a final result, in contrast the polymerase chain reaction (PCR) method and biosensor method need only 1～2 hours. Although the new techniques need to improve their sensitivity to the same level as the traditional method, the new methods provided a faster and more realistic approach for food pathogen detection (Olivier Lazckaa, F Javier Del Campob, and F Xavier Muñoza. Pathogen detection: A perspective of traditional methods and biosensors. Biosensors and Bioelectronics, 2007,22 (7): 1205-1217).

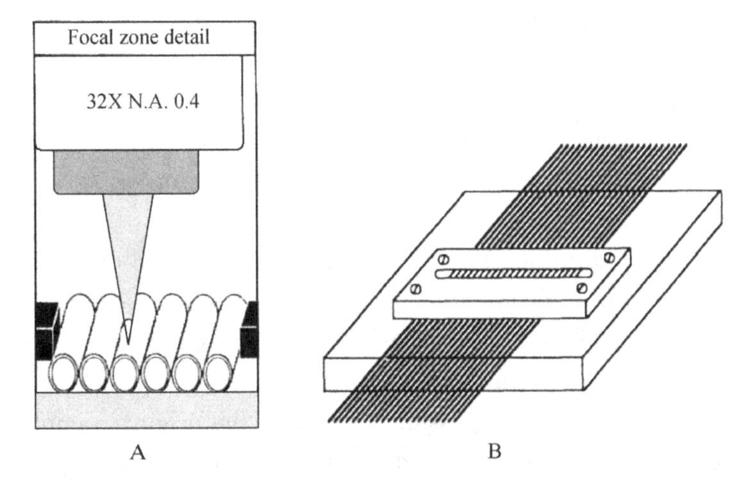

Figure 1.3 The exploded schematic view of focal zone (A) and the capillary array holder (B) of the laser-excited, confocal fluorescence capillary array scanner. From Huang X H C, Quesada M A, and Mathies R A. Capillary array electrophoresis using laser-excited confocal fluorescence detection. Anal Chem, 1992, 64(8): 967-972.

1.2 STEPS IN THE DEVELOPMENT OF AN ANALYTICAL METHOD

An **analyte** is a substance or chemical constituent of a sample that is to be measured by an analytical method. The development of an analytical method for an analyte always starts with a problem encountered, followed by choosing an analytical method, acquiring the sample, processing the sample, eliminating interferences, calibrating and measuring concentrations, calculating results, and evaluating results by estimating their reliability.

Choosing an analytical method is a crucial step. Many factors need to be considered in selecting an analytical method. Accuracy is usually the first factor to be considered. Cost in labor and time that combine to estimate the real cost often determine the selection of an analytical method. One always selects the method with the lowest real cost providing that the method meets the requirement of accuracy.

It is desirable to obtain a representative sample of less than one gram with the same composition as the bulk sample. Analysis of a non-representative sample will result in false information which in turn will result in wrong decision making.

The general sampling procedure for the solid sample is:

- Take small amount of samples from many sites within the bulk supply.
- Pool the samples together.
- Grind, sieve, and mix the sample to ensure homogeneity.

Sample preparation includes all necessary steps to obtain a final solution ready to be measured: drying, dissolution, eliminating interferences, enrichment and concentration. Sample preparation is usually the most difficult step and the major source of analytical error.

Absorption and desorption of water may occur during each of the sampling and storage steps that result in changes in the chemical composition of the solid sample. It is better that the sample is stored in a container located in an environment with controlled temperature and humidity. To best maintain the water content of the prepared solid sample as close as possible to that of the original sample, drying at selected temperature is usually conducted just before analysis starts.

Most analyses are performed on solutions of the analyte. Therefore, a dissolution step is necessary to dissolve the solid sample into a solution. The conditions of dissolution should be sufficiently mild to avoid loss of the analyte. The solvent should be able to dissolve the analyte rapidly and completely.

Once a sample is dissolved in a suitable solvent, it may be possible to proceed directly to the measurement step. However, in most cases, interferences must be eliminated before measurements. An **interference** is a species that coexists with the analyte and may affect the final measurement by enhancing or attenuating the signal. Few chemical and physical properties of importance in chemical analysis are unique to an individual chemical species, thus masking or separating the interfering chemical species from the analyte will be necessary. Separation methods used for analytical purpose will be introduced in Chapter 9.

Once the interferences are eliminated, the sample is ready for measurement step. After acquiring the sample concentration, a statistical report must be provided to estimate the reliability of the results.

1.3 CLASSIFICATION OF QUANTITATIVE ANALYTICAL METHODS

1.3.1 Chemical Analysis

Quantitative analytical methods are traditionally classified as **chemical analysis** and **instrumental analysis**. Chemical analysis, based on the chemical reactions, includes gravimetric and volumetric analyses and is the major focus of this text.

Gravimetry, based on mass measurement, includes **precipitation gravimetry**, volatilization gravimetry, and electrogravimetry. In **precipitation gravimetric analysis**, the mass of the product of the chemical reaction related to the analyte, which is the sparingly soluble and pure precipitate, is measured with an analytical balance. For example, a barium sulfate gravimetric method for determining sulfur content in iron ores is recommended by the International Organization for Standardization (ISO 4689:1986, Iron ores—determination of sulfur content—barium sulfate gravimetric method). After sample preparation, the sulfur in the sample is converted to sulfate, an excess of barium chloride ($BaCl_2$) is added to an aqueous solution of the sample to cause the precipitation of the sulfate as barium sulfate ($BaSO_4$). The precipitate ($BaSO_4$) is then filtered, washed to remove impurities, heated, and weighed to obtain the final mass of $BaSO_4$. Precipitation gravimetry is usually suitable for samples with an analyte greater than 1%.

Volumetric analysis (also **volumetric titration**) is a quantitative chemical analysis that is used to determine the unknown concentration of a known reactant. A standard solution is added from a buret to react with the analyte until the titration is complete, i.e., the endpoint is reached as determined by an indicator. In volumetric analysis, the volume measured is used to calculate the concentration of the analyte. For example, a hydrochloric acid (HCl) standard solution added from a buret can be used to determine the concentration of sodium hydroxide (NaOH) in a solution using methyl orange (MO) as the indicator. After the neutralization reaction reaches a point when all the NaOH has just reacted with HCl, an additional very small amount (about one half drop) of HCl changes the final solution from basic to acidic and the methyl orange indicator changes its color from yellow to orange.

Volumetric analysis are classified by the type of reactions occurring, e.g., acid-base titration (neutralization titration), complexometric titration, redox titration and precipitation titration.

1.3.2 Instrumental Analysis

In **instrumental analysis**, the physical and physicochemical properties of analytes, such as conductivity, electrode potential, light absorption or emission, mass-to-charge ratio, and magnetic resonance, can be used to detect as well as to measure the analyte. The properties and related instrumental methodologies are summarized in Table 1.1.

Table 1.1 Examples of Properties of Materials and Related Instrumental Methods

Property	Related Methods
Radiation emission	Emission spectroscopy—fluorescence, phosphorescence, luminescence, Raman, atomic emission
Radiation absorption	Absorption spectroscopy—spectrophotometry, nuclear magnetic resonance, electron spin resonance, atomic absorption
Electrical properties	Potential—potentiometry Charge—coulometry Current—voltammetry Resistance—conductometry
Mass-to-charge ratio	Mass spectrometry
Thermal	Thermal gravimetry, calorimetry
Radioactivity	Activation, isotope dilution

Quantitative analytical methods can also be classified based on sample size, e.g., **macro**, **semimicro**, **micro** and **ultramicro analyses**. The **macro analysis** is used for samples with mass more than 0.1 grams or volume more than 10 mL. The **semimicro analysis** is used for samples with mass in the range of 0.01~0.1 grams or volume in the range of 1~10 mL. The **micro analysis** is used for sample with mass less than 0.01 grams or volume less than 1 mL. Sometimes, the **ultramicro analysis** is used for sample with mass lower than 10^{-4} grams.

Classical analytical methods, including precipitation gravimetry, volumetric titration, spectrophotometry, and analytical data treatment as well as separation methods in analytical chemistry are introduced in this book.

1.4 PRINCIPLES OF VOLUMETRIC TITRATION

In this chapter, the principles of volumetric titration are described, including the basic terms, requirements of titration reactions and the classifications of titration, the primary standards and standard solution, and the calculations involved in titration.

1.4.1 Basic Terms

Titrimetry refers to the group of analytical techniques that are based on measuring the amount (the volume in volumetric titration) of a reagent of known concentration to determine the quantity of an analyte. The reagent may be a standard solution of a chemical as in volumetric and gravimetric titrimetry or an electric current of known magnitude. Other than visible color endpoint detection, instrumental endpoint detection can also be used, including electrochemical, thermoanalytical, optical and radiochemical detections. This chapter introduces general terms of volumetric titrimetry, which use color indicators whose color changes are visible to the eye to detect the endpoint(ep).

In a volumetric titration, the analyte (**titrand**) is put in an Erlenmeyer flask or a beaker and the standard solution (**titrant**) is added from a buret. Theoretically, the titration stops at the chemical **stoichiometric point** (SP) also called **equivalence point** (EP) which is the theoretical point reached when the amount of added titrant is equivalent to the amount of the analyte being titrated. However, one can only experimentally detect the endpoint by using a color **indicator**. In most cases, the endpoint is different than stoichiometric point because it is almost impossible to find an indicator that changes color exactly at the chemical equivalence point. This type of systematic error is called **titration error** (E_t). Titration error, the major source of error in volumetric titration, is affected by completeness of titration reaction and the response of the indicator.

1.4.2 Requirements of Titration Reactions

The volumetric titration is based on the chemical reaction between the analyte (titrand) and titrant. A titration reaction must meet the following requirements:

- The titrant must react completely with the analyte with a defined stoichiometry.
- The titrant should react rapidly with the analyte.
- The completeness of the reaction must be greater than 99.9% to ensure a reasonable titration break so that a color indicator is available for locating endpoint within the titration break.

1.4.3 Classification of Titration Processes

With **direct titration**, the analyte is directly titrated with titrant, for example, the titration of hydrochloric acid (HCl) with sodium hydroxide ($NaOH$) solution. Sometimes, an excess of the standard titrant is added to react with the analyte and then **back titration** is done using a second standard titrant. Back titration is carried out when: (1) the analyte is in solid form; (2) the reaction rate between the analyte and the standard solution is slow; (3) the standard solution is not adequately stable to complete the titration; and (4) no indicator is available for the direct-titration. For example, in the determination of calcium carbonate ($CaCO_3$) using acid-base titration, an excess of HCl standard solution is added, and the mixture is heated to complete the reaction. The excess amount of HCl is then titrated using NaOH standard solution. Other examples are aluminum (Al^{3+}) or nickel (Ni^{2+}) that cannot be directly titrated using ethlylenediaminetetraacetic acid (EDTA), because the metal ions block the indicator making the endpoint detection impossible. In this situation, an excess of EDTA is added to react with the metal ion. The excess EDTA is then back-titrated using the metal ion standard titrant. One should be aware that a back-titration cannot be applied to a reaction with completeness less than 99.9%, because the back titration approaches the endpoint from the opposite direction without changing the completeness of the reaction.

Indirect titration is applied when there is no defined stoichiometry between the analyte and the titrant. Thiosulfate ion ($S_2O_3^{2-}$) has been widely used to determine oxidizing agents by indirect titration that involves iodine (I_2) as an intermediate. Two moles of $S_2O_6^{2-}$ quantitatively produce one mole of tetrathionate ion ($S_4O_6^{2-}$) by reacting with liberated iodine (I_2). Indirect titration approach can be applied to determine an analyte that can not directly react with the titrant. Redox titration of calcium ion (Ca^{2+}) can be achieved by forming calcium oxalate (CaC_2O_4) precipitate first and then dissolving the precipitate with sulfuric acid (H_2SO_4) to quantitatively release oxalic acid ($H_2C_2O_4$). The oxalic acid can then be titrated with potassium permanganate ($KMnO_4$).

1.4.4 Primary Standards and Standard Solutions

A **primary standard** is a substance of known high purity which can serve as a reference material in volumetric and mass titrimetry. A primary standard must meet the following criteria:

- Purity greater than 99.9%.
- Reasonably stable to atmospheric conditions.
- Free from water of hydration.
- Reasonably soluble in the titration medium.
- Large molecular weight preferred to minimize the relative error associated with weighing.

Examples of primary standards are (1) arsenic trioxide (As_2O_3) for preparing sodium arsenite solution to standardize sodium periodate ($NaIO_4$) solution; (2) potassium bromate ($KBrO_3$) to standardize sodium thiosulfate ($Na_2S_2O_3$) solution; (3) potassium hydrogen phthalate ($KHC_8H_4O_4$, usually called KHP) to standardize an aqueous base and perchloric acid ($HClO_4$) in glacial acetic acid; (4) potassium dichromate ($K_2Cr_2O_7$) to standardize sodium thiosulfate solution.

However, very few compounds meet all the criteria for primary standards, and only a limited number of primary standards are available commercially, thus **secondary standards** are often used in titrimetry.

Standard solution is a solution of known concentration which is often used as the titrant. Standard solutions can be prepared by two basic methods:

- Direct preparation: a carefully weighed amount of a primary standard is dissolved in a suitable solvent and diluted to a known volume in a volumetric flask.
- Standardization: when a reagent does not meet the requirement of a primary standard, its standard solution can be prepared by standardization. First, a weighed amount of the titrant is dissolved in a volume of solvent close to the desired volume. Then, the solution to be standardized is used to titrate a weighed quantity of a primary standard, or a measured volume of another standard solution. Standard solutions of sodium hydroxide (NaOH), hydrochloric acid (HCl), ethylenediaminetetraacetic acid (EDTA), potassium permanganate ($KMnO_4$), and sodium thiosulfate ($Na_2S_2O_3$) are prepared by standardization.

1.4.5 Basic Apparatus in Chemical Analyses

1. Instruments for Measuring Mass

Weighing the analytical sample is often the first step of a quantitative analysis. Usually, analytical balances (macrobalances) have a maximum capacity of 160 to 200 g with a standard deviation of ± 0.1 mg. Semimicroanalytical balances have a maximum capacity of 10 to 30 g with a standard deviation of ± 0.01 mg. A microanalytical balance has a maximum capacity of 1 to 3 g with the precision of ± 0.001 mg. Based on the sample size and the accuracy required for the analysis, one needs to choose an appropriate analytical balance for mass measurement.

Historically, an equal-arm balance was used for mass measurement. The analytical sample is placed on one pan of the balance, and the standard weighs of equal mass of the objective are placed on the other pan to restore the light weight beam to the original position. The single-arm mechanical analytical balance (Figure 1.4) was commercialized in 1946. It vastly improved the speed and convenience of weighing compared with the equal-arm balance. With this mechanical balance, the balance pan and a set of removable standard weights on one side of a beam are balanced against a fixed counterweight on the other side of the beam. The beam itself is balanced on a fulcrum consisting of a sharp knife edge. Adding a sample to the balance pan makes the beam tilted away from the balance point. Selected standard weights are then removed until the beam is brought into balance. The combined mass of the removed weights equals the mass of the sample. Nowadays, electronic analytical balances are replacing the single-pan balance, because they provide better speed, ruggedness, convenience, and accuracy.

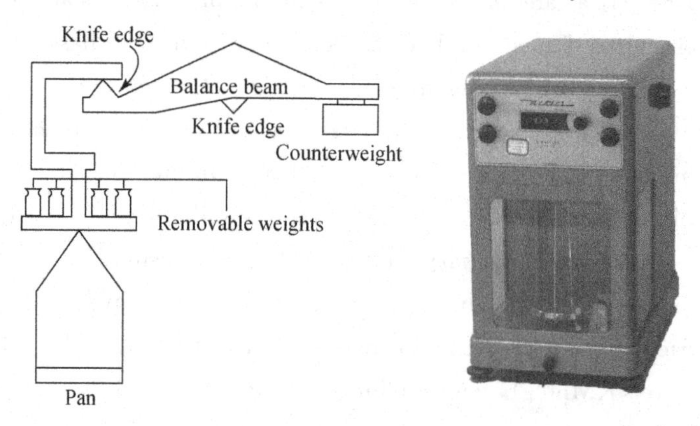

Figure 1.4 Schematic diagram and photo of the single-arm mechanical balance.

A typical electronic balance has the balance pan placed above an electromagnet as shown in Figure 1.5. When the pan is empty, the current is adjusted so that the level of the indicator arm is in the null position. Placing an object or a sample on the pan causes the pan and the indicator arm to move downward, the balance detects this downward movement and the electromagnet generates a counterbalancing force, thus the pan returns to its original null position. This device in which a small electric current causes a mechanical system to return to a null position is called a servo system. The current needed to produce this force is proportional to the mass of the object.

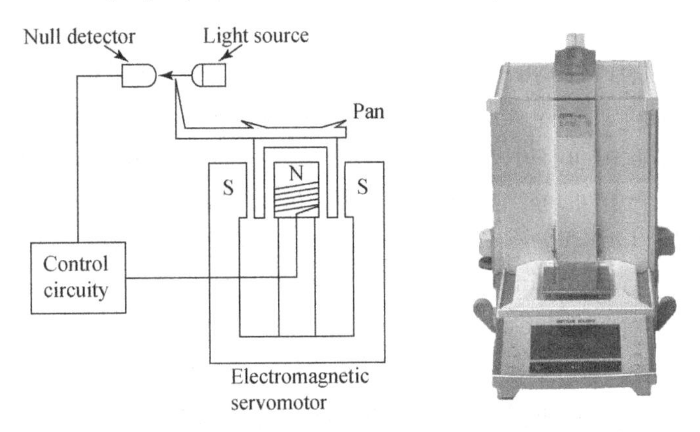

Figure 1.5 Schematic diagram and photo of the electronic balance.

2. Instruments for Measuring Volumes

The precise measurement of volume is of equal importance as is the precise measurement of mass.

A variety of glassware can be used to measure volume (Figure 1.6). The type

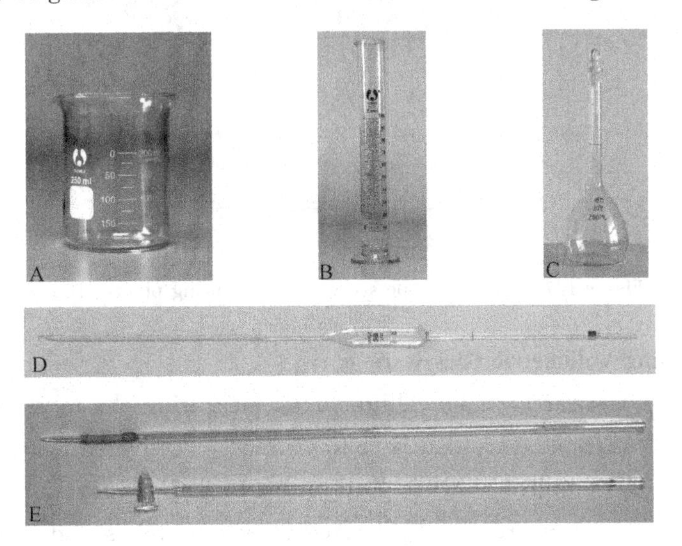

Figure 1.6 Common examples of glassware used to measure volume: A, beaker; B, graduated cylinder; C, volumetric flask; D, pipet; E, burets.

of glassware used depends on the accuracy of the volume needed. Beakers, dropping pipets, and graduated cylinders are used to measure volumes approximately with errors of several percent. Volumetric pipets and flasks provide more accurate measurement of volume. Volumetric flasks are designed **to contain** an accurate volume at the specified temperature (20℃ or 25℃) and are marked with "TC". Volumetric flasks are used in the dilution of a sample or solution to a specific volume. The flasks contain the exactly listed volume when the bottom of the meniscus of the solution just touches the line mark across the neck of the flask. A 10-mL volumetric flask contains 10.00 mL, and a 250-mL volumetric flask holds 250.0 mL. Graduated cylinders that are labeled "to contain" are less accurate containers to deliver volumes. Volumetric pipets are designed **to deliver** a specific volume at a specific temperature and are marked with "TD". There are two common types of pipets, the volumetric or transfer pipet, and the measuring pipet (Figure 1.7). A 100 mL transfer pipet delivers 100.0 mL solution, and a 25 mL transfer pipet delivers 25.00 mL solution.

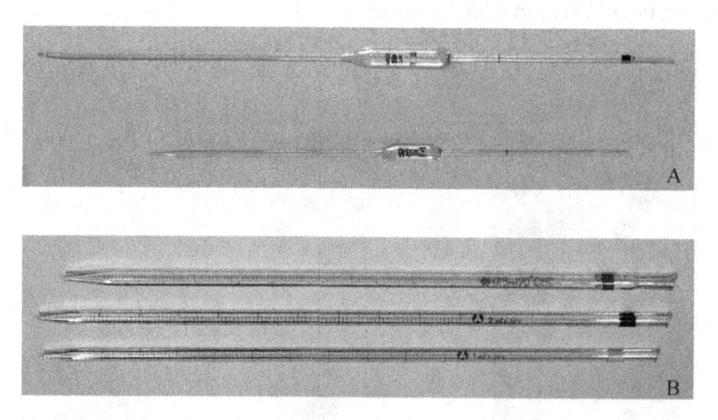

Figure 1.7 Transfer pipets (A) and measuring pipets (B).

3. Calibrating Volumetric Glassware

Volumetric Glassware can be calibrated by measuring the mass of a liquid of known density and temperature that is contained in (or delivered by) the volumetric glassware. Usually, distilled water is used for calibration purposes.

All volumetric glassware should be free of water breaks and droplets on the glassware walls before calibration. Burets and pipets are to deliver liquids and need not be dry. Volumetric flasks are to contain liquids and should be dry at room temperature. The water used for calibration should be in the thermal equilibrium with its surroundings.

The calibration of a volumetric flask relative to a pipet is to ensure that there is no systematic error caused by the inaccurate partition of a sample into aliquots. For example, to calibrate a 100-mL volumetric flask relative to a 25-mL transfer pipet, four 25-mL aliquots of water was transferred from the pipet to a dry 100-mL volumetric flask. The lower edge of meniscus is then marked. Dilution to the label permits the same pipet to deliver precisely a one fourth aliquot of the solution in the flask.

1.5 CALCULATIONS IN VOLUMETRIC TITRATION

1.5.1 Preparation of Standard Solutions

To prepare a standard solution from a primary standard, one needs to weigh accurately, dissolve and dilute the primary standard solution to an accurate volume in a volumetric flask.

For a primary standard (A), amount of A in mass, $m(A)$(g) can be calculated from the amount of A, $n(A)$(mol), and the molar mass, $M(A)$(g/mol). $n(A)$(mol) can be calculated from the volume of the solution, V(L), and the molar concentration, $c(A)$.

$$m(A) = n(A) \cdot M(A) = c(A) \cdot V(A) \cdot M(A) \qquad (1\text{-}1)$$

【Example 1.1】 Describe the preparation of 250.0 mL of 0.01000 mol · L^{-1} potassium dichromate ($K_2Cr_2O_7$) from the primary standard grade reagent.

Answer: For $K_2Cr_2O_7$, we have

$$\begin{aligned} m(A) &= n(K_2Cr_2O_7) \cdot M(K_2Cr_2O_7) \\ &= c(K_2Cr_2O_7) \cdot V(K_2Cr_2O_7) \cdot M(K_2Cr_2O_7) \\ &= 0.01000 \times (250.0/1000) \times 294.18 = 0.7354 \text{ (g)} \end{aligned}$$

Therefore, the solution can be prepared by dissolving 0.7354 g of $K_2Cr_2O_7$ in water and diluting to exactly 250.0 mL in a volumetric flask. In practice, the amount of primary standard weighed is acceptable in the range of $\pm 10\%$ of expected mass. For this example, the mass of $K_2Cr_2O_7$ in the range of 0.66~0.81 grams is acceptable.

A standard solution is often prepared by standardization, because the number of primary standard reagents such as potassium dichromate ($K_2Cr_2O_7$) is limited.

【Example 1.2】 Describe the preparation of 500 mL of 0.20 mol · L^{-1} HCl from commercially available concentrated hydrochloric acid (HCl) which is approximately

12 mol • L^{-1}.

Answer: The amount of the concentrated HCl needed is the same as that of the HCl standard solution to be prepared. Hence,

$$n(\mathrm{HCl}) = c(\mathrm{HCl}) \cdot V(\mathrm{HCl}) = (500/1000) \times 0.20 = 0.10\ (\mathrm{mol})$$

The volume of concentrated HCl needed is

$$V(\mathrm{HCl})_{\mathrm{concent}} = n(\mathrm{HCl})/c(\mathrm{HCl})_{\mathrm{concent}} = 0.10/12$$
$$= 0.0083(\mathrm{L}) = 8.3\ (\mathrm{mL})$$

Therefore, 8.3 mL of concentrated HCl as measured by a graduated cylinder is mixed with 490 mL water in a storage bottle. The diluted HCl can then be standardized and used as a HCl standard solution.

The amount of primary standard reagent for standardization can be calculated from the stoichiometry of the standardization reaction. The volume of titrant used for calculation must be greater than 20 mL (generally 25 mL) to assure a less than 0.1% volumetric error.

【Example 1.3】 To standardize 0.10 mol • L^{-1} NaOH using potassium hydrogen phthalate ($M_r(KHC_8H_4O_4)$=204.22), how many grams of $KHC_8H_4O_4$ are needed?

Answer: The titration reaction is

$$\mathrm{KHC_8H_4O_4 + NaOH \rightleftharpoons KNaC_8H_4O_4 + H_2O}$$

Therefore, the stoichiometric ratio is 1 : 1, and the amount of $KHC_8H_4O_4$ needed can be calculated from the amount of 25 mL of 0.10 mol • L^{-1} NaOH.

$$m(\mathrm{KHC_8H_4O_4}) = n(\mathrm{KHC_8H_4O_4}) \cdot M(\mathrm{KHC_8H_4O_4})$$
$$= c(\mathrm{NaOH}) \cdot V(\mathrm{NaOH}) \cdot M(\mathrm{KHC_8H_4O_4})$$
$$= 0.10 \times 25 \times 10^{-3} \times 204.22 = 0.51(\mathrm{g})$$

In practice, about 0.51 g (±10%) $KHC_8H_4O_4$ is weighed into an Erlenmeyer flask, dissolved in 25 mL water, and titrated with NaOH standard solution. Usually, three replicates weighing of $KHC_8H_4O_4$ are titrated to obtain the mean of three determinations.

【Example 1.4】 $Na_2S_2O_3$ solution is standardized using primary standard reagent $K_2Cr_2O_7$(M_r = 294.18). This is an example of indirect titration. $K_2Cr_2O_7$ can liberate stoichiometric amount of iodine when mixed with excess potassium iodide in acidic solution. Two moles of $S_2O_6^{2-}$ quantitatively produce one mole of tetrathionate ion ($S_4O_6^{2-}$) by reaction with one mole of iodine released. How many grams of $K_2Cr_2O_7$ are needed to standardize 0.020 mol • L^{-1} $Na_2S_2O_3$?

Answer: The reactions involved in standardization process are

$$\mathrm{Cr_2O_7^{2-} + 6I^- + 14H^+ \rightleftharpoons 2Cr^{3+} + 3I_2 + 7H_2O}$$

$$I_2 + 2S_2O_3^{2-} \rightleftharpoons 2I^- + S_4O_6^{2-}$$

According to reaction stoichiometry

$$1Cr_2O_7^{2-} \triangleq 3I_2 \triangleq 6S_2O_3^{2-}$$

For $aA \triangleq bB$, we have

$$n(A) = \frac{a}{b}n(B) \tag{1-2}$$

That is, with the reactant molecule as the reaction unit, calculation is carried out based on the stoichiometry factor for this reaction.

For the specific standardization reactions, 1 mole $K_2Cr_2O_7$ is equivalent to 6 moles of $Na_2S_2O_3$. That is, the amount of $K_2Cr_2O_7$, $n(K_2Cr_2O_7)$, is 1/6 of that of $Na_2S_2O_3$, $n(Na_2S_2O_3)$. Therefore,

$$m(K_2Cr_2O_7) = n(K_2Cr_2O_7) \cdot M(K_2Cr_2O_7) = \frac{1}{6}n(Na_2S_2O_3) \cdot M(K_2Cr_2O_7)$$

$$= \frac{1}{6}c(Na_2S_2O_3) \cdot V(Na_2S_2O_3) \cdot M(K_2Cr_2O_7)$$

$$= 0.020 \times 0.025 \times 294.18/6 = 0.025(g)$$

If 0.025 g $K_2Cr_2O_7$ is weighed to standardize $Na_2S_2O_3$, the weighing error will be $\pm(0.0002/0.025) = 1\%$. It is recommended to avoid potential weighing error that 10 times the calculated amount of $K_2Cr_2O_7$ (about 0.25 g) is weighed, dissolved, and diluted to 250.0 mL. Three aliquots of 25.00 mL $K_2Cr_2O_7$ are retrieved by a pipet for replicate titration. This approach reduces weighing error that occurs by direct weighing small amount of $K_2Cr_2O_7$ for individual titrations in this example. In Example 1.3, the weighing error for potassium hydrogen phthalate in one single titration is less than 0.1%, because the mass for one single weighing is 0.51 g, much more than 0.2 g. By multiple individual weighing, one can avoid the bias from a mistake in weighing, because one can easily tell from the individual titration results whether a weighing mistake occurred.

【Example 1.5】 To standardize ethylenediaminetetraacetic acid (EDTA) using zinc ion (Zn^{2+}) (at 1 : 1 stoichiometric ratio), 0.3126 g zinc flakes (99.99%) are dissolved in 10 mL concentrated HCl and diluted to 250.0 mL. One aliquot of 25.00 mL Zn^{2+} standard solution is titrated with EDTA. The volume of EDTA used for three replicate titrations is 24.32, 24.35, and 24.33 mL, respectively. Calculate the concentration of EDTA standard solution.

Answer: Concentration of zinc standard solution is

$$c(\text{Zn}) = \frac{m(\text{Zn})}{M(\text{Zn}) \cdot V} = \frac{0.3126}{65.38 \times 0.2500} = 0.01913(\text{mol} \cdot \text{L}^{-1})$$

The mean volume of EDTA from three replicate titrations is

$$\overline{V}(\text{EDTA}) = (24.32 + 24.35 + 24.33)/3 = 24.33(\text{mL})$$

Hence,

$$c(\text{EDTA}) = 25.00 \times 0.01913 / 24.33 = 0.01966(\text{mol} \cdot \text{L}^{-1})$$

1.5.2 Titration Results

Generally, the analytical results are presented in mass percentage (w) for a solid sample, and molar concentration (c) or mass concentration (ρ, at unit $\text{g} \cdot \text{L}^{-1}$ or $\text{mg} \cdot \text{L}^{-1}$) for a liquid sample.

For a species A in a sample,

$$w(\text{A}) = \frac{m(\text{A})}{m_s} \times 100\% \tag{1-3}$$

【Example 1.6】 A 0.2356 g (m_s) of iron ore sample is dissolved in acid and all ferric ion (Fe^{3+}) is reduced to ferrous (Fe^{2+}) before titration. A 23.42 mL of 0.01324 $\text{mol} \cdot \text{L}^{-1}$ $K_2Cr_2O_7$ is used for titrating Fe^{2+}. Calculate the mass percentage of iron in the sample as $w(\text{Fe})$ and $w(Fe_2O_3)$.

Answer: The titration reaction is

$$6Fe^{2+} + Cr_2O_7^{2-} + 14H^+ \rightleftharpoons 6Fe^{3+} + 2Cr^{3+} + 7H_2O$$

Therefore,

$$Cr_2O_7^{2-} \triangleq 6Fe^{2+} \triangleq 3Fe_2O_3$$

The stoichiometry is

$$n(\text{Fe}) = 6n(Cr_2O_7^{2-}) \quad \text{and} \quad n(Fe_2O_3) = 3n(Cr_2O_7^{2-})$$

Therefore,

$$w(\text{Fe}) = \frac{m(\text{Fe})}{m_s} = \frac{6c(K_2Cr_2O_7) \cdot V(K_2Cr_2O_7) \cdot M(\text{Fe})}{m_s} \times 100\%$$

$$= \frac{6 \times 0.01324 \times 23.42 \times 55.85}{0.2356 \times 1000} \times 100\%$$

$$= 44.10\%$$

$$w(Fe_2O_3) = \frac{m(Fe_2O_3)}{m_s} = \frac{3 \times 0.01324 \times 23.42 \times 159.7}{0.2356 \times 1000} \times 100\% = 63.06\%$$

【Example 1.7】 The calcium in human serum can be determined by titration of oxalate with potassium permanganate ($KMnO_4$) after quantitatively precipitating calcium as calcium oxalate (CaC_2O_4). The calcium oxalate precipitate (CaC_2O_4) can be prepared with the addition of ammonium hydroxide ($NH_3 \cdot H_2O$) in the presence of oxalic acid ($H_2C_2O_4$). CaC_2O_4 precipitate is then dissolved in hot and diluted H_2SO_4 and titrated with potassium permanganate ($KMnO_4$). If 1.35 mL of 0.01000 mol · L^{-1} $KMnO_4$ is used for the determination of 5.00 mL human serum, calculate the mass percentage of calcium in serum.

Answer:

Approach #1: The reactions involved in the determination of Ca^{2+} are

$$Ca^{2+} + (NH_4)_2C_2O_4 \rightleftharpoons CaC_2O_4 \downarrow + 2NH_4^+$$

$$CaC_2O_4 + 2H^+ \rightleftharpoons H_2C_2O_4 + Ca^{2+}$$

$$5H_2C_2O_4 + 2MnO_4^- + 6H^+ \rightleftharpoons 2Mn^{2+} + 10CO_2 + 8H_2O$$

Therefore, the stoichiometry is

$$Ca \triangleq C_2O_4^{2-} \quad \text{and} \quad 5C_2O_4^{2-} \triangleq 2MnO_4^-$$

Then we have $5Ca \triangleq 2MnO_4^-$. Hence,

$$\rho(Ca) = \frac{c(KMnO_4) \cdot V(KMnO_4) \cdot 5M(Ca)}{2V_s}$$

$$= \frac{0.0100 \times 1.35 \times 5 \times 40.08}{2 \times 5.00} = 0.271(g \cdot L^{-1})$$

Approach #2: Conservation law on proton transfer, electron transfer, complexation number, and charge balance is followed in acid-base, redox, complexation, and precipitation titrations. The number of protons that an acid gives equals the number of protons that a base accepts in an acid-base titration. For redox titration, the number of electrons that an oxidizing agent accepts equals the number of electrons that a reducing agent donates. This is also true for complexometric titration and precipitation titration.

If we define the reaction unit as the chemical formula associated with one electron transfer in redox titration, or one proton transfer in acid-base titration, we can write an equation which the number of reaction units that accepts one electron equals the number of reaction unit that provides one electron in a redox titration. From this approach, one can carry out the calculation based on the number of electrons that the reactant involves in the reaction without writing and balancing the reaction equation. For example, if molecule A accepts Z_A electrons and molecule B gives away Z_B electrons in a redox titration, we have

$$n\left(\frac{1}{Z_A}A\right) = n\left(\frac{1}{Z_B}B\right) \tag{1-4}$$

For redox titration in acidic pH, we have the reaction unit of 1/6 $K_2Cr_2O_7$ for $K_2Cr_2O_7$ and 1/5 $KMnO_4$ for $KMnO_4$, respectively. For titration of $H_2C_2O_4$ with $KMnO_4$ in acidic pH, 1/2 $H_2C_2O_4$ can be taken as the reaction unit because $C_2O_4^{2-} \xrightarrow{-2e} 2CO_2$. Therefore, we have

$$n\left(\frac{1}{5}KMnO_4\right) = n\left(\frac{1}{2}H_2C_2O_4\right)$$

And because $n(Ca) = n(H_2C_2O_4)$, we have $n\left(\frac{1}{2}Ca\right) = n\left(\frac{1}{5}KMnO_4\right)$. Then,

$$\rho(Ca) = \frac{c\left(\frac{1}{5}KMnO_4\right) \cdot V(KMnO_4) \cdot M\left(\frac{1}{2}Ca\right)}{V_s}$$

$$= \frac{0.0500 \times 1.35 \times 40.08/2}{5.00} = 0.271(g \cdot L^{-1})$$

Chapter 1 Questions and Problems

1.1 How many grams of the following primary standards are needed to standardize 25 mL of 0.05 mol · L^{-1} NaOH (sodium hydroxide)?

(1) $H_2C_2O_4 \cdot 2H_2O$ (oxalic acid);

(2) $KHC_8H_4O_4$ (potassium acid phthalate, KHP);

(3) State the advantages and disadvantages of standardization using the above primary standards.

(4) How to make sure that weighing error is within ±0.1%?

1.2 Describe the preparation of 500 mL of 0.1 mol · L^{-1} hydrogen chloride (HCl) solution from commercial concentrated HCl that has a specific gravity of 1.18 g · mL^{-1} and is 37% (w/w).

1.3 To standardize the concentration of NaOH using $H_2C_2O_4 \cdot 2H_2O$ that has been stored in a desiccator for longer time than it should be, will there be error in the result of standardization? If yes, will the result be high or low? Explain your answer.

1.4 Calculate the stoichiometry of the following reactants to H^+ when the equation for the titration is:

$$B_4O_7^{2-} + 2H^+ + 5H_2O \rightleftharpoons 4H_3BO_3$$

(1) $Na_2B_4O_7 \cdot 10H_2O$; (2) B; (3) B_2O_3; (4) $NaBO_2 \cdot 4H_2O$.

Note: $B_4O_7^{2-} + 5H_2O \rightleftharpoons 2H_3BO_3 + 2H_2BO_3^-$.

1.5 Assuming the volumes for both solutions used in titration are close enough, calculate the concentration of $Na_2C_2O_4$ standard solution used to standardize $KMnO_4$ ($c(\frac{1}{5}KMnO_4) \approx$ 0.10 mol · L^{-1}) in acidic solution. How many grams of $Na_2C_2O_4$ are needed for preparation of 100 mL such $Na_2C_2O_4$ standard solution?

1.6 A 0.8835 g sodium biphosphate ($Na_2HPO_4 \cdot 12H_2O$) is titrated using 27.30 mL of 0.1012 mol · L^{-1} HCl to the stoichiometric point of $H_2PO_4^-$. Calculate the percentage (w/w) of $Na_2HPO_4 \cdot 12H_2O$. Interpret the result, assuming that concentration of HCl is correctly prepared and standardized, the endpoint is properly judged, and volumetric glassware is correctly calibrated.

1.7 A mixed solution of tartaric acid ($H_2C_4H_4O_6$) and formic acid (HCOOH) is titrated according to the following procedure to determinate both acids in the solution. A 15.00 mL of 0.1000 mol · L^{-1} NaOH is used to titrate 10.00 mL of the solution to the endpoints of $C_4H_4O_6^{2-}$ and $HCOO^-$. Another 10.00 mL of the tested solution is acidified, and 30.00 mL of 0.2000 mol · L^{-1} Ce(Ⅳ) is added to oxidize the organic acids to CO_2. A 10.00 mL of 0.1000 mol · L^{-1} Fe(Ⅱ) is used to back titrate excess Ce(Ⅳ). Calculate concentration of tartaric acid and formic acid in the solution.

1.8 If exactly 25.00 mL of 0.1500 mol · L^{-1} NaOH was used to neutralize 25.00 mL KHC_2O_4-$H_2C_2O_4$, how many mL of 0.04000 mol · L^{-1} $KMnO_4$ is needed to titration 20.00 mL of above KHC_2O_4-$H_2C_2O_4$ solution?

CHAPTER 2

DATA ANALYSIS

2.1 Error and Classification

2.1.1 Accuracy and Precision

2.1.2 Errors and Deviation

2.1.3 Systematic and Random Errors

2.2 Distribution of Random Errors

2.2.1 Frequency Distribution

2.2.2 Normal Distribution

2.2.3 Predicting the Probability of Random Errors—Area under Gaussian Curve

2.3 Statistical Data Treatment

2.3.1 Estimation of Population Mean (μ) and Population Standard Deviation (σ)

2.3.2 Confidence Interval for Population Mean

2.3.3 Statistical Aids to Hypothesis Testing

2.3.4 Detection of Gross Errors

2.4 Propagation of Error

2.4.1 Systematic Errors

2.4.2 Random Errors (Standard Deviation)

2.4.3 Maximum Errors (E_R)

2.4.4 Distribution of Errors

2.5 Significant Figure Convention

2.5.1 Significant Figures

2.5.2 Numerical Rounding in Calculations

Measurement invariably involves errors, which come from mistakes of the analyst, faulty calibrations and inaccurate standardizations. Frequent calibration and standardization with replicate analyses of defined sample specimens can lessen all but the random error. Even so, it is still impossible to perform a chemical analysis that is totally free of error. Every measurement is influenced by many uncertainties. Even if the same measurement was performed by the same person a number of times with the same experimental conditions, scattered results can still occur. Because measurement uncertainties can never be eliminated, the experimental result gives only an estimate of the "true" value. By exploring the nature of experimental errors and the effects on the chemical analysis, one can evaluate the quality of data and define limits within which the true value of a measured quantity lies with a defined level of probability. Data analysis can help with selecting appropriate experimental design and instruments to better develop and validate a cost effective analytical method.

2.1 ERROR AND CLASSIFICATION

2.1.1 Accuracy and Precision

Accuracy indicates the closeness of the measurement to the true or accepted value. As the difference between the measured value and the true value decreases, the accuracy increases. **Precision** describes the agreement among several results obtained in the same way. To obtain optimal quality analytical results, replicates should be performed with the same size samples and experimental conditions through the entire analytical procedure. If the results of the replicates are close, one can say that the precision is good.

How can one judge the analytical results based on accuracy and precision? Let us discuss one example. The analytical results of four analysts for an iron sample are illustrated in Figure 2.1. The results reveal the following: Analyst A results show good precision and accuracy; Analyst B results show good precision but poor accuracy, indicating systematic error in the measurement; thus, it would be dangerous to assume that precise results are also accurate; Analyst C results show poor precision and accuracy; Analyst D gives a result close to the true value but with poor precision. If two replicates were taken randomly from D's set of data to estimate the iron content in the sample, the mean values could be scattered and distinctively different from the true value.

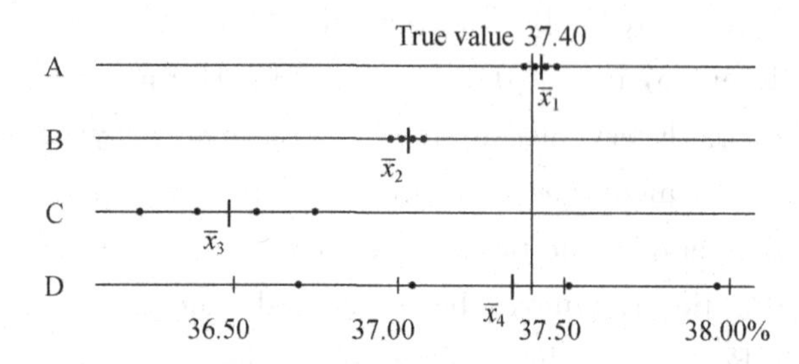

Figure 2.1 Iron sample analytical results from four analysts (A, B, C, D).

Precision is a pre-requirement to assure the accuracy. If precision is poor, it is not possible to decide whether the results are accurate. Results with reasonable precision do not necessarily mean that the accuracy will be acceptable.

2.1.2 Errors and Deviation

Accuracy is expressed in the term of error, which means the difference between the measured value and the true value (T). The error of one determination (x_i) in an assay is given as follows:

$$x_1 - T,\ x_2 - T,\ \cdots,\ x_i - T,\ \cdots,\ x_n - T$$

In reality, the result of an analysis is given as the mean value and the error, expressed as $\bar{x}$ and E. Errors can be expressed as either the **absolute error** (E_a) or the **relative error** (E_r) with the equations as follows:

- Absolute error $$E_a = \bar{x} - T \quad (2\text{-}1a)$$
- Relative error $$E_r = \frac{E_a}{T} \times 100\% \quad (2\text{-}1b)$$

A small error means the measured value is close to the true value, thus good accuracy; or vice versa, large error means poor accuracy. The sign of the error tells that whether the result is higher or lower than the true value. If the measurement result is lower than the true value, the sign is negative; or vice versa.

Relative error indicates the percentage of error as to the true value in the result; thus, the relative error often is a more useful quantity than the absolute error.

However, the true value can never be "known" exactly. In practice, "a reference value or standard value" is often employed to judge the accuracy of an analytical result. The reference value refers to the results of experienced analysts after repeating the same experiment for a number of times.

【Example 2.1】 The $w(\mathrm{Cl})$ in a pure NaCl sample determined by precipitation titration was 60.53%. Calculate E_a and E_r.

Answer: The theoretical $w(\mathrm{Cl})$ in pure NaCl is

$$w(\mathrm{Cl}) = \frac{M(\mathrm{Cl})}{M(\mathrm{NaCl})} \times 100\% = \frac{35.45}{35.45 + 22.99} \times 100\% = 60.66\%$$

$$E_a = 60.53\% - 60.66\% = -0.13\%$$

$$E_r = \frac{-0.13\%}{60.66\%} \times 100\% = -0.2\%$$

2.1.3 Systematic and Random Errors

In Figure 2.1, the results from all analysts showed errors toward the true value and the data of each of the analysts were scattered. These observations suggest that analytical results can be affected by at least two types of errors—**systematic error** and **random error**.

1. Systematic Errors

Systematic errors have a definite value, an assignable cause and are of the same order of magnitude for replicate measurements made in the same way. By examining the sources, systematic errors, also called determinate errors, can be determined experimentally and thus may be reduced.

The sources of systematic errors are:

(1) **Method errors** arise from the analytical method itself. In precipitation gravimetric analysis, method errors are caused by co-precipitation of unexpected entities or dissolution of the precipitate. In volumetric analysis, incompleteness of the reaction, the difference between the endpoint and stoichiometric point, and side reactions are all the possible sources of method errors that may make the results either lower or higher than the true value. Method errors are the most difficult to identify and correct. Selecting an alternative method and calibration can help with determining and reducing method errors.

(2) **Instrumental errors** are from the measurement devices including volumetric glassware and electronic instruments. Using volumetric glassware at a temperature significantly different from the calibration temperature makes the volume differ from its calibrated volume. When an equal-arm balance has two arms that are not equal as calibrated or the weights used have not been calibrated for a long time, the mass measured by the balance may differ from the true mass. Frequent calibration can

reduce most of the instrumental errors.

(3) **Reagent errors** come from impurities in reagents used in experiments. In volumetric analysis, poor water quality is one example of reagent errors. This type of error can be eliminated by conducting reagent blank determinations and using high purity water.

(4) **Personal errors** come from personal judgment in the experiments. For example, endpoint judgment in a volumetric analysis, when the analyst is insensitive to color changes at the endpoint, excess reagent is used. In this situation, analytical procedures should be adjusted so that any known physical limitation of the analyst will not cause errors.

One should be aware that systematic errors cannot be eliminated by increasing the number of replicates.

2. Random Errors

Random errors, also called indeterminate errors, are caused by many uncontrollable variables that are an inevitable part of every analysis. Random errors bear no sign and can never be totally determined. The uncertainties in experiments cause replicate measurements to fluctuate randomly around the mean of the data set.

Random errors cannot be reduced or eliminated by calibration. Random errors in analytical results follow a Gaussian or normal distribution. We can use statistical methods to evaluate the random errors. By increasing the number of replicates, random errors can be reduced to better describe the analytical result.

A human mistake, also called gross error, is not included in the above two types of errors. For example, partial loss of precipitate during filtration, spill of the sample solution during dissolution, and false reading will all lead to gross error. These results differ markedly from all other data in a set of replicate measurements. Various statistical tests can be performed to determine if a result is an **outlier.** Outliers can or should be excluded from the data analysis.

2.2 DISTRIBUTION OF RANDOM ERRORS

Although random errors come from uncertainties and are not directionally determinate, one may still want to know "Do random errors follow any particular pattern of distribution?".

2.2.1 Frequency Distribution

As an example, consider results for determining the purity of $BaCl_2 \cdot H_2O$ by precipitation gravimetric method in a quantitative laboratory class. If results from 173 students are listed one by one, the data may appear scattered without any pattern. The information will be easier to visualize when the data are rearranged into frequency distribution groups as in Table 2.1. By sorting the data, we know that all data lie in 98.9%~100.2%. The data are then divided into 14 groups with each group including data points within an adjacent 0.1% range. Then we tabulate the number of data points falling into a series of adjacent 0.1% ranges. The number of data falling in one group is the number in range (n_i); ratio of n_i to total number of measurements (n) is the frequency. The ratio of frequency to the spread of group (0.1%) is the frequency density. The frequency distribution data in Table 2.1 are plotted as a **histogram** (labeled A in Figure 2.2).

Table 2.1 Frequency Distribution Illustration

Group	/%	Number in Range (n_i)	Frequency (n_i/n)	Frequency Density
1	98.85~98.95	1	0.006	0.06
2	98.95~99.05	2	0.012	0.12
3	99.05~99.15	2	0.012	0.12
4	99.15~99.25	5	0.029	0.29
5	99.25~99.35	9	0.052	0.52
6	99.35~99.45	21	0.121	1.21
7	99.45~99.55	30	0.173	1.73
8	99.55~99.65	50	0.289	2.89
9	99.65~99.75	26	0.150	1.50
10	99.75~99.85	15	0.087	0.87
11	99.85~99.95	8	0.046	0.46
12	99.95~100.05	2	0.012	0.12
13	100.05~100.15	1	0.006	0.06
14	100.15~100.25	1	0.006	0.06
Total		173	1.001	

A few conclusions can be drawn from Table 2.1 and Figure 2.2: (1) most frequent occurrence is zero deviation from the mean (99.6%); (2) about 87% of the data points are within ±0.3% of this mean with a few having a larger deviation from the mean. As the number of measurements increases, the spread for group is narrower, and the histogram approaches the shape of the continuous curve shown as plot B in Figure 2. 2, known as a **frequency density plot.**

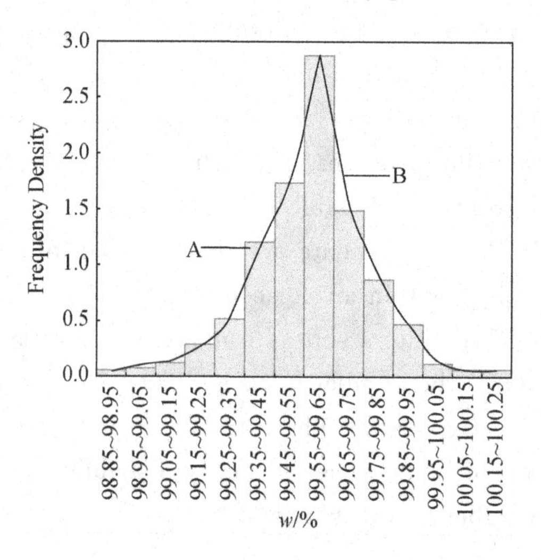

Figure 2. 2 Histogram (A) showing distribution of the 173 results in Table 2. 1 and the frequency density plot (B) for the same data in the histogram.

2.2.2 Normal Distribution

In most cases, analytical results follow a normal or Gaussian distribution. Figure 2. 3 shows two Gaussian curves with frequency density of the data points or the deviations from the mean plotted against the data points or the deviations from the mean. The normal distribution equation is

$$y = f(x) = \frac{1}{\sigma\sqrt{2\pi}} e^{-\frac{(x-\mu)^2}{2\sigma^2}} \tag{2-2}$$

in which $f(x)$ denotes the frequency density and x denotes the data values as the variable. μ denoted as the **population mean** and σ as the **population standard deviation** are the two parameters in the distribution named as $N(\mu,\sigma)$.

The population mean (μ) is the mean of the collection of all measurements and is correspondent to the x-axis value of the peak. When there is no systematic error, μ is the true value of the measured quantity.

The population standard deviation (σ) describes the precision of a population of data. σ is one half of the distance between two reflexion points of the normal distribution curve. For example, σ_A and σ_B in Figure 2. 3 are for Gaussian curves A and B, respectively. The precision of data set A is greater than that of B, which makes σ_A smaller than σ_B, thus curve A is narrower than curve B. σ is given by Equation (2-3)

$$\sigma = \sqrt{\frac{\sum_{i=1}^{n}(x_i - \mu)^2}{n}} \tag{2-3}$$

where n is the number of data points of the population.

$x-\mu$ in Equation (2-2) is a measure of random error. Plotting frequency density of a data set versus $x-\mu$ gives a normal distribution of random error with a peak value at 0 in the x-axis.

From this normal distribution, one can see that: (1) small random errors are observed much more frequently than very large ones; (2) there is a symmetrical distribution of the positive and negative random errors about the maximum.

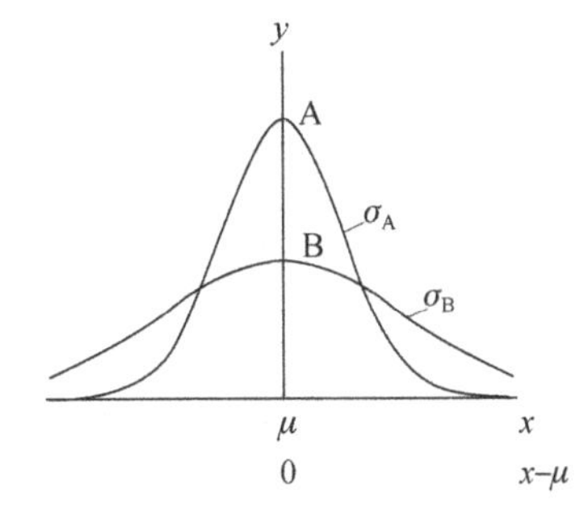

Figure 2.3 Normal distribution curves.

The shape of a normal distribution curve changes as σ changes. If the x-axis is replaced by a new variable u, defined as $u=(x-\mu)/\sigma$, a plot of frequent density versus u yields a single Gaussian curve, which describes all data populations regardless of the population standard deviation. u is the ratio of the deviation or difference between a data point and the population mean to one population standard deviation.

The Gaussian curve is then described in the following equation

$$y = f(x) = \frac{1}{\sigma\sqrt{2\pi}}e^{-u^2/2}$$

Because $dx=\sigma du$, then

$$f(x)dx = \frac{1}{\sqrt{2\pi}}e^{-u^2/2}du = \phi(u)du$$

That is,

$$y = \phi(u)du = \frac{1}{\sqrt{2\pi}}e^{-u^2/2}du \tag{2-4}$$

The Gaussian distribution described by Equation (2-4) is called standardized normal distribution noted as N (0,1), and is illustrated in Figure 2.4.

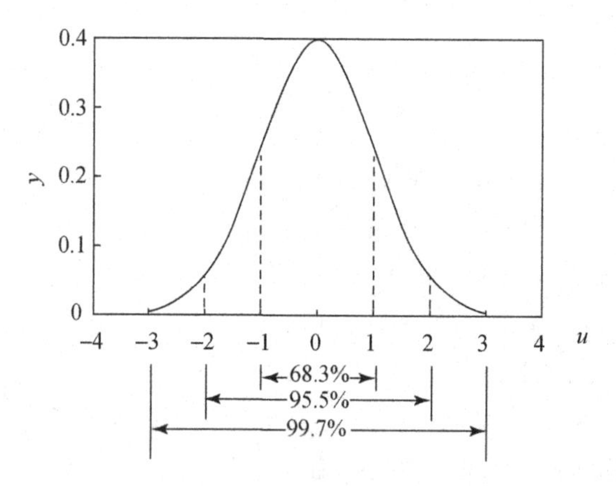

Figure 2.4 The standardized normal distribution curve.

2.2.3 Predicting the Probability of Random Errors—Area under Gaussian Curve

The total area under the Gaussian curve represents the probability that random errors in the full data set. Integration of Equation (2-4) will give a value of 1 as illustrated in the following equation

$$\int_{-\infty}^{\infty} \phi(u)\,du = \frac{1}{\sqrt{2\pi}}\int_{-\infty}^{\infty} e^{-u^2/2}\,du = 1$$

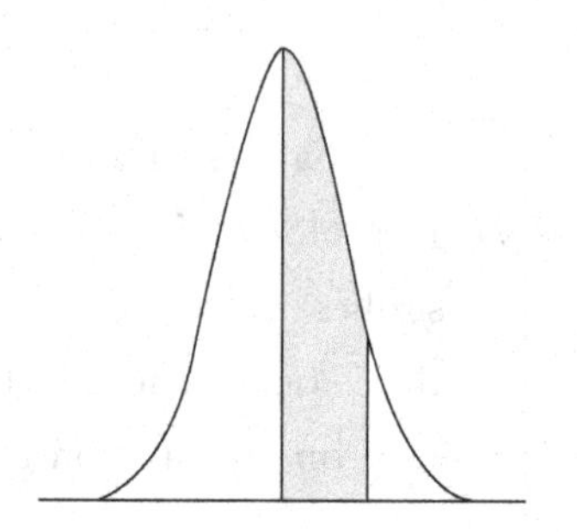

Figure 2.5 Areas under the Gaussian curve.

Filled area $= \int_0^{\mu} \phi(u)\,du = \frac{1}{\sqrt{2\pi}}\int_0^{\mu} e^{-u^2/2}\,du$

The area under the curve between a pair of limits gives the probability of a measured value occurring between those limits and can be obtained by integration of Equation (2-4) in a defined interval. Table 2.2 gives a few integrated results as illustrated in Figure 2.4. The probability of $\pm u$ is 2 fold of the filled area in Figure 2.5.

Table 2.2 Representative Integrated Results

$\|u\|$	0.674	1.000	1.645	1.960	2.000	2.576	3.000	∞
Area	0.2500	0.3413	0.4500	0.4750	0.4773	0.4950	0.4987	0.5000

It is easy to calculate the probability that a random error or a measured value lies within the defined limits. For example, about 68.3% of the values making up the population will lie within one standard deviation ($\pm 1\sigma$) of the population mean

(μ). The other representative probabilities are given in Table 2.3. There is less than 0.3% probability that a random error will exceed $\pm 3\sigma$.

Table 2.3 Probability of a Measured Value between Varied Limits

Limits for Random Errors Relative to σ	Limits for Measured Values	Probabilities
(−1, +1)	($\mu-1\sigma$, $\mu+1\sigma$)	68.3%
(−1.96, +1.96)	($\mu-1.96\sigma$, $\mu+1.96\sigma$)	95.0%
(−2, +2)	($\mu-2\sigma$, $\mu+2\sigma$)	95.5%
(−2.58, +2.58)	($\mu-2.58\sigma$, $\mu+2.58\sigma$)	99.0%
(−3, +3)	($\mu-3\sigma$, $\mu+3\sigma$)	99.7%

2.3 STATISTICAL DATA TREATMENT

Statistical distribution of random errors has been derived from the populations. A population is the collection of all measurements of interest to the analyst. In many of the examples in analytical chemistry, the population is conceptual, very large, and close to infinite; therefore, in most situations, the population mean (μ) cannot be determined because a huge number of measurements (approaching infinity) would be required. In reality, a limited number of measurements are performed to predict the characteristics of the population. A sample is a limited number of measurements which is a subset of the population. Based on statistical analysis, we can scientifically judge the quality of experimental measurements, including prediction of population parameters, μ and σ, as well as conducting statistical tests.

2.3.1 Estimation of Population Mean (μ) and Population Standard Deviation (σ)

Population mean and population standard deviation (μ and σ), known as parameters for the central value and precision of population, cannot be determined but can be estimated from a limited number of measurements.

1. The Mean and Median

The average value of n measurements is called the **mean**, or **average**. It is calculated as

$$x = \frac{x_1 + x_2 + \cdots + x_n}{n} = \frac{1}{n}\sum_{i=1}^{n} x_i \tag{2-5}$$

where x_i represents the individual values of x making up the set of n replicate measurements. $\overline{x}$ is an estimate of μ. For a limited number of measurements, replicate values distribute around $\overline{x}$. When $n \to \infty$, $\overline{x} \to \mu$.

The **median** ($\tilde{x}$) is the middle result when replicate data are arranged in increasing or decreasing sequence. For an odd number of results, the median can be selected directly. For an even number of results, median is the mean of the middle pair. In ideal cases, the mean and median are identical, but in some cases they differ. The median is used to an advantage when a set of data contains an outlier. An outlier can have a significant effect on the mean of the data set but has no effect on the median.

2. Deviation

(1) Spread or range (R)

Spread or **range** is the difference between the largest value (x_{max}) and the smallest (x_{min}) in the data set.

$$R = x_{max} - x_{min} \tag{2-6}$$

Range is a simple way of describing precision, especially when the number of replicates is small. In volumetric titration, the precision of replicate titrations is often estimated by the range expressed in a relative value $\frac{R}{x} \times 100\%$.

(2) Average deviation ($\overline{d}$)

Deviation from the mean (d_i) tells how much an individual result, x_i, differs from the mean.

$$d_i = x_i - \overline{x} \quad (i = 1, 2, \cdots, n)$$

Average deviation is the arithmetic average of the absolute value of individual deviation in the data set.

$$\overline{d} = \frac{|d_1| + |d_2| + \cdots + |d_n|}{n} = \frac{1}{n}\sum_{i=1}^{n} |d_i| \tag{2-7}$$

If the sign of deviation is considered (no absolute value is taken), the average deviation would have been close to zero; therefore, the average deviation in this way gives a false estimate of precision.

(3) Sample standard deviation (s)

Sample standard deviation is given by the equation

$$s = \sqrt{\frac{\sum_{i=1}^{n}(x_i - \overline{x})^2}{n-1}} = \sqrt{\frac{\sum_{i=1}^{n} d_i^2}{n-1}} \tag{2-8a}$$

where $n-1$ is the number of degrees of freedom (denoted as f).

Relative standard deviation (RSD), also called the **coefficient of variation** (CV), is calculated according to Equation (2-8b).

$$CV = \frac{s}{\bar{x}} \times 100\% \tag{2-8b}$$

s is an unbiased estimate of the population standard deviation (σ), but one has to remember that s differs from σ in two ways: (1) population mean (μ) is replaced by sample mean ($\bar{x}$); (2) n is replaced by $n-1$.

3. Standard Error of the Mean

If a series of replicate results, each containing n measurements, are taken randomly from a population, the mean of each set will be less scattered as n increases. The standard deviation of each mean is known as **standard error of the mean** and is inversely proportional to the square root of the number of data points n used to calculate the mean as given by Equation (2-9) and (2-10).

$$\sigma_{\bar{x}} = \frac{\sigma}{\sqrt{n}} \tag{2-9}$$

$$s_{\bar{x}} = \frac{s}{\sqrt{n}} \tag{2-10}$$

The above two equations tell us that precision can be improved by increasing the number of replicate measurements; however, the improvement to be gained in this way is limited because of the square root dependence. Figure 2.6 gives the trend of

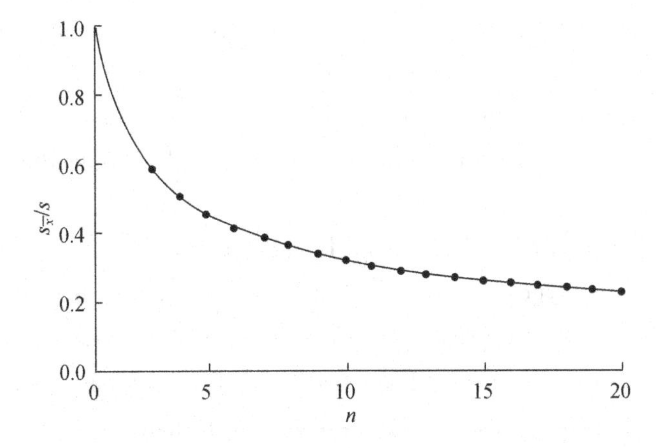

Figure 2.6 Standard error of the mean as a function of number of replicate measurements.

$s_{\bar{x}}/s$ as n increases. Initially, the pace of improvement is fast as n increases. When $n > 5$, the improvement is slower. When $n > 10$, the improvement is very minor and not worth the effort.

【Example 2.2】 Five replicate determinations of the iron content of an ore sample give: 37.45, 37.20, 37.50, 37.30, and 37.25(%). Calculate mean, median, range, average deviation, sample standard deviation, coefficient of variation, and standard error of the mean.

Answer:

$$\bar{x} = \frac{37.45 + 37.20 + 37.50 + 37.30 + 37.25}{5}\% = 37.34\%$$

$$\tilde{x} = 37.30\%$$

d_i: +0.11, −0.14, −0.04, +0.16, −0.09(%), thus

$$\bar{d} = \frac{\sum_{i=1}^{n} |d_i|}{n} = \frac{0.11 + 0.14 + 0.04 + 0.16 + 0.09}{5}\% = 0.11\%$$

$$s = \sqrt{\frac{\sum_{i=1}^{n} d_i^2}{n-1}} = \sqrt{\frac{0.11^2 + 0.14^2 + 0.04^2 + 0.16^2 + 0.09^2}{5-1}} \times 100\% = 0.13\%$$

$$CV = \frac{s}{\bar{x}} \times 100\% = \frac{0.13\%}{37.34\%} \times 100\% = 0.35\%$$

$$s_{\bar{x}} = \frac{s}{\bar{x}} = \frac{0.13\%}{\sqrt{5}} = 0.058\% \approx 0.06\%$$

The final results can be reported as $n, \bar{x}, s$. In this example, the result is

$$n = 5, \quad \bar{x} = 37.34\%, \quad s = 0.13\%$$

With n, $\bar{x}$, and s, one can make an estimate of μ and σ. The μ and σ are used to evaluate the quality of the results and can be used for subsequent statistical analysis.

2.3.2 Confidence Interval for Population Mean

As stated above, the true value of μ cannot be determined because a very large number of measurements (approaching infinity) would be required and uncertainties exist in measurements. An interval around $\bar{x}$ can be established within which the population mean (μ) lies in with a certain degree of probability.

1. *t* Distribution

In limited replicate measurements, we obtain $\bar{x}$ and s. An English chemist and statistician, W. S. Gosset, discovered the statistical treatment of small data sets. To evaluate the variability of s, t statistic (often called Student's t) is used as defined in exactly the same way as u.

For a single measurement with the result x, t is defined as $t=(x-\mu)/s$. For the mean of n measurements, t is defined as

$$t=\frac{\bar{x}-\mu}{s_{\bar{x}}}=\frac{\bar{x}-\mu}{s/\sqrt{n}} \tag{2-11}$$

Random errors of limited replicate measurements follow t distribution as shown in Figure 2.7. t distribution curve varies as number of degree of freedom (f) differs. When $n \rightarrow \infty$, t curve is the same as the Gaussian curve. Area under the curve between a pair of limits is the probability of the random error (Figure 2.8).

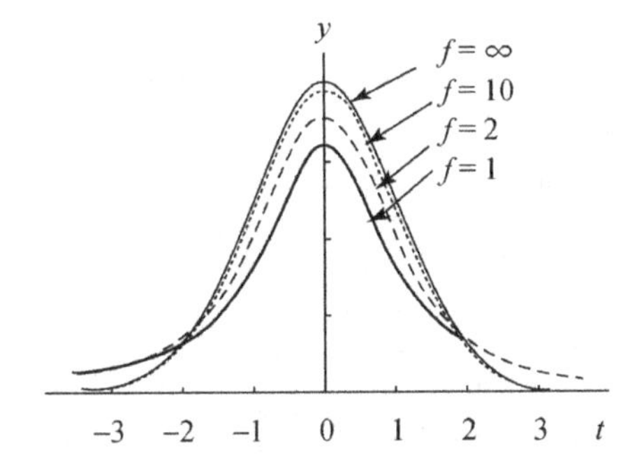

Figure 2.7 The t distribution curve.

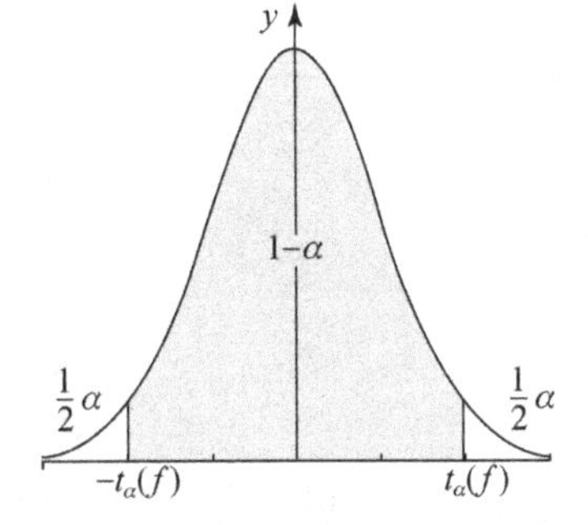

Figure 2.8 Probability for interval $[-t_\alpha(f), t_\alpha(f)]$.

Test statistic (t) depends on both the confidence level and the number of degree of freedom. t values are calculated and provided in various mathematical and statistical handbooks. Table 2.4 provides examples of t values associated with the respective degrees of freedom and **significance level.** α in Table 2.4 represents the probability when $t>t_\alpha(f)$ and $t<-t_\alpha(f)$ and is called **significance level.** The significant level in clinical studies is often referred to as the p value. The probability for t in $[-t_\alpha(f), t_\alpha(f)]$ equals to $(1-\alpha)\times 100\%$, called confidance level. As $f \rightarrow \infty$, $s \rightarrow \sigma$, then t equals to u. When $f=20$, t is very close to u.

2. Confidence Interval (CI)

According to Figure 2.7, it is with the probability of $(1-\alpha)$ that $-t_\alpha(f)<t<t_\alpha(f)$. Substitution of t with Equation (2-11) yields Equation (2-12).

Table 2.4 Values of t for Various Significance Levels

f \ $t_\alpha(f)$ \ α	0.50	0.10	0.05	0.01
1	1.00	6.31	12.71	63.66
2	0.82	2.92	4.30	9.93
3	0.77	2.35	3.18	5.84
4	0.74	2.13	2.78	4.60
5	0.73	2.02	2.57	4.03
6	0.72	1.94	2.45	3.71
7	0.71	1.90	2.37	3.50
8	0.71	1.86	2.31	3.36
9	0.70	1.83	2.26	3.25
10	0.70	1.81	2.23	3.17
20	0.69	1.73	2.09	2.85
∞	0.67	1.65	1.96	2.58

$$-t_\alpha(f) < \frac{\overline{x}-\mu}{s/\sqrt{n}} < t_\alpha(f) \tag{2-12}$$

Then,

$$\overline{x} - t_\alpha(f)\frac{s}{\sqrt{n}} < \mu < \overline{x} + t_\alpha(f)\frac{s}{\sqrt{n}} \tag{2-13}$$

Thus **confidence interval** (CI) at confidence level of $(1-\alpha)\times 100\%$ is

$$\overline{x} \pm t_\alpha(f)\frac{s}{\sqrt{n}} \tag{2-14}$$

Confidence interval is symmetrical to the sample mean with $(1-\alpha)\times 100\%$ probability such that μ lies in this interval as defined by Equation (2-14).

【Example 2.3】 The content of Fe (%) in a sample is $n=4$, $\overline{x}=35.21\%$, $s=0.06\%$. Calculate: (1) confidence interval (CI) for the population mean with 95% probability; (2) confidence interval (CI) for the population mean with the same probability at $n=9$.

Answer:

(1) For 95% confidence level, significant level is 0.05. From Table 2.4, $t_{0.05}(3)=3.18$. Hence, $t_\alpha(f)\frac{s}{\sqrt{n}}=0.10\%$.

CI is 35.21± 0.10(%). That is [35.11%, 35.31%].

(2) For $n=9$ at 95% confidence level, $t_{0.05}(8)=2.31$, $t_\alpha(f)\frac{s}{\sqrt{n}}=0.05\%$.

CI is 35.21 ± 0.05 (%). That is [35.16%, 35.26%].

The confidence interval changes with confidence level (probability). The number of replicates in an analysis affects the confidence interval. Four measurements will narrow the CI by a factor of 2, and 9 measurements will narrow the CI by a factor of 3.

In practice, a large number of measurements can be pooled in routine analysis for one sample. In this situation, σ can be considered to be known or s is a good estimate of σ. CI for μ can be obtained from the Gaussian distribution (Equation (2-15) & Figure 2.9).

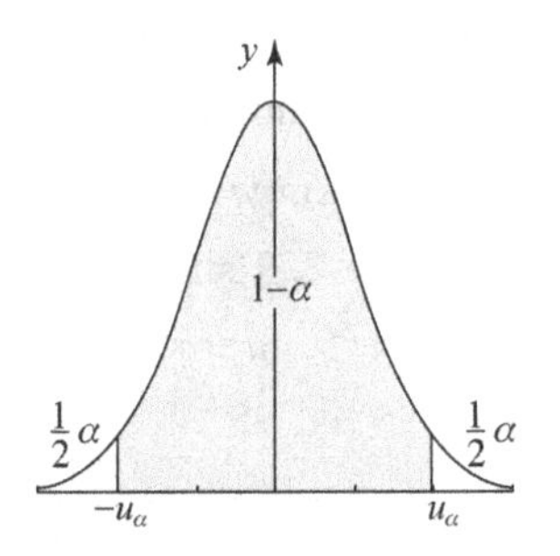

Figure 2.9 Probability for interval $[-u_\alpha, u_\alpha]$ when s is a good estimate of σ.

$$u = \frac{\bar{x} - \mu}{\sigma/\sqrt{n}} \tag{2-15}$$

CI can be obtained with a $(1-\alpha)$ confidence level

$$\bar{x} \pm u\frac{\sigma}{\sqrt{n}} \tag{2-16}$$

【Example 2.4】 The content of Fe (%) in a sample is $n=4, \bar{x}=35.21\%, \sigma=0.06\%$. Calculate confidence interval for the population mean with 95% probability.

Answer: $u=1.96$ with $\alpha=0.05$.

$$u\frac{\sigma}{\sqrt{n}} = 1.96 \times \frac{0.06\%}{\sqrt{4}} = 0.06\%$$

Then CI with 95% confidence is 35.21±0.06(%), that is [35.15%, 35.27%].

By comparing Examples 2.3 and 2.4, with the same value for standard deviation (s or σ), the CI is narrower when s is a good estimate of σ at the same confidence level.

2.3.3 Statistical Aids to Hypothesis Testing

The mean ($\bar{x}$) of experimental results seldom agrees with the true value (μ) predicted from theoretical model. The difference can be the result of random errors inevitable in all measurements or the result of systematic errors. Statistical tests can be helpful in sharpening these judgments.

We use a **statistical hypothesis** test to draw conclusion about the population mean (μ) and its nearness to the known value (μ_0). A **null hypothesis** assumes that the mean of the two populations are the same.

1. Comparing an Experimental Mean with a Known Value—*u* Test (Known σ or *s* is a Good Estimate of σ)

If σ is known or s is a good estimate of σ, u test is appropriate.

Figure 2.9 shows that there is a probability of α that random error will lead to a value of $u < -u_\alpha$ or $u > u_\alpha$. This can be expressed as

$$\frac{\bar{x}-\mu_0}{\sigma/\sqrt{n}} < -u_\alpha \quad \text{or} \quad \frac{\bar{x}-\mu_0}{\sigma/\sqrt{n}} > u_\alpha$$

where μ_0 is the known value.

The above expression can be rewritten as

$$\bar{x} < \mu_0 - u_\alpha \frac{\sigma}{\sqrt{n}} \quad \text{or} \quad \bar{x} > \mu_0 + u_\alpha \frac{\sigma}{\sqrt{n}}$$

The probability of this result lying in the regions as indicated in the above expressions as well as Figure 2.10 is the significance level (α) and the regions are called **rejection region.** If the test statistic u lies in the rejection region, the null hypothesis is rejected at the given significance level. Usually a significance level at 0.05 is chosen for the hypothesis test. Other significance levels, such as 0.01 or 0.001 may also be chosen, depending on the judgment certainty desired. The confidence level is related to the significance level (α) on a percentage basis by $(1-\alpha)\times 100$.

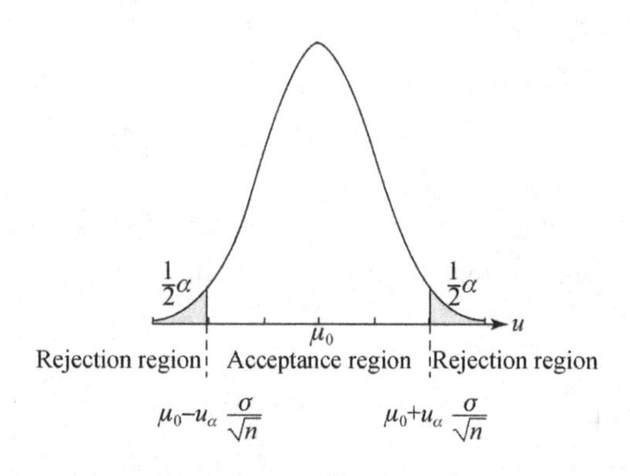

Figure 2.10 Rejection and acceptance regions for population mean.

The procedure that is used for u test is summarized as follows:

(1) State the null hypothesis, $H_0: \mu = \mu_0$.

(2) Calculate the test statistic u at the given significance level, α: $u_{cal} = \frac{\bar{x}-\mu_0}{\sigma/\sqrt{n}}$.

(3) State the alternative hypothesis, H_a, and determine the rejection regions (Figure 2.11).

For H_a: $\mu \neq \mu_0$, reject H_0 if $|u_{cal}| \geqslant u_{crit}$ (Figure 2.11A);

For H_a: $\mu > \mu_0$, reject H_0 if $u_{cal} \geqslant u_{crit}$ (Figure 2.11B);

For H_a: $\mu < \mu_0$, reject H_0 if $u_{cal} \leqslant u_{crit}$ (Figure 2.11C).

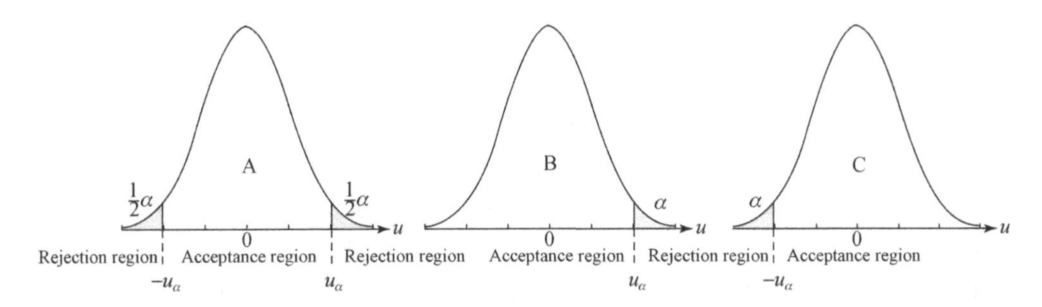

Figure 2.11 Rejection and acceptance regions. A, Two tailed test for $H_a: \mu = \mu_0$; B, one tailed test for $H_a: \mu > \mu_0$; C, one tailed test for $H_a: \mu < \mu_0$.

【Example 2.5】 In a routine test of the carbon content in molten iron, results obtained over a long period of time showed that the data followed Gaussian distribution with $\mu_0 = 4.55\%$ and $\sigma = 0.08\%$. Five more determinations were: 4.28%, 4.40%, 4.42%, 4.35%, and 4.37%. Was the carbon content changed at 0.05 significance level?

Answer:

$H_0: \mu = \mu_0 = 4.55\%$, the mean calculated from the data was 4.36%,

$$u_{cal} = \frac{\overline{x} - \mu}{\sigma / \sqrt{n}} = \frac{4.36\% - 4.55\%}{0.08\% / \sqrt{5}} = -5.3$$

At 0.05 significance level, $u_{crit} = 1.96$. As $|u_{cal}| \geqslant u_{crit}$, H_0 is rejected. The carbon content changed, that is, $u_{cal} \leqslant u_{crit}$, $\mu < \mu_0$. The carbon content is now lower.

2. Comparing an Experimental Mean with a Known Value μ_0—t Test (Unknown σ)

When σ is unknown, t test is chosen. The procedure is:

(1) State the null hypothesis, $H_0: \mu = \mu_0$.

(2) Construct the test statistic t at given significant level, $\alpha: t_{cal} = \frac{\overline{x} - \mu_0}{s/\sqrt{n}}$.

(3) State the alternative hypothesis, H_a, and determine the rejection region.

【Example 2.6】 The CaO content of a sample is 30.43%. One analysis ($n=6$) of this sample is $\overline{x} = 30.51\%$, $s = 0.05\%$. Is there any systematic error at 0.05 significance level?

Answer:

$H_0: \mu = \mu_0 = 30.43\%$,

$$t_{cal} = \frac{\overline{x} - \mu_0}{s / \sqrt{n}} = \frac{30.51\% - 30.43\%}{0.05\% / \sqrt{6}} = 3.9$$

$t_{crit} = t_{0.05}(5) = 2.57$

$|t_{cal}| \geqslant t_{crit}$

H_0 was rejected and there is systematic error with the experimental results.

3. The Test for Difference in Means

(1) Comparison of precision—F test

Sometimes, we need to compare the standard deviation of two populations. For example, the normal t test requires that the standard deviations of the data sets being compared are statistically equivalent. To illustrate, let us assume that n_1 replicate analyses by Analyst A yield a mean value of $\overline{x}_1$ and that n_2 analyses of Analyst B by the same method give $\overline{x}_2$. The F test compares two populations with standard deviations of σ_1 and σ_2 under the provision that the two populations follow the normal (Gaussian) distribution. The test statistic F is defined as the ratio of the two sample variances (s^2) with the larger variance always appearing in the numerator:

$$F = s_1^2 / s_2^2 \tag{2-17}$$

The test statistic F follows $F((n_1 - 1), (n_2 - 2))$ distribution with degrees of freedom, $(n_1 - 1)$ and $(n_2 - 2)$.

The null hypothesis for the F test is that the variances of the two populations under consideration are equal, $H_0 : \sigma_1^2 = \sigma_2^2$. Then the test statistic F is calculated and compared with the critical value at the desired significance level. Acceptance and rejection regions for F distribution are illustrated in Figure 2.12. Critical values of F at the 0.05 significance level are shown in Table 2.5 with two degrees of freedom, one for the numerator and the other for the denominator.

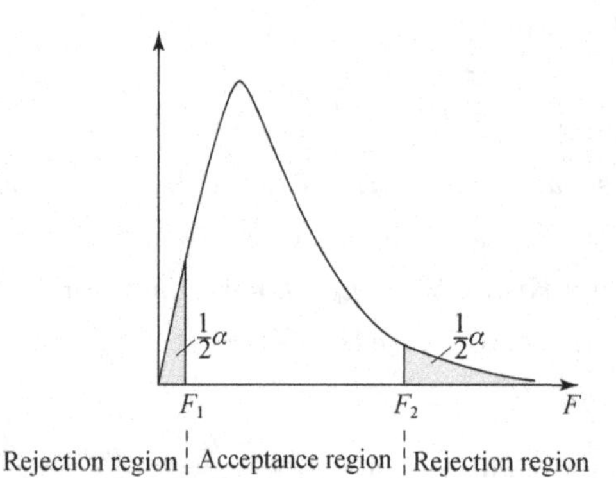

Figure 2.12 Acceptance and rejection regions for F distribution.

The F test can be used in either a one-tailed mode or a two-tailed mode. For a one-tailed test as described above, we test the alternative hypothesis that one variance is greater than the other. The alternative hypothesis is $H_a : \sigma_1^2 > \sigma_2^2$. For a two-tailed test, we test whether the variances are different, $H_0 : \sigma_1^2 \neq \sigma_2^2$. The significance level of the F values is doubled from 5% to 10%, but $F_{\alpha/2}$ table is still used.

Table 2.5 Critical Values of F at the 0.05 Significance Level (95% Confidence Level)

Degrees of Freedom (Denominator)	Degrees of Freedom (Numerator)									
	2	3	4	5	6	7	8	9	10	∞
2	19.00	19.16	19.25	19.30	19.33	19.35	19.37	19.38	19.40	19.50
3	9.55	9.28	9.12	9.01	8.94	8.89	8.85	8.81	8.79	8.53
4	6.94	6.59	6.39	6.26	6.16	6.09	6.04	6.00	5.96	5.63
5	5.79	5.41	5.19	5.05	4.95	4.88	4.82	4.77	4.74	4.37
6	5.14	4.76	4.53	4.39	4.28	4.21	4.15	4.10	4.06	3.67
7	4.74	4.35	4.12	3.97	3.87	3.79	3.73	3.68	3.64	3.23
8	4.46	4.07	3.84	3.69	3.58	3.50	3.44	3.39	3.35	2.93
9	4.26	3.86	3.63	3.48	3.37	3.29	3.23	3.18	3.14	2.71
10	4.10	3.71	3.48	3.33	3.22	3.14	3.07	3.02	2.98	2.54
∞	3.00	2.60	2.37	2.21	2.10	2.01	1.94	1.88	1.83	1.00

(2) Comparison of means—t test

Based on the provision that there is no significant difference between the standard deviation of the two data sets, t test is used to compare the means. The null hypothesis states that the two means are identical and that any difference is due to random error, thus, we can state $H_0: \mu_1 = \mu_2$.

$$t_{cal} = \frac{\bar{x}_1 - \bar{x}_2}{s_p}\sqrt{\frac{n_1 n_2}{n_1 + n_2}} \tag{2-18}$$

t is the test statistic of t distribution with degrees of freedom, $n_1 + n_2 - 2$. The s_p is the pooled standard deviation and calculated as follows:

$$s_p = \sqrt{\frac{(n_1 - 1)s_1^2 + (n_2 - 1)s_2^2}{n_1 + n_2 - 2}} \tag{2-19}$$

t_{cal} is then compared with critical value of t in Table 2.4 for the desired confidence level. If $|t_{cal}| > t_{crit}$, then reject H_0, there is a significant difference between these two means. If $|t_{cal}| < t_{crit}$, accept H_0, there is no significant difference between these two means.

【Example 2.7】 The results of two analytical methods for Na_2CO_3 content in an alkali lime sample are: $n_1 = 5, \bar{x}_1 = 42.34\%, s_1 = 0.10\%$ and $n_2 = 4$, $\bar{x}_2 = 42.44\%, s_2 = 0.12\%$.

Does the data indicate difference between the two methods at a 0.05 significance level?

Answer:

(1) Precision of two methods needs to be compared first with the one-tailed F test, thus the null hypothesis is $H_0: \sigma_1^2 = \sigma_2^2$.

$$F_{cal} = s_2^2/s_2^1 = 0.12^2/0.10^2 = 1.44$$

From Table 2.5, $F_{crit}(n_2-1, n_1-1) = F_{0.05}(3,4) = 6.59$, $F_{cal} < F_{crit}$; therefore, there is no significant difference on precision between these two methods.

(2) The population means are compared with two-tailed t test with the null hypothesis as $H_0: \mu_1 = \mu_2$.

$$s_p = \sqrt{\frac{(n_1-1)s_1^2 + (n_2-1)s_2^2}{n_1+n_2-2}} = \sqrt{\frac{(5-1)\times 0.10^2 + (4-1)\times 0.12^2}{5+4-2}}$$

$$= 0.109\%$$

$$t_{cal} = \frac{\bar{x}_1 - \bar{x}_2}{s_p}\sqrt{\frac{n_1 n_2}{n_1+n_2}} = \frac{42.34-42.44}{0.109}\sqrt{\frac{5\times 4}{5+4}} = -1.37$$

From Table 2.4, $t_{crit}(n_1+n_2-2) = t_{0.05}(7) = 2.37$. Then $|t_{cal}| < t_{crit}$, accept H_0. There is no difference between these two methods.

2.3.4 Detection of Gross Errors

When a set of data contains a result that appears to be outside the range of random errors in the procedure, then one should suspect an outlier. An outlier will produce a bias estimation of the population mean and the statistical test. The outlier can be excluded, if the cause is an individual mistake. Among the methods to judge an outlier, the Q test is a simple, widely used one to retain or reject a suspected result.

The procedure for a Q test:

(1) Arrange the data in ascending or descending order.

(2) Calculate the range of the set of data, R.

(3) Calculate the absolute value of difference between the suspected result and its nearest value, d. Q is calculated as

$$Q_{cal} = d/R \tag{2-20}$$

Q_{cal} is then compared with Q_{crit} in Table 2.6. If $Q_{cal} > Q_{crit}$, then the suspected result can be excluded at the desired confidence level.

Table 2.6 Critical Values for the Rejection Quotient, Q

Number of Observations	3	4	5	6	7	8	9	10
90% Confidence level	0.941	0.765	0.642	0.560	0.507	0.468	0.437	0.412
95% Confidence level	0.970	0.829	0.710	0.625	0.568	0.526	0.493	0.466
99% Confidence level	0.994	0.926	0.821	0.740	0.680	0.634	0.598	0.568

【Example 2.8】 The results of a volumetric assay of a solute are: 0.1014, 0.1012, 0.1016, and 0.1025 mol • L^{-1}. The last value appears suspicious; can this value be retained or rejected at the 90% confidence level?

Answer:

$R = 0.1025 - 0.1012 = 0.0013$ and $d = 0.1025 - 0.1016 = 0.0009$

Then $Q_{cal} = 0.0009/0.0013 = 0.691$

From Table 2.6, $Q_{crit} = 0.765$ ($n=4$ at 90% confidence level), thus $Q_{cal} < Q_{crit}$, 0.1025 can be retained at the 90% confidence level.

2.4 PROPAGATION OF ERROR

Generally, analytical results are calculated from two or more experimentally measured parameters, such as mass, volume, potential, and absorption, each of which may have a systematic error and some uncertainty. Errors from each measurement step can effect the analytical results. The effect of sequential multiple errors is called propagation of errors.

2.4.1 Systematic Errors

The propagation of system errors is summarized in Table 2.7.

Table 2.7 System Error Propagation in Arithmetic Calculation

Type of Calculation	Example	System Error of Result
Addition or subtraction	$R=A+B-C$ $R=A+mB-nC$	$E_{a(R)}=E_{a(A)}+E_{a(B)}-E_{a(C)}$ $E_{a(R)}=E_{a(A)}+m E_{a(B)}-nE_{a(C)}$
Multiplication or division	$R=AB/C$ $R=m(AB/C)$	$E_{r(R)}=E_{r(A)}+E_{r(B)}-E_{r(C)}$
Logarithm	$R=k+n\lg A$	$E_{a(R)}=0.434n(E_{a(A)}/A)$
Exponentiation	$R=k+A^n$	$E_{r(R)}=n E_{r(A)}$

Note: E_a denotes the absolute error and E_r denotes the relative error.

2.4.2 Random Errors (Standard Deviation)

The propagation of random errors is summarized in Table 2.8.

Table 2.8 Random Error Propagation in Arithmetic Calculation

Type of Calculations	Examples	System Error of Result
Addition or subtraction	$R=aA+bB-cC$	$s_R=\sqrt{a^2s_A^2+b^2s_B^2+c^2s_C^2}$
Multiplication or division	$R=AB/C$ $R=m(AB/C)$	$s_R=R\sqrt{\frac{s_A^2}{A^2}+\frac{s_B^2}{B^2}+\frac{s_C^2}{C^2}}$
Logarithm	$R=k+n\lg A$	$s_R=0.434n(s_A/A)$
Exponentiation	$R=k+A^n$	$s_R=n(s_A/A)$

2.4.3 Maximum Errors (E_R)

If the calculation of the propagation systematic errors has the same sign, the error can be as large as the sum of the individual errors, range error (E_R). On the other hand, it is possible that the individual errors could combine to give a collective value of less than E_R. E_R and its relative value are calculated as follows:

$$E_R = |E_A|+|E_B|+|E_C| \qquad (2\text{-}21)$$

$$\frac{E_R}{R} = \left|\frac{E_A}{A}\right|+\left|\frac{E_B}{B}\right|+\left|\frac{E_C}{C}\right| \qquad (2\text{-}22)$$

2.4.4 Distribution of Errors

In an analysis, the error of each analytical step shall be within the desired range. In a volumetric analysis, the error should be less than 0.2%, then the error in each of the weighing and volumetric steps should be less than 0.1%. When using an analytical balance, the variability of the delivered mass is 0.2 mg, therefore, the weighed sample mass shall not be smaller than 200 mg to ensure that weighing error is less than 0.1%. For titrations, the variability of the delivered volume is 0.01 mL, therefore, the titrant used shall not be smaller than 20 mL.

【Example 2.9】 Determine the barium content (%) in a sample using precipitation gravimetry. A 0.4503 g sample was weighed to give a 0.4291 g of $BaSO_4$ precipitate. If the standard deviation of weighing is $s=0.1$ mg, calculate standard deviation for the analytical result (s_w).

Answer:

$$w(\text{Ba}) = \frac{m_2 \cdot \frac{A_r(\text{Ba})}{M_r(\text{BaSO}_4)}}{m_1}\times 100\% = \frac{0.4291\times\frac{137.33}{233.39}}{0.4503}\times 100\% = 56.07\%$$

Calculate the error of analytical result using

$$\frac{s_w^2}{w^2} = \frac{s_{m_2}^2}{m_2^2}+\frac{s_{m_1}^2}{m_1^2}$$

m_1 and m_2 were from weighing the sample and the precipitate. The precipitate needs to be heated and weighed at least two time intervals to obtain a constant weight. Hence,

$$s_{m_1}^2 = s_{m_{1(2)}}^2 + s_{m_{1(1)}}^2 = 2s^2 \quad \text{and} \quad s_{m_2}^2 = s_{m_{2(4)}}^2 + s_{m_{2(3)}}^2 + s_{m_{2(2)}}^2 + s_{m_{2(1)}}^2 = 4s^2$$

Therefore,

$$\frac{s_w^2}{w^2} = \frac{4s^2}{m_2^2} + \frac{2s^2}{m_1^2} = 4\left(\frac{s}{m_2}\right)^2 + 2\left(\frac{s}{m_1}\right)^2 = 4\left(\frac{0.1}{429.1}\right)^2 + 2\left(\frac{0.1}{450.3}\right)^2$$

$$= 3.16 \times 10^{-7}$$

$$\frac{s_w}{w} = 5.62 \times 10^{-4}$$

$$s_w = 5.62 \times 10^{-4} \times 56.07\% = 0.032\%$$

2.5 SIGNIFICANT FIGURE CONVENTION

2.5.1 Significant Figures

To ensure the quality of analytical results, one needs to use the number of digits that reflect the precision and accuracy of the measurements. The number of digits indicates the uncertainty in the experimental measurements. Data must contain only the number of significant figures. By definition, significant figures are the number of digits that can be obtained from the measurements.

The number of significant figures is all the digits known with certainty plus the first uncertain digit associated with analytical methods and instrumental accuracy. For example, if one sample mass weighed using an analytical balance is 0.5000 g, the forth digit after decimal point is uncertain. The relative error for weighing is

$$\frac{\pm 0.0002 \text{ g}}{0.5000 \text{ g}} \times 100\% = \pm 40\%$$

If a sample mass weighed on a balance is 0.5 g with an uncertainty of 0.1 g, then relative error for weighing is

$$\frac{\pm 0.2 \text{ g}}{0.5 \text{ g}} \times 100\% = \pm 40\%$$

If the volume of a solution is reported as 24 mL, then measuring with a graduated cylinder would be sufficiently accurate. The same volume measured with a buret has two more significant digits and is reported as 24.00 mL.

【Example 2.10】 Back titration was used to determine an acid sample. To obtain necessary accuracy, 40.00 mL of 0.1000 mol · L^{-1} NaOH was added, and the excess NaOH was back titrated with 39.10 mL of HCl with the same concentration as NaOH. Do you think reporting the result as 10.12% appropriate?

Answer: The number of significant figures is not correct. The amount of NaOH reacting with the acid is only (40.00−39.10) mL=0.90 mL. The relative error by volume is $\frac{\pm 0.02\ \text{mL}}{0.90\ \text{mL}} \times 100\% = \pm 2\%$. Therefore, the correct reporting of the result is 10.1%.

A zero may or may not be significant, depending on the location in a number. A zero without any other number before the decimal point is not significant, e.g., 0.1. On the other hand, a zero surrounded by other digits is always significant. For example, the two zeros in 20.30 mL obtained from titration with a buret are both significant. If the volume is rewritten as 0.02030 L, the only function of the first two zeros in the number before the 2 is to locate the decimal point. The number of significant figures is still four. Terminal zeros may or may not be significant. For example, a mass reported as 25.0 mg means the mass has three significant figures. If this mass is written in micrograms, the mass must be reported as 2.50×10^4 g. Reporting the same mass as 25000 μg is not correct.

Factors in analytical calculations are considered as numbers with unlimited significant figures. When you take the logarithm of a number, such as pH, pM, and lg*K*, the number of significant figures is the number of digits to the right of the decimal point. The numbers to the left of the decimal point is only an indication of the power of exponent. From pH 11.02, we can calculate $[H^+]=9.6\times10^{-12}$ mol · L^{-1}. In this case, pH 11.02 has two significant figures instead of four.

In a calculation, a number starting with a digit equal to or greater than eight can be considered one more digit of its actual significant figures. For example, 0.0985 can be considered as a number of four significant figures.

The rounding convention:

- A number can be rounded up when ending in a digit greater than 5 and can be rounded down when ending in a digit smaller than five.
- A number ending in 5 shall be rounded up to a result that ends with an even number.

For example, the following numbers are rounded to have four significant figures:

0.52664→0.5266, 0.36266→0.3627, 10.2350→10.24,
2.50650→2.506, and 18.0852→18.09

2.5.2 Numerical Rounding in Calculations

In calculating an analytical result, errors from each measurement are propagated to the reported result. The calculation must be carried out in a way that no significant figure is lost. It is important to postpone rounding until calculation is complete. During calculation, at least one extra digit beyond the significant figures should be carried through all of the computations to avoid a rounding error. Rounding in each step is not necessary in calculations using calculator or computer. Rounding is performed on the final result according to error propagation convention.

1. Rounding in Addition or Subtraction

The result of addition or subtraction obtains the error from the absolute error of each term in the calculation. The number of digits after decimal point for the calculated result shall be taken according to the one with largest absolute error, i. e. , the term with least number of digits after decimal point. For example, in calculating 50.1+1.45+0.5812, 50.1 has the largest absolute error; therefore, the result shall have only one digit after decimal point. First, the individual terms can be rounded to one digit after decimal point. Then the addition can be made and the final result is 52.1.

2. Rounding in Multiplication or Division

The result of multiplication or division obtains the relative error of each term in the calculation. The number of digits for the calculated result shall be the same as the one with the largest relative error. For example, in calculating 0.0121×25.64×1.05782, 0.0121 has the least number of significant figures; therefore, the result shall have three significant figures. First, the individual terms can be rounded to three significant figures. Then the multiplication can be completed and the final result is 0.328.

In a calculation of a chemical equilibrium, two significant figures are enough to report the final results because equilibrium constant determinations often have two significant figures. In gravimetric or volumetric analysis, the relative error is about 0.1% and four significant figures are reported. For entities with content greater than 80%, three significant digits are reported to be consistent with 0.1% relative error.

Chapter 2 Questions and Problems

2.1 Differentiate between

(1) accuracy and precision;

(2) random error and systematic error.

2.2 Name the types of systematic errors.

2.3 What are the probability, confidence level, and confidence interval?

2.4 What is the difference between Gaussian distribution and the t-distribution?

2.5 The color change of a chemical indicator requires an extra 0.02 mL of titrant. Calculate the relative error, if the stoichiometric volume of titrant is as listed in the table below assuming that the concentrations of titrant and titrand are the same.

V/mL	5.0	10.0	20.0	30.0
E_r/%				

2.6 A loss of 0.2 mg of Zn occurs in the course of an analysis for the zinc element. Calculate the relative error due to this loss, if the mass of Zn in the sample is as listed in the table below.

m/mg	20	100	200	400
E_r/%				

2.7 Consider the following sets of replicate measurements in the table below:

Samples	A	B	C	D	E
x_i	20.48	0.792	2.6	70.21	0.472
	20.55	0.794	2.7	70.63	0.486
	20.58	0.812	2.8	70.64	0.497
	20.60	0.900	3.0	70.65	0.503
	20.53		3.2		
	20.50				

For each set, calculate the (1) mean; (2) median; (3) range, or spread; (4) standard deviation; (5) coefficient of variation; and (6) the 95% confidence interval.

2.8 The percentage of manganese (Mn) present in an ore was found to be 9.56% with the standard deviation s = 0.12%. If s is a good estimate of σ, calculate the 95% confidence intervals for the result based on (1) single analysis; (2) the mean of 4 analyses, (3) the mean of 9 analyses.

2.9 The analytical results of calcium (Ca) in a sample based on colorimetry and atomic absorption spectrometry are given in the table below.

Replicates	1	2	3	4	5
Colorimetry / ppm	3.92	3.28	4.18	3.53	3.35
Atomic Absorption / ppm	4.40	4.92	3.51	3.97	4.59

Determine whether there are differences in the two methods at the 95% confidence levels.

2.10 Apply the Q test to the following data sets to determine whether the outlying results should be retained or rejected at the 95% confidence level.

(1) 41.27, 41.61, 41.84, 41.70; (2) 7.295, 7.284, 7.388, 7.292.

2.11 How many grams of Na_2CO_3 should be weighed to standardize 25 mL of 0.1 mol · L^{-1} HCl? Estimate the accuracy from sample weighing error standpoint. What will the relative error be, if sodium borate decahydrate ($Na_2B_4O_7 \cdot 10H_2O$) is used for the standardization?

2.12 A back complexometric titration is carried out to determine aluminum (Al, in w/w%) in a sample. A 25.00 mL of 0.02002 mol · L^{-1} EDTA is added to a 0.2000 g test sample. A 23.12 mL of 0.02012 mol · L^{-1} Zn^{2+} is used to back titrate excess EDTA. Calculate percentage of Al in the sample (%, w/w). How many significant digits are correct for description of the result? How could one improve accuracy of this determination?

CHAPTER 3

ACID-BASE EQUILIBRIUM

3.1 Equilibrium Constants and Effect of Electrolytes

3.2 Acid-base Reactions and Equilibria

3.2.1 Acid and Base—Brønsted Concept

3.2.2 Dissociation of Acid or Base and Acid-base Equilibria

3.2.3 Magnitude of Dissociating Species at a Given pH: x-values

3.3 Solving Equilibrium Calculations Using pH Calculations as an Example

3.3.1 General Approaches (Systematic Approaches)

3.3.2 pH Calculations

3.4 Buffer Solutions

3.4.1 pH Calculations of Buffer Solutions

3.4.2 Buffer Capacity

3.4.3 Preparation of Buffers

Most analytical techniques are based on chemical reactions and require the state of chemical equilibrium. All reactions will reach a state of dynamic equilibrium when the state of the reaction is in its forward direction. In most cases, an analyst must know where the equilibrium point lies to develop a successful analytical process to assure that the reaction proceeds toward >99.9% completion (or conversely remains >99.9% as reactants). This chapter provides a fundamental approach to chemical equilibria, starting with acid-base equilibrium because pH calculation and pH buffers are extremely important in many areas of science. Details of the equilibrium state for particular reactions will be discussed in the subsequent four chapters, which deal with quantitation methods.

3.1 EQUILIBRIUM CONSTANTS AND EFFECT OF ELECTROLYTES

When a reaction, $a\mathrm{A}+b\mathrm{B} \rightleftharpoons c\mathrm{C}+d\mathrm{D}$, reaches equilibrium, the equilibrium state can be represented mathematically by

$$K^{\mathrm{C}} = \frac{[\mathrm{C}]^c \cdot [\mathrm{D}]^d}{[\mathrm{A}]^a \cdot [\mathrm{B}]^b} \tag{3-1}$$

where K^{C} is the equilibrium constant with the concentrations in the expression as the equilibrium concentration of A, B, C, and D. Equation (3-1) is only an approximate form of a thermodynamic equilibrium constant expression. The K^{C} value is also affected by ionic strength in addition to temperature. The rigorous form is given by Equation (3-2)

$$K^{\ominus} = \frac{a^c(\mathrm{C}) \cdot a^d(\mathrm{D})}{a^a(\mathrm{A}) \cdot a^b(\mathrm{B})} \tag{3-2}$$

$K^{\ominus}$ is the thermodynamic constant which is independent of ionic strength and dependent on temperature and pressure. $a(\mathrm{A})$, $a(\mathrm{B})$, $a(\mathrm{C})$, and $a(\mathrm{D})$ are the activity (effective concentration) of the species in the reaction. For example, the activity of the species X, $a(\mathrm{X})$, depends on the ionic strength in the medium and is defined by

$$a(\mathrm{X}) = \gamma(\mathrm{X}) \cdot [\mathrm{X}] \tag{3-3}$$

where $\gamma(\mathrm{X})$ is activity coefficient, and $[\mathrm{X}]$ is its molar concentration. The activity coefficient and thus the activity vary with concentration of electrolytes in the medium such that substitution of activity for equilibrium concentration in Equation (3-1) makes the equilibrium constant $K^{\ominus}$ independent of the concentration of electrolytes.

In a very dilute solution, the concentration of electrolytes is very small and the activity coefficient of the species is close to unity. In this situation the activity is approximately equal to the concentration of the solute. The activity coefficient of the neutral solute without net charges is considered to be unity.

In 1923, P. Debye and Hückel derived an equation using ionic atmosphere model to calculate the activity coefficients of ions based on their charge and average size. This equation, known as Debye-Hückel equation, is

$$-\lg\gamma(\mathrm{X}) = \frac{0.509z^2(\mathrm{X})\sqrt{I}}{1 + B\mathring{a}\sqrt{I}} \quad (I < 0.1\ \mathrm{mol \cdot L^{-1}}) \tag{3-4}$$

where $\gamma(\mathrm{X})$ is activity of the species X, $z(\mathrm{X})$ is charge on the species X, I is ionic strength of the solution, $\mathring{a}$ is effective diameter of the hydrated ion X in nanometers. At 25℃, $B=3.28$. $\mathring{a}$ and γ values of selected species are given in Appendix C.

The ionic strength (I) is a measure of total electrolyte concentration and charges, and is defined by

$$I = \frac{1}{2}(c_1 z_1^2 + c_2 z_2^2 + \cdots) = \frac{1}{2}\sum(c_i z_i^2) \tag{3-5}$$

where $c_1, c_2, \cdots$, represent the concentration of individual ions in the solution and $z_1, z_2, \cdots$, are their charges.

For I in the range of 0.1~0.6 mol · L^{-1}, the activity coefficient is calculated by Davis equation

$$-\lg\gamma(\mathrm{X}) = 0.509z^2(\mathrm{X})\left(\frac{\sqrt{I}}{1+\sqrt{I}} - 0.2I\right) \tag{3-6}$$

Figure 3.1 shows the activity coefficient of ions as a function of square root of ionic strength (I) with same effective diameter of the hydrated ion. From Figure 3.1, one can calculate: (1) as the solution approaches infinite dilution, $I \to 0, \gamma \to 1$; (2) when I increases, γ decreases, and γ of an ion with greater number of charges decreases more rapidly than an ion with smaller number of charges; (3) for a given I, the activity coefficient of an ion departs farther from unity as the number charges carried by the species increases; (4) the change of γ is much less significant when $I > 0.1$ mol · L^{-1} than when $I < 0.1$ mol · L^{-1}; thus, γ at $I = 0.1$ mol · L^{-1} is often used for calculation when $I > 0.1$ mol · L^{-1}.

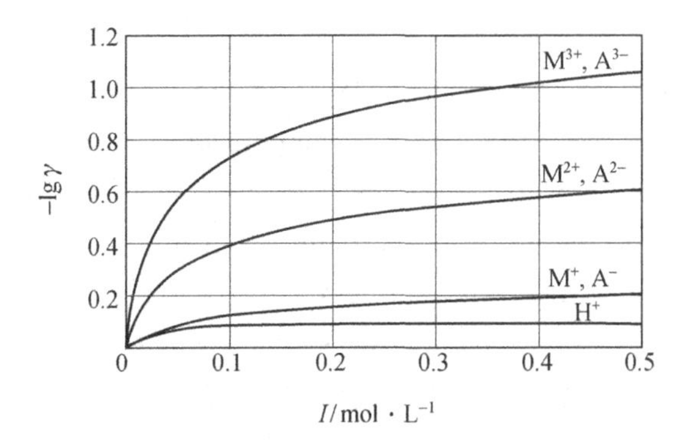

Figure 3.1 Effect of ionic strength on activity coefficients.

3.2 ACID-BASE REACTIONS AND EQUILIBRIA

3.2.1 Acid and Base—Brønsted Concept

In 1923, J. N. Brønsted in Denmark and J. M. Lowry in England, proposed independently a theory of acidity and alkalinity, which is valid in all solvents and is particularly useful in analytical chemistry. In Brønsted-Lowry theory, an **acid** is a substance capable of donating a proton to another substance and a **base** is a substance that accepts a proton. Acids can be cationic, anionic, or electrically neutral. The same is true for bases. The product formed when an acid gives a proton is called the **conjugate base** of the parent acid. Similarly, the base produces a conjugate acid when accepting a proton. Examples in aqueous solution are:

Acid	$\rightleftharpoons$	Base	+	Proton
HAc	$\rightleftharpoons$	Ac^-	+	H^+
NH_4^+	$\rightleftharpoons$	NH_3	+	H^+
H_2CO_3	$\rightleftharpoons$	HCO_3^-	+	H^+
HCO_3^-	$\rightleftharpoons$	CO_3^{2-}	+	H^+
$(CH_2)_6N_4H^+$	$\rightleftharpoons$	$(CH_2)_6N_4$	+	H^+

In aqueous solution, H_2O can act as both the proton donor and acceptor to induce basic or acidic behavior in solutes. The proton in the above examples is present in solvated form (hydrated form). The conjugate acid of water is written as H_3O^+, and called **hydronium ion** or **hydrogen ion**. The simplified form of the hydronium ion, H^+, is used in writing chemical equations in which the proton is involved.

Species, such as HSO_3^-, HCO_3^-, HS^-, $H_2PO_4^-$, HPO_4^{2-}, and HSO_3^-, that possess both acidic and basic properties are **amphiprotic.**

Water is the classic example of an amphiprotic solvent that can act either as an acid or as a base depending on the solute. Amphiprotic solvent undergoes **autoprotolysis** to form a pair of ionic species:

$Base_1$	+	$Acid_2$	$\rightleftharpoons$	$Acid_1$	+	$Base_2$
H_2O	+	H_2O	$\rightleftharpoons$	H_3O^+	+	OH^-
CH_3OH	+	CH_3OH	$\rightleftharpoons$	$CH_3OH_2^+$	+	CH_3O^-
CH_3COOH	+	CH_3COOH	$\rightleftharpoons$	$CH_3COOH_2^+$	+	CH_3COO^-
NH_3	+	NH_3	$\rightleftharpoons$	NH_4^+	+	NH_2^-

The dissociate constant of water, K_w, is an autoprotolysis constant, also called **ion-product constant.**

$$K_w = \frac{a(H_3O^+) \cdot a(OH^-)}{a^2(H_2O)} \tag{3-7}$$

The activity of water is constant because this solution is very dilute for hydrogen and hydroxide ions and can be considered as to be very close to the standard state of pure water. Therefore,

$$K_w = a(H_3O^+) \cdot a(OH^-) = 1.0 \times 10^{-14} \quad (25℃) \tag{3-8}$$

Because $pH = -\lg a(H^+)$ and $pOH = -\lg a(OH^-)$, we have

$$pH + pOH = pK_w = 14.00 \quad (25℃) \tag{3-9}$$

The autoprotolysis of H_2O is of utmost importance in understanding the behavior of aqueous solutions. The tendency of a solvent to accept or donate protons determines the strength of a solute acid or base dissolved in the solvent. Figure 3.2 shows the conjugate acid-base pairs arranged according to the strength of the acid. The strength of their conjugate bases is in the reversed order. HCl and $HClO_4$ are strong acids, because these acids are completely dissociated in water. These two acids show no difference in strength in water; thus, water is a **leveling solvent** for $HClO_4$ and HCl. The remainder in Figure 3.2 are weak acids, which dissociate partially to yield solutions containing significant quantities of both the parent acid and its conjugate base. Water is a **differentiating solvent** for them. If glacial acetic acid is used as a solvent, neither $HClO_4$ nor HCl undergo complete dissociation. $HClO_4$ ($pK_a = 4.87$, 25℃) is considerably stronger than HCl ($pK_a = 8.55$, 25℃). Glacial acetic acid acts as a differentiating solvent toward these two acids by revealing the inherent differences in their acidities.

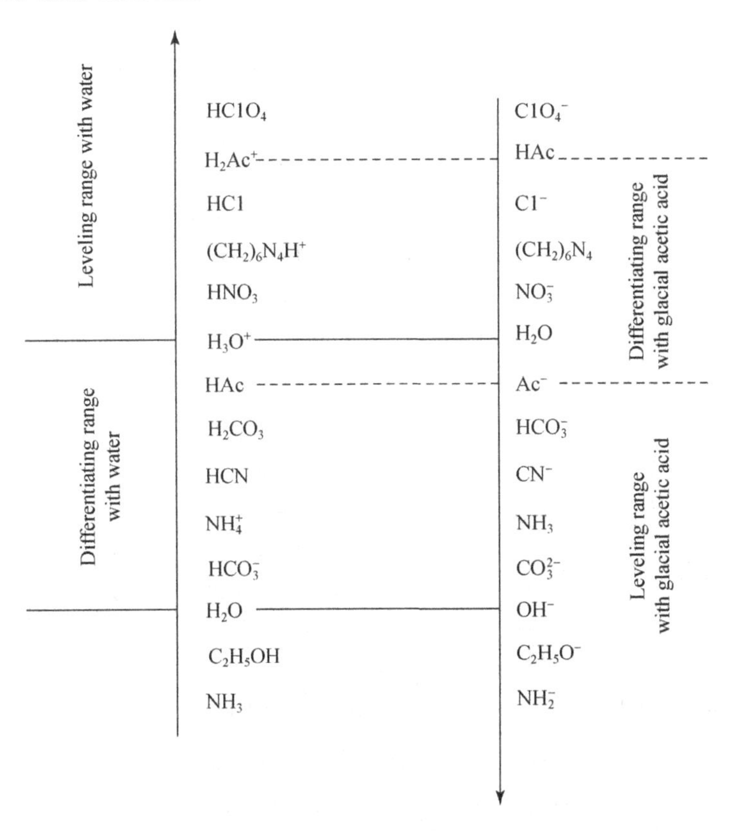

Figure 3. 2 The conjugate acid-base pairs and the leveling and differentiating range of water and glacial acetic acid.

3.2.2 Dissociation of Acid or Base and Acid-base Equilibria

An acid can not dissociate by itself. A substance acts as an acid to dissociate only in the presence of a base, and vice versa.

For example, the dissociation reaction of acetic acid (HAc) in aqueous solution is

$$HAc + H_2O \rightleftharpoons H_3O^+ + Ac^-$$

In this reaction, the solvent water accepts one proton from HAc to become its conjugate acid, H_3O^+. HAc gives one proton to become its conjugate base, Ac^-.

The dissociation constant of weak acid (HA), K_a, is calculated as

$$K_a = \frac{a(H^+) \cdot a(A^-)}{a(HA)} \tag{3-10}$$

K_a value indicates the strength of an acid. The larger the K_a is, the stronger the acid is.

The dissociation reaction of a weak base, acetate (Ac^-) in aqueous solution is

$$Ac^- + H_2O \rightleftharpoons HAc + OH^-$$

H_2O gives one proton to become its conjugate base, OH^-. Ac^- accepts the proton released from H_2O to become its conjugate acid, HAc.

The dissociation constant of weak base (A^-), K_b, is calculated as

$$K_b = \frac{a(HA) \cdot a(OH^-)}{a(A^-)} \tag{3-11}$$

K_a and K_b are thermodynamic constants and depend on only temperature. In practical applications, concentrations are used instead of activities. The equilibrium constant based on concentration (K_a^C) will be affected not only by temperature but also by the ionic strength of the solution.

For HA, the dissociation constant (K_a^C) is calculated by

$$K_a^C = \frac{[H^+] \cdot [A^-]}{[HA]} = \frac{a(H^+) \cdot a(A^-)}{a(HA)} \cdot \frac{\gamma(HA)}{\gamma(H^+) \cdot \gamma(A^-)}$$

$$= K_a^{\ominus} \cdot \frac{\gamma(HA)}{\gamma(H^+) \cdot \gamma(A^-)} \tag{3-12}$$

If $a(H^+)$ (or $a(OH^-)$) is used to substitute for $[H^+]$ (or $[OH^-]$), we get a constant that is called a mixed constant (K_a^M).

$$K_a^M = \frac{a(H^+) \cdot [A^-]}{[HA]} = K_a/\gamma(A^-) \tag{3-13}$$

In practical applications, $a(H^+)$ is widely measured with pH meter, thus, the equilibrium constant measured is a mixed constant that is convenient for calculations.

The thermodynamic constants (activity constant, $I = 0 \text{ mol} \cdot L^{-1}$) and mixed constants ($I = 0.1 \text{ mol} \cdot L^{-1}$) of dissociation of weak acids and bases are listed in Appendix D1. For diluted solutions, the activity constant is used for calculation. In acid-base titration as well as the solubility calculations of precipitate in pure water, the activity constants are used for calculation. In most other calculations, especially when a pH buffer is involved, mixed constants are often used. For rigorous results in calculating the pH of standard buffer solution, the effect of ionic strength must be considered. For a conjugate acid-base pair,

$$K_a \cdot K_b = \frac{a(H^+) \cdot a(A^-)}{a(HA)} \cdot \frac{a(HA) \cdot a(OH^-)}{a(A^-)} = a(H^+) \cdot a(OH^-)$$

$$= K_w = 1.0 \times 10^{-14} \quad (25^\circ C) \tag{3-14}$$

or $$pK_a + pK_b = pK_w = 14.00 \quad (25℃) \tag{3-15}$$

Many acids or bases are polyfunctional, which means they can give away or accept more than one protons in a stepwise manner. For H_3A, there are three dissociation reactions, thus there are three dissociation constants:

$$H_3A \underset{K_{b3}}{\overset{K_{a1}}{\rightleftharpoons}} H_2A^- \underset{K_{b2}}{\overset{K_{a2}}{\rightleftharpoons}} HA^{2-} \underset{K_{b1}}{\overset{K_{a3}}{\rightleftharpoons}} A^{3-}$$

The dissociation constants, K_{a1}, K_{a2}, and K_{a3}, are generally $K_{a1} > K_{a2} > K_{a3}$.

For the polyfunctional base (A^{3-}) the dissociation constants are K_{b1}, K_{b2}, and K_{b3} that are generally $K_{b1} > K_{b2} > K_{b3}$.

According to Equation (3-15), we have

$$K_{b1} = K_w/K_{a3}, \quad pK_{b1} = 14.00 - pK_{a3}$$
$$K_{b2} = K_w/K_{a2}, \quad pK_{b2} = 14.00 - pK_{a2}$$
$$K_{b3} = K_w/K_{a1}, \quad pK_{b3} = 14.00 - pK_{a1}$$

For amphiprotic species in aqueous solution, such as HA^-, the dissociation reactions are

$$HA^- + H_2O \overset{K_{a2}}{\rightleftharpoons} H_3O^+ + A^{2-}$$
$$HA^- + H_2O \overset{K_{b2}}{\rightleftharpoons} H_2A + OH^-$$

Whether a solution of HA^- will be acidic or basic depends on the relative magnitude of the equilibrium constants for the above dissociations.

3.2.3 Magnitude of Dissociating Species at a Given pH: *x*-values

Most reagents used in analytical chemistry (precipitation, complexation and extraction agents) are weak acids or bases. pH changes result in the change of ratio of the acid/base forms. Therefore, the completeness of a specific reaction depends on the concentration of a specific form of reactant that is a function of pH.

For a monoprotic acid (HA) if we let c be the analytical concentration (total concentration) of HA, then $c = [HA] + [A^-]$. If we define x_0 as the fraction of A^- and x_1 as the fraction of HA in the solution at the given pH, then we have

$$x_1 = x(HA) = \frac{[HA]}{c} = \frac{[HA]}{[HA] + [A^-]}$$
$$= \frac{[H^+]}{[HA] + \dfrac{[HA] \cdot K_a}{[H^+]}} = \frac{[H^+]}{[H^+] + K_a}$$

$$x_0 = x(A^-) = \frac{[A^-]}{c} = \frac{K_a}{[H^+]+K_a} \tag{3-16}$$

$$x_1 + x_0 = 1$$

x values depend only on $[H^+]$ and K_a, and are independent of c. As an example, the plot of x_1 and x_0 values as a function of pH for acetic acid and its conjugate base are given in Figure 3. 3: The two curves cross at $pH = pK_a$, where $[HA]=[A^-]=0.5$. When $pH < pK_a$, $x_1 > x_0$, acid form is predominates. When $pH > pK_a$, $x_1 < x_0$, base form is predominates. A buffer solution contains an acid and its conjugate base. The effective pH for a buffer is usually $pK_a \pm 1$.

A ladder diagram (also predominance-area diagram) is another convenient graphical tool to illustrate the distribution of acid/base forms changing with pH. For a better page layout, it is presented horizontally instead of vertically. For HAc, the ladder diagram is illustrated in Figure 3. 3.

F^- and CN^- are widely applied to masking of metal ions in analytical methods. To obtain an effective masking result, pH must be in a range where F^- and CN^- are the predominant forms. By referring to the fraction diagram or ladder diagram, we know that for masking with HF ($pK_a = 3.17$), the pH needs to be greater than 3. 17, and for masking Cu^{2+} or Zn^{2+} with hydrogen cyanide ($HCN, pK_a = 9.31$), the pH must be greater than 10 to avoid volatilization of the very toxic HCN.

【Example 3. 1】 For a HAc($pK_a = 4.76$) solution with $c = 0.10\ mol \cdot L^{-1}$, calculate [HAc] and $[Ac^-]$ at $pH = pK_a - 2.00$, $pK_a - 1.30$, $pK_a - 1.00$, pK_a, $pK_a + 1.00$, $pK_a + 1.30$, and $pK_a + 2.00$, respectively.

Answer:

At $pK_a - 2.00 = 2.76$, $x(HAc) = \frac{10^{-2.76}}{10^{-2.76} + 10^{-4.76}} = 0.99$, and $x(Ac^-) = 1.00 - 0.99 = 0.01$. Hence,

$$[HAc] = x(HAc) \cdot c(HAc) = 0.99 \times 0.10 = 0.099\ (mol \cdot L^{-1})$$

$$[Ac^-] = 0.01 \times 0.10 = 0.001\ (mol \cdot L^{-1})$$

The x values and the concentrations of HAc and Ac^- at selected pH are summarized in the table below:

pH	$pK_a-2.00$	$pK_a-1.30$	$pK_a-1.00$	pK_a	$pK_a+1.00$	$pK_a+1.30$	$pK_a+2.00$
	2.76	3.46	3.76	4.76	5.76	6.06	6.76
x_1	0.99	0.95	0.91	0.50	0.09	0.05	0.01
x_0	0.01	0.05	0.09	0.50	0.92	0.95	0.99
[HAc]/mol · L^{-1}	0.099	0.095	0.091	0.050	0.009	0.005	0.001
$[Ac^-]$/mol · L^{-1}	0.001	0.005	0.009	0.050	0.091	0.095	0.099

The x-pH function and ladder diagram for acetic acid and its conjugate base are illustrated in Figure 3.3.

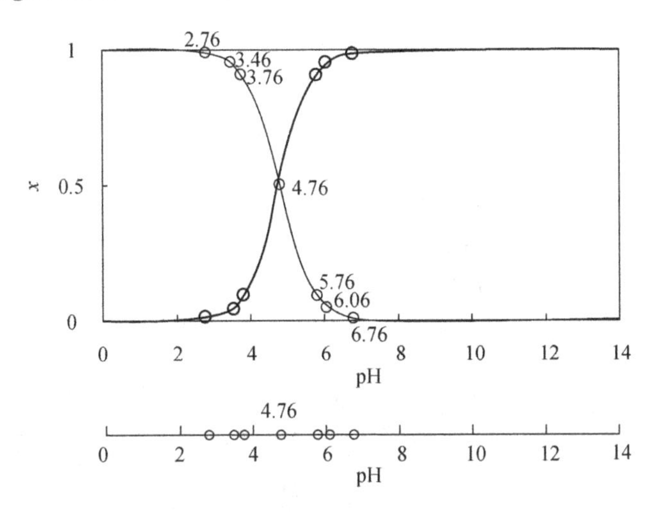

Figure 3.3 x-pH curves and ladder diagram of acetic acid (HAc) and its conjugate base (Ac^-).

For any weak monoprotic acid and its conjugate base, the profile of x-pH is the same as acetic acid and its salt by adjusting the cross point to the pK_a value of the specific acid.

For a diprotic acid (H_2A), x_0, x_1, and x_2 stand for the molar fraction of A^{2-}, HA^-, H_2A in a solution at a given pH with analytical concentration of c, we have

$$c = [H_2A] + [HA^-] + [A^{2-}] = [H_2A] \cdot \left(1 + \frac{K_{a1}}{[H^+]} + \frac{K_{a1} \cdot K_{a2}}{[H^+]^2}\right)$$

Then,

$$x_2 = \frac{[H_2A]}{c} = \frac{1}{1 + \frac{K_{a1}}{[H^+]} + \frac{K_{a1} \cdot K_{a2}}{[H^+]^2}} = \frac{[H^+]^2}{[H^+]^2 + [H^+]K_{a1} + K_{a1} \cdot K_{a2}}$$

$$x_1 = \frac{[HA^-]}{c} = \frac{[H^+]K_{a1}}{[H^+]^2 + [H^+]K_{a1} + K_{a1} \cdot K_{a2}}$$

$$x_0 = \frac{[A^{2-}]}{c} = \frac{K_{a1} \cdot K_{a2}}{[H^+]^2 + [H^+]K_{a1} + K_{a1} \cdot K_{a2}}$$

$$x_2 + x_1 + x_0 = 1$$

At $pH = \frac{1}{2}(pK_{a1} + pK_{a2})$, $[HA^-]$ reaches maximum, and $[H_2A] = [A^{2-}]$. With x values, the concentration of the acid and its salt can be calculated as $[H_2A] = c \cdot x_2$, $[HA^-] = c \cdot x_1$, and $[A^{2-}] = c \cdot x_0$.

For tartaric acid ($pK_{a1} = 3.04$ and $pK_{a2} = 4.37$), the x-pH profile and ladder diagram are illustrated in Figure 3.4. The difference between pK_{a1} and pK_{a2} is only 1.3 pH unit. When $[HA^-]$ reaches maximum, H_2A and A^{2-} each is 15% of the total amount. For tartaric acid, it is impossible to have a pH where $x_1 > 0.99$.

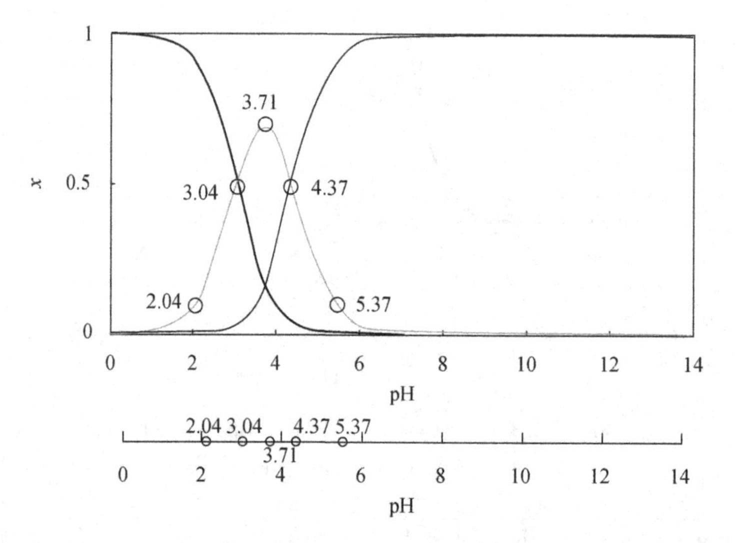

Figure 3.4 x-pH curves and ladder diagram of tartaric acid and its salts.

For a triprotic acid (H_3A), $c = [H_3A] + [H_2A^-] + [HA^{2-}] + [A^{3-}]$. We have

$$x_3 = \frac{[H_3A]}{c} = \frac{[H^+]^3}{[H^+]^3 + [H^+]^2 \cdot K_{a1} + [H^+] \cdot K_{a1} \cdot K_{a2} + K_{a1} \cdot K_{a2} \cdot K_{a3}}$$

$$x_2 = \frac{[H_2A^-]}{c} = \frac{[H^+]^2 \cdot K_{a1}}{[H^+]^3 + [H^+]^2 \cdot K_{a1} + [H^+] \cdot K_{a1} \cdot K_{a2} + K_{a1} \cdot K_{a2} \cdot K_{a3}}$$

$$x_1 = \frac{[HA^{2-}]}{c} = \frac{[H^+] \cdot K_{a1} \cdot K_{a2}}{[H^+]^3 + [H^+]^2 \cdot K_{a1} + [H^+] \cdot K_{a1} \cdot K_{a2} + K_{a1} \cdot K_{a2} \cdot K_{a3}}$$

$$x_0 = \frac{[A^{3-}]}{c} = \frac{K_{a1} \cdot K_{a2} \cdot K_{a3}}{[H^+]^3 + [H^+]^2 \cdot K_{a1} + [H^+] \cdot K_{a1} \cdot K_{a2} + K_{a1} \cdot K_{a2} \cdot K_{a3}}$$

For phosphoric acid ($pK_{a1} = 2.16$, $pK_{a2} = 7.21$, and $pK_{a3} = 12.32$), the x-pH profile and ladder diagram are illustrated in Figure 3.5. The difference between adjacent pK_a values is about 5, which is much larger than that for tartaric acid. This makes stepwise titration of H_3PO_4 feasible because at $pH=\frac{1}{2}(pK_{a1}+pK_{a2})$ $[H_2PO_4^-]$ is almost equal to the analytical concentration. The same is true for $[HPO_4^{2-}]$ at $pH=\frac{1}{2}(pK_{a2}+pK_{a3})$.

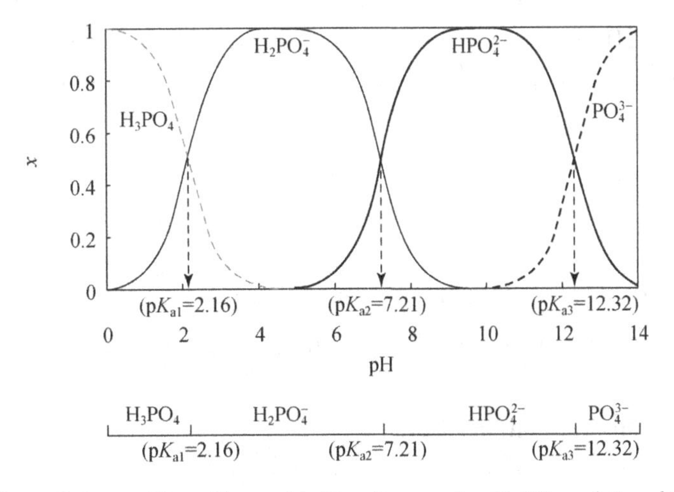

Figure 3.5 x-pH profiles and ladder diagram for H_3PO_4 and its salts.

3.3 SOLVING EQUILIBRIUM CALCULATIONS USING pH CALCULATIONS AS AN EXAMPLE

3.3.1 General Approaches (Systematic Approaches)

The equilibrium problems can be solved by a systematic approach: a number of independent equations need to be developed to cover all the participants in the system. Three types of algebraic equations are used: (1) equilibrium constant expressions, (2) mass-balance equations, and (3) a charge-balance equation. The equilibrium constant expressions for acid-base equilibria are introduced in this chapter as acid-base titration is introduced as the first titration method. Other equilibria will be introduced with the titration methods.

1. Mass-balance Equation (MBE)

Mass-balance equations (MBEs) state by the law of conservation of mass that the total concentration of each reactant must be constant; thus, the analytical concentration of one species in a solution is the sum of the concentrations of all its forms. Because all species are in the same volume of solvent, the volume factor can be canceled from the equation. Therefore, equating the mass to concentration does not create a problem. The same is for charge-balance equation.

For example, the mass-balance equations for NaH_2PO_4 would be $c(Na^+) = c(H_3PO_4) = [H_3PO_4] + [H_2PO_4^-] + [HPO_4^{2-}] + [PO_4^{3-}]$, and the mass-balance equations for Na_2HPO_4 would be $c(Na^+) = 2c(H_3PO_4) = [H_3PO_4] + [H_2PO_4^-] + [HPO_4^{2-}] + [PO_4^{3-}]$. For the system consisting 0.1 mol · L^{-1} NH_3 at pH 10.0, the mass-balance expression is $c(NH_3) = [NH_3] + [NH_4^+]$. When there is 0.01 mol · L^{-1} $ZnCl_2$ in the solution, the mass-balance expressions are $c(NH_3) = [NH_3] + [NH_4^+] + [Zn(NH_3)^{2+}] + [Zn(NH_3)_2^{2+}] + [Zn(NH_3)_3^{2+}] + [Zn(NH_3)_4^{2+}]$, $c(Zn^{2+}) = [Zn^{2+}] + [Zn(NH_3)^{2+}] + [Zn(NH_3)_2^{2+}] + [Zn(NH_3)_3^{2+}] + [Zn(NH_3)_4^{2+}] + [Zn(OH)^+] + [Zn(OH)_2] + [Zn(OH)_3^-] + [Zn(OH)_4^{2-}]$, and $2c(Zn^{2+}) = c(Cl^-)$. The mass-balance equation for $NaHCO_3$ with concentration of c is $c = [H_2CO_3] + [HCO_3^-] + [CO_3^{2-}]$.

2. Charge-balance Equation (CBE)

Charge-balance equation (CBE or electroneutrality) states that there are as many positive charges as negative charges in solution. The total molar concentration of cations is always equal to that of the total anions. To calculate the charge concentration of an ion, one must multiply the molar concentration of the ion by its charge. For example, the charge-balance equation for a 0.100 mol · L^{-1} solution of NaCl is $[Na^+] + [H^+] = [Cl^-] + [OH^-]$. For a 0.100 mol · L^{-1} solution of $MgCl_2$, the CBE is $2[Mg^{2+}] + [H^+] = [Cl^-] + [OH^-]$. The CBE for a 0.100 mol · L^{-1} solution of NaH_2PO_4 is $c + [H^+] = [H_2PO_4^-] + 2[HPO_4^{2-}] + 3[PO_4^{3-}] + [OH^-]$. The CBE for a 0.100 mol · L^{-1} solution of Na_2HPO_4 is $2c + [H^+] = [H_2PO_4^-] + 2[HPO_4^{2-}] + 3[PO_4^{3-}] + [OH^-]$.

With the chemical equilibrium constant expressions, the mass-balance equations, and the charge-balance equation, one can solve problems involving several equilibria using the following systematic steps:

- Step 1 list: (1) all pertinent chemical reactions; (2) all equilibrium constant expressions; (3) mass-balance equations; (4) charge-balance equation.
- Step 2 count: (1) the number of unknown concentrations; (2) compare this number with the number of equations developed in Step 1.

The number of equations developed must equal the number of unknowns.

There are two ways to calculate the concentration.

Approach #1: Solve the simultaneous equations exactly for the concentrations using a computer program.

Approach #2: Make suitable approximations to reduce the number of unknowns and thus the number of equations needed to provide an answer. Solve manually the simplified algebraic equations to give provisional concentrations for the species in the solution. Check the validity of the approximation.

The computer method can provide a quick solution for solving multiple, nonlinear, simultaneous equations rigorously and is very helpful in study of chemical equilibria. However, nearly all equation-solving software requires an initial estimate of the concentration. One should consider the reality of the conclusion before solving the equations to be sure that the answer conforms to the chemical dynamics of the equilibria. One also needs to be sensitive to the limitations of the software.

In solving chemical equilibria problems, suitable approximations simplify calculations. One should be aware that approximations can be made only in charge-balance and mass-balance equations because in these equations the concentration terms appear as sums or differences rather than as products or quotients. The decision to neglect a term in a mass-balance or charge-balance equation is based on the chemistry of the system. For example, in an acid solution, OH^- concentration is often neglected without introducing significant error to such a calculation.

3. Proton-condition Equation (PCE)

The **proton-condition equation** (PCE) is a special mass balance of protons, and is an essential part of solving for proton-transfer in the equilibria. The proton mass balance is established with reference to a "zero level" or a "reference level" for protons. The "zero level" is established as the species when the solution was prepared, including the solvent molecules and the species participating in proton transfer. With "zero level" as a reference, the molar concentration of species gaining protons are equated with species losing protons. For HA in aqueous solution, the "zero level" species are HA and H_2O, then the proton condition equation is written as $[H^+]=[A^-]+[OH^-]$. For NaA in aqueous solution, the "zero level" species are A^- and H_2O, then the PCE is $[HA]+[H^+]=[OH^-]$. To obtain the proton concentration that one species gains, one must multiply the molar concentration of the species by the number of protons this species gains or loses. For $NH_4H_2PO_4$

aqueous solution, the "zero level" species are NH_4^+, $H_2PO_4^-$, and H_2O, and the PCE is $[H^+]+[H_3PO_4]=[NH_3]+[HPO_4^{2-}]+2[PO_4^{3-}]+[OH^-]$. Again, as all species are in the same volume of solvent, equating the mass to concentration does not create a problem.

The proton balance can be always obtained from the mass- and charge-balance relations. PCE tells the essence of proton transfer. In pH calculations, chemical equilibrium expressions and PCE are used in most cases.

Now we go to pH calculations for aqueous solution, in buffer preparation and titrations. The pH calculation for the following solutions may encounter: strong acid and base, monoprotic acid and base, polyprotic acid and base, buffer (acid and its conjugated base), mixed acid or base, and amphiprotic species.

3.3.2 pH Calculations

1. Weak Monoprotic Acids and Bases

(1) Weak monoprotic acids

For a c_a ($mol \cdot L^{-1}$) of HA aqueous solution, the PCE is $[H^+]=[A^-]+[OH^-]$. Substitute equilibrium expression into the equation to obtain an expression in terms of $[H^+]$, we obtain

$$[H^+]=\frac{K_a[HA]}{[H^+]}+\frac{K_w}{[H^+]}$$

On arranging terms, we have

$$[H^+]=\sqrt{K_a[HA]+K_w} \tag{3-17a}$$

Equation (3-17a) is the rigorous expression to calculate $[H^+]$. As we do not know $[HA]$, an approximation needs to be made to solve the equation. Equation (3-17a) can be simplified by making the assumption that $c_aK_a>20K_w$; thus, the dissociation of H_2O as compared to dissociation of HA can be ignored. In this case, the error introduced by this assumption is about 5%. Equation (3-17a) reduces to

$$[H^+]=\sqrt{K_a[HA]}$$

With MBE as $[HA]=c_a-[A^-]$ and PCE as $[A^-]=[H^+]-[OH^-]$, we have $[HA]=c_a-[H^+]+[OH^-]\approx c_a-[H^+]$. $[OH^-]$ is neglected because pH of the solution is acidic. Then Equation (3-17a) is simplified to be

$$[H^+]=\sqrt{K_a[HA]}=\sqrt{K_a(c_a-[H^+])} \tag{3-17b}$$

$[H^+]$ can be obtained by solving the quadratic equation.

Equation (3-17b) can be simplified by the additional assumption that the dissociation does not decrease appreciably the HA equilibrium concentration. If the dissociation is less than 5%, i. e., dissociation error is less than 5%, thus $K_a/c_a \leqslant 2.5\times10^{-3}$, Equation (3-17b) reduces to

$$[H^+] = \sqrt{K_a c_a} \quad (3\text{-}17c)$$

【Example 3.2】 Calculate pH of 0.10 mol • L^{-1} HAc solution ($pK_a=4.76$).

Answer: Because $K_a c_a > 20K_w$, $[H^+] = \sqrt{K_a[HA]}$ is used for calculation.

Because $K_a/c_a < 2.5\times10^{-3}$, $[H^+] = \sqrt{K_a c_a}$. Hence,

$$[H^+] = \sqrt{K_a c_a} = \sqrt{10^{-4.76-1.00}} = 10^{-2.88}(\text{mol} \cdot L^{-1}), \quad pH = 2.88$$

【Example 3.3】 Calculate pH of 0.20 mol • L^{-1} dichloroacetic acid ($pK_a=1.26$).

Answer: Because $c_a K_a \gg K_w$, $[H^+] = \sqrt{K_a[HA]}$ is used for calculation.

Because $K_a/c_a \gg 2.5\times10^{-3}$, the amount of acid dissociated can not be ignored. Hence,

$$[H^+] = \sqrt{K_a[HA]} = \sqrt{K_a(c_a - [H^+])}$$

By solving the quadratic equation, $[H^+]=10^{-1.09}(\text{mol} \cdot L^{-1})$; therefore, pH= 1.09.

(2) Weak monobases

The PCE for a monobase A^- is $[H^+]+[HA]=[OH^-]$. pH can be calculated from the K_b value of A^-,

$$[OH^-] = \sqrt{K_b[A^-] + K_w} \quad (3\text{-}18a)$$

If the base is not very weak, i. e., $c_b K_b > 20K_w$,

$$[OH^-] = \sqrt{K_b[A^-]} = \sqrt{K_b(c_b - [OH^-])} \quad (3\text{-}18b)$$

If $K_b/c_b < 2.5\times10^{-3}$, i. e. the dissociation of A is less than 5%,

$$[OH^-] = \sqrt{K_b c_b} \quad (3\text{-}18c)$$

$$pH = pK_w - pOH$$

【Example 3.4】 Calculate pH of 0.10 mol • L^{-1} NaAc solution (for HAc, $pK_a=4.76$).

Answer:

$$pK_b = pK_w - pK_a = 14.00 - 4.76 = 9.24$$

Because $K_b c_b > 20K_w$, $[H^+] = \sqrt{K_b[Ac^-]}$ is used for calculation.

Because $K_b/c_b < 2.5\times10^{-3}$, $[OH^-] = \sqrt{K_b c_b}$. Hence,

$$[OH^-] = \sqrt{K_b c_b} = \sqrt{10^{-9.24-1.00}} = 1.88\ (mol \cdot L^{-1})$$

$$pH = pK_w - pOH = 14.00 - 1.88 = 12.12$$

2. Polyprotic Acids

The general approach for calculating the pH of a polyprotic acid solution will be illustrated using succinic acid (H_2A, $pK_{a1}=4.21$, $pK_{a2}=5.64$) as a specific diprotic acid. The dissociation equilibria and the constant expressions are

$$H_2A \rightleftharpoons HA^- + H^+ \quad K_{a1} = \frac{[HA^-][H^+]}{[H_2A]}$$

$$HA^- \rightleftharpoons A^{2-} + H^+ \quad K_{a2} = \frac{[A^{2-}][H^+]}{[HA^-]}$$

Choosing H_2A and H_2O as the reference proton level, the PCE is $[H^+]=[HA^-]+2[A^{2-}]+[OH^-]$. Because the solution is acidic, $[OH^-]$ can be immediately eliminated from the PCE. Substitution of $[HA^-]$ and $[A^{2-}]$ with constant expression into this PCE yields

$$[H^+] = \frac{K_{a1}[H_2A]}{[H^+]} + 2\frac{K_{a1}K_{a2}[H_2A]}{[H^+]^2} = \frac{K_{a1}[H_2A]}{[H^+]}\left(1+\frac{2K_{a2}}{[H^+]}\right)$$

Thus,

$$[H^+] = \sqrt{K_{a1}[H_2A]\left(1+\frac{2K_{a2}}{[H^+]}\right)} \tag{3-19}$$

This is a cubic equation for $[H^+]$. One approach is to solve this equation with a computer program that may not be convenient or necessary. However, the equation can be simplified by suitable approximation to obtain a provisional $[H^+]$. If $2K_{a2}/[H^+] \ll 1$, Equation (3-19) can be simplified as $[H^+]=\sqrt{K_{a1}[HA]}$; thus, the provisional hydronium ion concentration of this diprotic acid solution is calculated by considering this acid as a monoprotic acid. $[H^+]$ can be calculated according to Equation (3-17c),

$$[H^+] = \sqrt{K_{a1} \cdot c(H_2A)} = \sqrt{10^{-4.21-1.00}} = 10^{-2.61}\ (mol \cdot L^{-1})$$

When $2K_{a2}/[H^+] \ll 1$, the assumption is valid. Then calculation of pH can be simplified by considering only the first dissociation of H_2A. The general steps for a monoprotic acid can be followed. In most cases, calculation of hydronium ion concentration of H_2A can be simplified as discussed above. Hydronium ion concentration of polyfunctional base solution can be calculated by following the same approach by using the K_b values.

【Example 3.5】 Calculate the pH of a 0.10 mol · L^{-1} oxalic acid ($H_2C_2O_4$) solution. ($K_{a1}=5.6\times10^{-2}$, $K_{a2}=5.1\times10^{-5}$)

Answer: The value of $2K_{a2}/(c_a K_{a1})^{1/2} = 2\times 5.1\times 10^{-5}/(0.1\times 5.6\times 10^{-2})^{1/2}$ is 0.0014; thus the second dissociation can be neglected. Oxalic acid solution can be treated as the acid were a monoprotic acid with $K_a = 5.6\times 10^{-2}$. Because $K_a c > 20K_w$ and $K_a/c > 2.5\times 10^{-3}$, $[H^+] = (-K_a + \sqrt{K_a^2 + 4K_a \cdot c})/2 = 0.052\ \text{mol} \cdot L^{-1}$; then, pH=1.28.

3. Amphiprotic Species

For an amphiprotic species (HA), the principal equilibria in the solution are:

$$HA \rightleftharpoons H^+ + A^- \qquad K_{a2} = \frac{[H^+][A^-]}{[HA]}$$

$$HA + H_2O \rightleftharpoons H_2A^+ + OH^- \qquad K_{b2} = \frac{[H_2A^+][OH^-]}{[HA]} = \frac{K_w}{K_{a1}}$$

The reference proton levels are H_2O and HA; PCE is $[H^+]+[H_2A^+]=[A^-]+[OH^-]$. Substitution of H_2A^+ and A^- with constant expressions into this equation gives

$$[H^+] + \frac{[H^+][HA]}{K_{a1}} = \frac{K_{a2}\cdot[HA]}{[H^+]} + \frac{K_w}{[H^+]}$$

Rearrangement of this equation gives

$$[H^+] = \sqrt{\frac{K_{a2}\cdot[HA] + K_w}{1 + [HA]/K_{a1}}} \tag{3-20a}$$

Under most circumstances, $K_{a1} \gg K_{a2}$; thus, we can make the approximation that $[HA] \approx c$. Then

$$[H^+] = \sqrt{\frac{K_{a2}c + K_w}{1 + c/K_{a1}}}$$

If $K_{a2}\cdot c > 20K_w$, K_w in the numerator can be neglected,

$$[H^+] = \sqrt{\frac{K_{a2}\cdot c(HA)}{1 + c(HA)/K_{a1}}} \tag{3-20b}$$

If $c/K_{a1} > 20$, the unity in the denominator can also be neglected,

$$[H^+] = \sqrt{\frac{K_{a2}c + K_w}{c/K_{a1}}}$$

If both K_w in the numerator and the unity in the denominator can be neglected at the same time,

$$[H^+] = \sqrt{K_{a1}\cdot K_{a2}} \tag{3-20c}$$

【Example 3.6】 Calculate pH of 0.050 mol · L^{-1} $NaHCO_3$ solution. ($K_{a1} = 4.2\times 10^{-7}$, $K_{a2} = 5.6\times 10^{-11}$)

Answer: Because $K_{a1} \gg K_{a2}$, $[HCO_3^-] \approx c$.

Because $cK_{a2} = 5.6\times10^{-11}\times0.050 = 2.8\times10^{-12} > 20K_w$ and $c/K_{a1} = 0.050/4.2\times10^{-7} = 1.2\times10^5 > 20$, $[H^+] = (K_{a1}\cdot K_{a2})^{1/2} = (4.2\times10^{-7}\times5.6\times10^{-11})^{1/2} = 4.8\times10^{-9}(mol\cdot L^{-1})$.

Therefore, pH=8.32.

【Example 3.7】 Calculate pH of 0.05 $mol\cdot L^{-1}$ NaH_2PO_4 solution.

Answer: $H_2PO_4^-$ is an amphiprotic species which involves dissociation equilibria as following

$$(a)\ H_2PO_4^- + H_2O \rightleftharpoons H_3PO_4 + OH^-$$

$$(b)\ H_2PO_4^- \rightleftharpoons HPO_4^{2-} + H^+$$

$$(c)\ HPO_4^{2-} \rightleftharpoons PO_4^{3-} + H^+$$

Taking $H_2PO_4^-$ and H_2O as the reference proton levels, the PCE is

$$[H^+]+[H_3PO_4] = [HPO_4^{2-}]+2[PO_4^{3-}]+[OH^-]$$

Because there are about five orders of magnitude difference between K_{a2} and K_{a3}, $2[PO_4^{3-}]$ can be neglected; therefore, Equations (a) and (b) are the principal equilibria in the solution.

Because $K_{a1} \gg K_{a2}$, $[H^+]=\sqrt{\dfrac{K_{a2}\cdot c+K_w}{1+c/K_{a1}}}$.

When $cK_{a2} = 10^{-7.21}\times0.050 = 10^{-8.51} \gg K_w$, K_w in the numerator can be neglected. Because $c/K_{a1}=0.01/10^{-2.16}=7.2<20\times1$, the unity in the denominator can not be neglected; therefore

$$[H^+]=\sqrt{\frac{K_{a2}\cdot c}{1+c/K_{a1}}}=\sqrt{\frac{10^{-7.21}\times0.05}{1+0.05/10^{-2.16}}} = 10^{-4.71}(mol\cdot L^{-1})$$

$$pH = 4.71$$

At pH 4.71, PO_4^{3-} is negligible, which means the approximation is valid.

【Example 3.8】 Calculate pH of 0.033 $mol\cdot L^{-1}$ Na_2HPO_4 solution.

Answer: HPO_4^{2-} is an amphiprotic species which involves dissociation equilibria as follows

$$(a)\ HPO_4^{2-} + H_2O \rightleftharpoons H_2PO_4^- + OH^-$$

$$(b)\ H_2PO_4^- + H_2O \rightleftharpoons H_3PO_4 + OH^-$$

$$(c)\ HPO_4^{2-} \rightleftharpoons PO_4^{3-} + H^+$$

Taking HPO_4^{2-} and H_2O as the reference proton levels, the PCE is

$$[H^+]+[H_2PO_4^-]+2[H_3PO_4] = [PO_4^{3-}]+[OH^-]$$

Because there are about five orders of magnitude difference between K_{a1} and K_{a2}, $2[H_3PO_4]$ can be neglected. Therefore, Equations (a) and (c) are the principal equilibria in solution.

Because $K_{a2} \gg K_{a3}$, $[H^+]=\sqrt{\frac{K_{a3} \cdot c+K_w}{1+c/K_{a2}}}$. Because $cK_{a3}=10^{-12.32}\times 0.033=1.6\times 10^{-14}$, K_w in the numerator can not be neglected. Because $c/K_{a2}>20$, the unity in the denominator can be neglected. Therefore,

$$[H^+]=\sqrt{\frac{K_{a3} \cdot c+K_w}{c/K_{a2}}}=\sqrt{\frac{10^{-12.32}\times 0.033+1.0\times 10^{-14}}{0.033/10^{-7.21}}}$$
$$=10^{-9.66}\,(\text{mol}\cdot \text{L}^{-1})$$
$$\text{pH}=9.66$$

At pH 9.66, H_3PO_4 is negligible, which means the approximation is valid.

4. Mixture of Strong Acid and Weak Acid (HA)

If a solution contains c_a ($\text{mol}\cdot\text{L}^{-1}$) strong acid (e.g. HCl) and $c(\text{HA})$ ($\text{mol}\cdot\text{L}^{-1}$) weak acid HA, the reference proton levels are H^+, HA, and H_2O, and the PCE is $[H^+]=c_a+[A^-]+[OH^-]$. The PCE tells us that the total hydronium ion concentration in the solution comes from HCl, HA, and H_2O. Because the solution is acidic, $[OH^-]$ can be neglected from the PCE. Substitution of $[A^-]=cx_0$ into PCE yields

$$[H^+]=c_a+\frac{c(\text{HA})\cdot K_a}{K_a+[H^+]} \tag{3-21a}$$

$[H^+]$ can be calculated by solving the quadratic equation.

Generally, the approximation is made by assuming that the dissociation of HA can be ignored in the presence of c_a ($\text{mol}\cdot\text{L}^{-1}$) strong acid. A provisional hydronium ion concentration is then calculated from $[H^+]=c_a$. If $[A^-]\ll c(\text{HCl})$, then the assumption is valid. If the assumption is not valid, then one must go back and solve the quadratic equation.

For a mixture of a strong base and a weak base, e.g., NaOH and A^-, the following equation is used for calculation of $[OH^-]$:

$$[OH^-]=c(\text{NaOH})+\frac{c(A^-)\cdot K_b}{K_b+[OH^-]} \tag{3-21b}$$

【Example 3.9】 Calculate the pH of the solution that results when 20.04 mL of 0.1000 $\text{mol}\cdot\text{L}^{-1}$ NaOH is mixed with 20.00 mL of 0.1000 $\text{mol}\cdot\text{L}^{-1}$ HA ($K_a=10^{-7.00}$).

Answer: After mixing, the concentrations are

$$c(A^-)=\frac{0.1000\times 20.00}{20.00+20.04}=10^{-1.30}\,(\text{mol}\cdot\text{L}^{-1})$$

$$c(\text{NaOH})=\frac{0.1000\times 0.04}{20.00+20.04}=10^{-4.00}\,(\text{mol}\cdot\text{L}^{-1})$$

Thus, we get $[OH^-]=c(NaOH)=10^{-4.00}(mol\cdot L^{-1})$

With this provisional result,

$$[HA]=\frac{c(A^-)\cdot K_b}{K_b+[OH^-]}=\frac{10^{-1.30}\times 10^{-7.00}}{10^{-7.00}+10^{-4.00}}=10^{-4.30}(mol\cdot L^{-1})$$

Because $[HA]\approx c(NaOH)$, the assumption is not valid. Then, $[OH^-]=\frac{10^{-1.30}\times 10^{-7.00}}{10^{-7.00}+[OH^-]}+10^{-4.00}$.

$$[OH^-]=10^{-3.86}(mol\cdot L^{-1}),\quad pH=10.14$$

5. Mixture of Weak Acid and Weak Base

If a solution contains a weak acid (HA) and a weak base (B^-) ($K_a(HB)>K_a(HA)$), there will be two major dissociation equilibria in addition to autoprotolysis of H_2O.

$$HA+H_2O \rightleftharpoons H_3O^+ + A^-$$
$$B^- + H_2O \rightleftharpoons HB + OH^-$$

With HA, B^- and H_2O as the reference proton levels, the PCE is

$$[H^+]+[HB]=[A^-]+[OH^-]$$

Substitution of equilibria constant expressions into the PCE yields

$$[H^+]+\frac{[B^-][H^+]}{K_a(HB)}=\frac{K_a(HA)\cdot[HA]}{[H^+]}+\frac{K_w}{[H^+]}$$

Rearrangement of the above equation produces

$$[H^+]=\sqrt{\frac{K_a(HA)\cdot[HA]+K_w}{1+[B^-]/K_a(HB)}} \tag{3-22a}$$

If $K_a(HB)\gg K_a(HA)$, then $[HA]\approx c(HA)$ and $[B^-]\approx c(B^-)$. Equation (3-22a) becomes

$$[H^+]=\sqrt{\frac{K_a(HA)\cdot c(HA)+K_w}{1+c(B^-)/K_a(HB)}} \tag{3-22b}$$

This is similar to an amphiprotic solution with amphiprotic species as both a weak acid and a weak base; therefore, a similar approximation to simplify the calculation can be used to obtain the hydronium ion concentration of the solution.

【Example 3. 10】 Calculate the pH of a solution containing 0. 10 mol · L^{-1} NH_4Cl and 0. 20 mol · L^{-1} NaAc. (for NH_3, $pK_b=4.75$; for HAc, $pK_a=4.76$)

Answer: Taking NH_4^+, Ac^-, and H_2O as the reference proton levels, the PCE is

$$[H^+]+[HAc]=[NH_3]+[OH^-]$$

Then we have $[H^+]=\sqrt{\frac{K_a(NH_4^+)\cdot[NH_4^+]+K_w}{1+[Ac^-]/K_a(HAc)}}$.

As $K_a(NH_4^+) \cdot c(NH_4^+) \gg K_w$, K_w in the numerator can be neglected; and because $c(HAc)/K_a(HAc) \gg 1$, the unity in the denominator can be neglected. Because $K_a(HAc) \gg K_a(NH_4^+)$, one can assume that $c(NH_4^+) \approx [NH_4^+]$ and $c(Ac^-) \approx [Ac^-]$. Thus,

$$[H^+] = \sqrt{\frac{K_a(NH_4^+) \cdot c(NH_4^+)}{c(Ac^-)/K_a(HAc)}} = \sqrt{\frac{10^{-9.25} \times 0.10}{0.20/10^{-4.76}}} = 10^{-7.16}(\text{mol} \cdot \text{L}^{-1})$$

$$\text{pH} = 7.16$$

At pH 7.16, NH_4^+ and Ac^- are the major form of NH_4^+ and Ac^-, respectively; therefore, the above calculation is valid.

3.4 BUFFER SOLUTIONS

3.4.1 pH Calculations of Buffer Solutions

A buffer solution contains a weak acid (HA), and its conjugate base (A^-). The dissociation equilibrium of HA is

$$HA \rightleftharpoons H^+ + A^- \qquad K_a = \frac{[H^+][A^-]}{[HA]}$$

For HA, the analytical concentration is c_a, and for A^-, the analytical concentration is c_b. The CBE and MBE are written as

$$[H^+] + c_b = [OH^-] + [A^-]$$

$$[HA] + [A^-] = c_a + c_b$$

Then we have

$$[A^-] = c_b + [H^+] - [OH^-] \tag{3-23}$$

$$[HA] = c_a - [H^+] + [OH^-] \tag{3-24}$$

Substitution of Equations (3-23) and (3-24) into dissociation constant expression and rearrangement yields

$$[H^+] = K_a \frac{[HA]}{[A^-]} = K_a \frac{c_a - [H^+] + [OH^-]}{c_b + [H^+] - [OH^-]} \tag{3-25}$$

If pK_a is less than 7, the buffering pH is acidic, Equation (3-25) becomes

$$[H^+] = K_a \frac{[HA]}{[A^-]} = K_a \frac{c_a - [H^+]}{c_b + [H^+]} \tag{3-26a}$$

If pK_a is greater than 7, the buffering pH is basic, Equation (3-25) becomes

$$[H^+] = K_a \frac{[HA]}{[A^-]} = K_a \frac{c_a + [OH^-]}{c_b - [OH^-]} \tag{3-26b}$$

If $c_a \gg [OH^-]-[H^+]$ and $c_b \gg [H^+]-[OH^-]$, Equation (3-25) becomes

$$[H^+] = \frac{c_a}{c_b} K_a \tag{3-26c}$$

Then we get the **Henderson-Hasselbalch equation**

$$pH = pK_a + \lg \frac{c_b}{c_a} \tag{3-26d}$$

Generally, Equation (3-26c) is used initially. The provisional values for $[H^+]$ and $[OH^-]$ will then be used to compare with c_a and c_b to test the assumptions.

【Example 3.11】 Calculate pH of a solution that is composed of 0.040 mol · L^{-1} HAc (pK_a=4.76) and 0.060 mol · L^{-1} NaAc.

Answer: According to Equation (3-26c),

$$[H^+] = \frac{c_a}{c_b} K_a = \frac{0.040}{0.060} \times 10^{-4.76} = 10^{-4.94} (\text{mol} \cdot \text{L}^{-1})$$

$[H^+] \ll c_a$, $[H^+] \ll c_b$, therefore one may write pH=4.94.

【Example 3.12】 Calculate pH of a solution resulted when 20.00 mL 0.1000 mol · L^{-1} HA ($K_a = 10^{-7.00}$) is mixed with 19.96 mL 0.1000 mol · L^{-1} NaOH.

Answer: After mixing,

$$c_a = \frac{0.1000 \times 0.04}{20.00 + 19.96} = 10^{-4.00} (\text{mol} \cdot \text{L}^{-1})$$

$$c_b = \frac{0.1000 \times 19.96}{20.00 + 19.96} = 10^{-1.30} (\text{mol} \cdot \text{L}^{-1})$$

According to Equation (3-26c),

$$[H^+] = \frac{c_a}{c_b} K_a = \frac{10^{-4.00}}{10^{-1.30}} \times 10^{-7.00} = 10^{-9.70} (\text{mol} \cdot \text{L}^{-1})$$

To check the validity of the approximation, it is found that $[OH^-] = 10^{-4.30}$ mol · L^{-1} is comparable with c_a, thus Equation (3-26b) is used for calculating

$$[OH^-] = \frac{c_b - [OH^-]}{c_a + [OH^-]} K_b = \frac{10^{-1.30} - [OH^-]}{10^{-4.00} + [OH^-]} \times 10^{-7.00}$$

The solution of this quadratic equation is $[OH^-] = 10^{-4.44}$ mol · L^{-1}, thus one may write pH=9.56.

3.4.2 Buffer Capacity

A buffer solution is able to prevent significant change in pH when a small amount of acid/base is introduced. **Buffer capacity** (β) is a quantitative measure of this resistance to pH change, and is defined as the number of moles of a strong acid or a strong base that causes 1.00 L of the buffer to undergo a 1.00 pH unit change.

Mathematically, β is given by

$$\beta = -\frac{da}{dpH} = \frac{db}{dpH} \tag{3-27}$$

where da and db is the small amount of acid and base added to the 1 L buffer. Note that the buffer capacity is always positive. Because the addition of acid will reduce pH, da/dpH is negative, thus a negative sign is put in front of da/dpH.

For a monoprotic acid and its conjugate base, such as acetic acid (HAc) and sodium acetate (NaAc), assuming that the buffer is made by addition of strong base, e.g., NaOH to HAc, then the CBE is $[H^+]+[Na^+]=[OH^-]+[Ac^-]$. It is possible to replace $[Na^+]$ with the analytical concentration of NaOH added, b. then we have

$$b = -[H^+]+[OH^-]+[Ac^-] = -[H^+]+\frac{K_w}{[H^+]}+\frac{cK_a}{[H^+]+K_a} \tag{3-28a}$$

Differentiating with respect to $[H^+]$, one gets

$$\frac{db}{d[H^+]} = -1-\frac{K_w}{[H^+]^2}-\frac{cK_a}{([H^+]+K_a)^2} \tag{3-28b}$$

Since $d[H^+]/d(pH) = -\ln 10[H^+] = -2.303[H^+]$, then

$$\frac{db}{dpH} = 2.303\times\left([H^+]+\frac{K_w}{[H^+]}+\frac{cK_a[H^+]}{([H^+]+K_a)^2}\right) \tag{3-28c}$$

The profile of Equation (3-27) is presented in Figure 3.6. The term $[H^+]$ becomes significant with pH approximately less than 2 and the term $K_w/[H^+]$ becomes significant with pH approximately greater than 11.5. Both situations are the results of the acid and base properties of H_2O and are independent of the weak acid and its salt. For the term $\frac{cK_a[H^+]}{([H^+]+K_a)^2}$,

- Buffer capacity reaches its maximum when $pH=pK_a$. $\beta_{max}=2.3c\times0.5\times0.5=0.575c$. Therefore, it is appropriate to select a buffer with pK_a close to the desired pH.
- At $pH=pK_a\pm1$ the buffer capacity falls to 1/3 of the maximum. This is the pH range within which a buffer is effective. At $pH=pK_a-1$, The Henderson-Hasselbalch equation shows that the ratio $[HA]:[A^-]$ is 10 : 1; and at $pH=pK_a+1$, $[HA]:[A^-]=1:10$.
- Buffer capacity is directly proportional to the analytical concentration of the acid. Therefore, an adequate concentration of the buffer system can effectively prevent pH from changes.

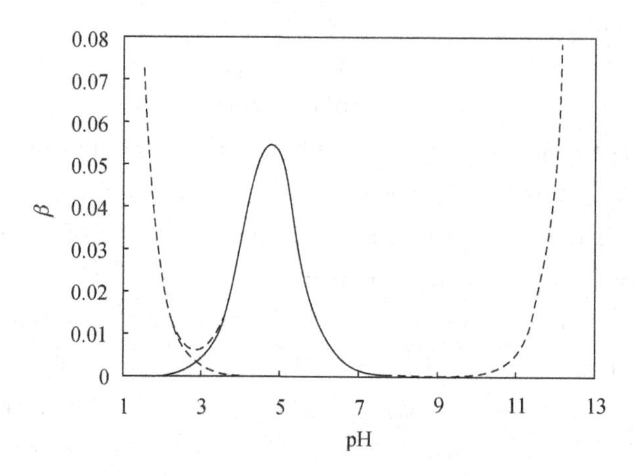

Figure 3.6 Buffer capacity as a function of pH.

The most commonly used buffers are listed in Table 3.1. A buffer widely used in the clinical laboratory and in biochemical studies in the physiological pH range is the one prepared from tris(hydroxymethyl)aminomethane ($NH_2C(CH_2OH)_3$, abbr. as Tris or THAM by adding hydrochloric acid to a solution of Tris to adjust the pH to the desired value. This buffer does not precipitate calcium salts, nor inhibits many enzymatic systems, and is compatible with biological fluids.

Table 3.1 List of Frequently Used Buffers

Buffer	pK_a	Buffer	pK_a
HCOOH-NaOH	3.77	$Na_2B_4O_7 \cdot 10H_2O$	9.24
CH_3COOH-CH_3COONa	4.76	$NH_3 \cdot H_2O$-NH_4Cl	9.25
$(CH_2)_6N_4$-HCl	5.13	$NaHCO_3$-Na_2CO_3	10.25
NaH_2PO_4-Na_2HPO_4	7.21	Na_2HPO_4-NaOH	12.32
$NH_2C(CH_2OH)_3$-HCl	8.21		

3.4.3 Preparation of Buffers

Theoretically, a buffer solution of a desired pH can be prepared by combining calculated quantities of a suitable conjugate acid/base pair. However, pH of buffer often differs from the calculated results because that effect of ionic strength and temperature are often neglected. Usually, a buffer solution is made by adjusting the buffer solution to a desired pH as indicated by a pH meter. For example, to prepare an one liter of pH 7.60, 0.10 $mol \cdot L^{-1}$ Tris buffer, the following steps can be followed:

- Dissolve calculated amount of Tris, 12.1 grams, in 800 mL water.
- Adjust pH with strong acid to pH 7.60.
- Add water to close to the final solution volume.
- Check the pH and adjust if necessary.
- Add additional water to one liter.

Alternatively, empirically derived recipes for preparation of pH buffers are available in handbooks. For example, **Britton-Robinson buffer** is a pH buffer that can be used in the range of pH 2 to 12. This buffer consists of a mixture of 0.04 mol · L^{-1} boric acid (H_3BO_3), 0.04 mol · L^{-1} phosphoric acid (H_3PO_4) and 0.04 mol · L^{-1} acetic acid (HAc) that has been titrated to the desired pH with 0.2 mol · L^{-1} NaOH.

【Example 3.13】 Calculate how much NH_4Cl in gram and how much 1.0 mol · L^{-1} ammonia (NH_3) in mL are needed for preparing 200 mL of pH 9.4 NH_4Cl-NH_3 buffer to ensure that pH change is less than 0.2 pH unit when 1.0 mmol HCl or NaOH is introduced to the buffer. ($pK_a(NH_4^+)=9.4$)

Answer: From the definition of buffer capacity,

$$\beta = \frac{1.0/200}{0.2} = 2.5 \times 10^{-2}\ (\text{mol} \cdot \text{L}^{-1})$$

and because

$$\beta = 2.3c \cdot x(\text{HA}) \cdot x(\text{A}) = 2.3c \cdot \frac{K_a \cdot [\text{H}^+]}{(K_a + [\text{H}^+])^2} = 0.575c$$

$$c = \frac{2.5 \times 10^{-2}}{0.575} = 0.043\ (\text{mol} \cdot \text{L}^{-1})$$

Thus,

$$m(\text{NH}_4\text{Cl}) = 0.043 \times 0.50 \times 0.200 \times 53.49 = 0.23(\text{g})$$

$$V(\text{NH}_3) = 0.043 \times 0.50 \times 200/1.0 = 4.3(\text{mL})$$

Chapter 3 Questions and Problems

3.1 Calculate x values (molar fraction, x_0, x_1, x_2, and x_3) for H_3PO_4 at pH 2.16, 4.69, 7.21, 9.77, and 12.32. ($pK_{a1} \sim pK_{a3}$: 2.16, 7.21, and 12.32)

pH	$[H^+]$	x_3	x_2	x_1	x_0
2.16					
4.69					
7.21					
9.77					
12.32					

3.2 Calculate pK_a^M and pK_a^C for NH_4^+ at $I=0.10$. (For NH_4^+, $pK_a=9.25$)

3.3 Write the MBE, CBE, and PCE of the following aqueous solutions:

(1) NH_3; (2) NH_4Cl; (3) Na_2CO_3; (4) KH_2PO_4; (5) $NaAc+H_3BO_3$.

3.4 Write expressions for the autoprotolysis of

(1) H_2O; (2) CH_3COOH; (3) CH_3NH_2; (4) CH_3OH.

3.5 Calculate pH of the following aqueous solution:

(1) 0.10 mol · L^{-1} $ClCH_2COOH$ (chloroacetic acid, $pK_a=2.86$);

(2) 0.10 mol · L^{-1} $ClCH_2COONa$ (sodium chloroacetate, $pK_a=2.86$);

(3) 0.10 mol · L^{-1} $(CH_2)_6N_4$ (hexamethylene tetramine, $pK_b=8.87$);

(4) 0.10 mol · L^{-1} Na_2S (sodium sulfide, $pK_{a1}=7.05$, $pK_{a2}=12.92$);

(5) 0.010 mol · L^{-1} H_2SO_4 (sulfuric acid, $pK_{a2}=1.92$);

(6) 0.010 mol · L^{-1} H_2NCH_2COOH (glycine)

$$H_2A^+ \xrightleftharpoons{pK_{a1} = 2.35} HA \xrightleftharpoons{pK_{a2} = 9.78} A^-$$

(7) pI (isoelectric point) for glycine.

3.6 Calculate the pH of C_6H_5COOH (benzoic acid, $K_a = 6.2 \times 10^{-5}$) solutions that have the concentrations as given in table.

c/mol · L^{-1}	K_a/c	Equation for Calculation	pH
1.00×10^{-1}			
1.00×10^{-2}			
1.00×10^{-4}			

3.7 Calculate the pH of C_6H_5COONa (sodium benzoate, $K_a = 6.2 \times 10^{-5}$) solutions that have the concentrations as given in table.

c/mol · L^{-1}	K_b/c	Equation for Calculation	[OH^-]	pOH	pH
1.00×10^{-1}					
1.00×10^{-2}					
1.00×10^{-4}					

3.8 Calculate the pH of NH_3 (ammonia, $K_b=1.75\times10^{-5}$) solutions that have the concentrations as given in table.

c/mol · L^{-1}	K_b/c	Equation for Calculation	[OH^-]	pOH	pH
1.00×10^{-1}					
1.00×10^{-2}					
1.00×10^{-4}					

3.9 Calculate the pH of NH_4^+ (ammonium ion, $K_b = 1.75\times10^{-5}$) solutions that have the concentrations as given in table.

c/mol · L^{-1}	K_a/c	Equation for Calculation	pH
1.00×10^{-1}			
1.00×10^{-2}			
1.00×10^{-4}			

3.10 Calculate the pH of the solution that results when 50 mL of 0.10 mol · L^{-1} H_3PO_4 (pK_{a1} = 2.16, pK_{a2} = 7.21, pK_{a3} = 12.32) is

(1) mixed with 25 mL of 0.10 mol · L^{-1} NaOH;

(2) mixed with 50 mL of 0.10 mol · L^{-1} NaOH;

(3) mixed with 75 mL of 0.10 mol · L^{-1} NaOH.

3.11 Prepare 100 mL buffer solutions of pH 2.00 and pH 10.00 by addition of glycine (H_2NCH_2COOH, M_r = 75.07) to HCl or NaOH solution, respectively. If the total concentration of the buffer is 0.10 mol · L^{-1}, how many grams of glycine is needed and how many mL of 1.0 mol · L^{-1} HCl or 1.0 mol · L^{-1} NaOH is needed, respectively? (pK_{a1} = 2.35, pK_{a2} = 9.78)

3.12 How many grams of $Na_2HPO_4\,2H_2O$ must be added to 400 mL of 0.200 mol · L^{-1} H_3PO_4 to give a buffer of pH 7.30? (H_3PO_4, $pK_{a1}\sim pK_{a3}$: 2.16, 7.21, 12.32)

3.13 How would you prepare 1.00 L of a buffer with a pH of 7.00 from 0.200 mol · L^{-1} H_3PO_4 and 0.160 mol · L^{-1} NaOH?

CHAPTER 4

ACID-BASE TITRATION

4.1 Acid/Base Indicators

4.1.1 Principle

4.1.2 Examples

4.1.3 Titration Errors

4.1.4 Factors Influencing Performance

4.2 Titration Curves and Selection of Indicators

4.2.1 Strong Acids (Bases)

4.2.2 Monoprotic Acids (Bases)

4.2.3 Strong and Weak Acids (Bases)

4.2.4 Polyfunctional Weak Acids (Bases)

4.2.5 Mixture of Weak Acids (Bases)

4.3 Titration Error Calculations

4.3.1 Strong Acids (Bases)

4.3.2 Monoprotic Weak Acids (Bases)

4.3.3 Polyfunctional Acids (Bases)

4.4 Preparation of Standard Solutions

4.4.1 Standard Acid Solutions

4.4.2 Standard Base Solutions

4.4.3 The Carbonate Error

4.5 Examples of Acid-base Titrations

4.5.1 Determination of Total Alkalinity

4.5.2 Determination of Nitrogen

4.5.3 Determination of Boric Acid

4.6 Acid-base Titrations in Non-aqueous Solvents

4.6.1 Non-aqueous Solvents

4.6.2 Examples of Non-aqueous Titrations

Acid-base titration is based on the reaction between an acid and a base that can be applied to the quantifying of assorted acids and bases as well as non-acidic and basic species. Acid-base titrations can also be used to determine the molar mass or the dissociation constant of an acid or a base. A titration curve can be prepared by calculating the pH after each addition of titrant and plotting the pH as a function of titrant volume. This information facilitates the selection of an indicator for endpoint detection in acid-base titration. Endpoint detection is the major source of titration error. The acid/base indicators will be introduced in this chapter. Titration of polyprotic acids, non-aqueous titrations, and some practical applications will also be discussed in this chapter.

4.1 ACID/BASE INDICATORS

4.1.1 Principle

An acid/base indicator is a weak organic acid or a weak organic base which dissociates or associates with a proton as the pH changes. The dissociation and association cause the internal structural changes between the acid/base and their conjugated forms, thus producing a color change. The color change of such an acid or base can be used to indicate pH change at the endpoint detection.

If HIn represents the acid form of an acid/base indicator and In^- represents the conjugated base, the dissociation equilibrium is

$$\underset{\text{(acid color)}}{HIn} \rightleftharpoons \underset{\text{(base color)}}{In^-} + H^+$$

The dissociation constant expression will be

$$K_{HIn} = \frac{[H^+][In^-]}{[HIn]} \quad \text{or} \quad \frac{[In^-]}{[HIn]} = \frac{K_{HIn}}{[H^+]} \tag{4-1}$$

The ratio of conjugated base form to acid form of an acid indicator or the ratio of base form to conjugated acid form of a base indicator determines the color of the solution and is dictated by the hydronium ion concentration of the solution.

The human eye usually perceives the acid form when $[In^-]/[HIn] \leqslant 1:10$, and the base form when $[In^-]/[HIn] \geqslant 10:1$. The color appears to be intermediate for ratios between these two values. If Equation (4-1) is put in the logarithmic form, we have

$$\lg[H^+] = \lg K_{HIn} - \lg \frac{[In^-]}{[HIn]} \tag{4-2}$$

or
$$\text{pH} = \text{p}K_{\text{HIn}} + \lg \frac{[\text{In}^-]}{[\text{HIn}]} \tag{4-3}$$

If the two concentration ratios are substituted into Equations (4-2) and (4-3), the color of acid form will be observed with $\text{pH}=\text{p}K_{\text{HIn}}+\lg(1/10)=\text{p}K_{\text{HIn}}-1$ (or lower), and the color of base form will be observed with $\text{pH}=\text{p}K_{\text{HIn}}+\lg(10/1)=\text{p}K_{\text{HIn}}+1$ (or higher). A transition from the acid color to base color will occur in the pH range

$$\text{pH} = \text{p}K_{\text{HIn}} \pm 1 \tag{4-4}$$

In practice, individuals can differ significantly in their ability to distinguish between colors. The color spectrum of room lighting can influence the sensitivity of the human eye to color change, e. g. , there must be a red component in lighting for the analyst to observe a purple color change to blue. Therefore, the actual transition range may deviate from the theoretically derived range and may vary considerably from indicator to indicator.

4.1.2 Examples

Methyl orange (MO, $\text{p}K_a = 3.4$) is a weak base indicator with the molecular structure and the associated color described in Figure 4.1.

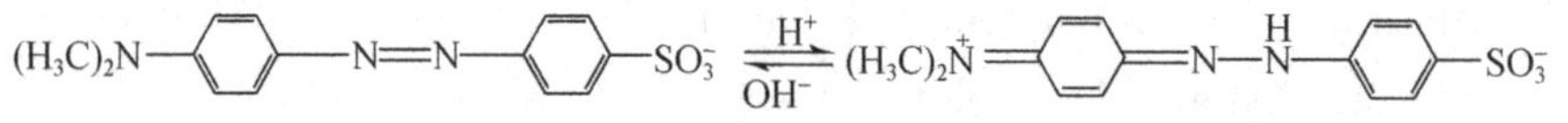

$$(H_3C)_2N{-}C_6H_4{-}N{=}N{-}C_6H_4{-}SO_3^- \underset{OH^-}{\overset{H^+}{\rightleftharpoons}} (H_3C)_2\overset{+}{N}{=}C_6H_4{=}N{-}NH{-}C_6H_4{-}SO_3^-$$

(yellow) (red)

Figure 4.1 Color change associated with dissociation/association of amine group in methyl orange.

The color transition range for methyl orange is pH 3.1～4.4. The acid color is distinctively observed at pH 3.1, although $[\text{HIn}] : [\text{In}^-]$ is only 2 : 1. The base color can be observed at $[\text{In}^-] : [\text{HIn}] = 10 : 1$. When the pH decreases in the transition range (pH 4.4→3.1), a very distinctive color change from yellow to orange can be easily observed at pH 4.0 (indicator transition point). When the pH increases within the transition range (pH 3.1→4.4), a color change from red to orange at pH 4.0 is difficult to see, but the yellow base color at pH 4.4 is easier to see. The color change is similar for methyl red (MR, $\text{p}K_a = 5.0$) whose color transition range is 4.4～6.2 with the distinctive color change from yellow to orange at pH 5.0.

Phenolphthalein (PP, $\text{p}K_a = 9.1$) is a weak organic acid with a colorless acid form and a purple base form (Figure 4.2). The color transition range for phenolphthalein is 8. 0~9. 8. The solution becomes a distinctively purple at 9. 0, but the purple color completely disappears at pH 8. 0.

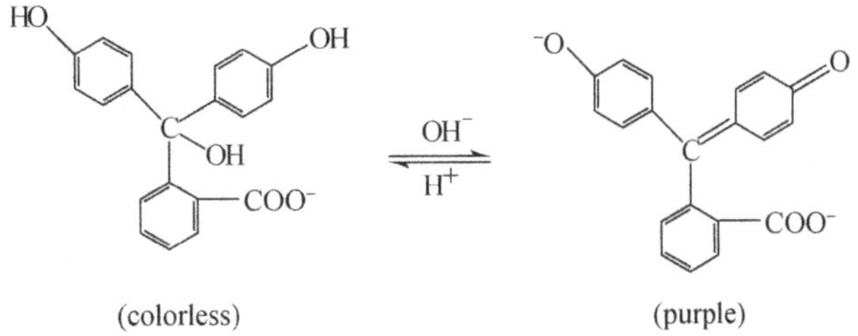

Figure 4. 2 Color change associated with dissociation/association of hydroxyl group of phenolphthalein.

The endpoint indication can be sharpened by using a mixed indicator. One approach is to simply mix two acid/base indicators. Mixed indicators of methyl red (MR, transition range 5. 0~6. 2 from red to yellow) and bromocresol green (BCG, transition range 3. 8~5. 4 from yellow to green blue) have a sharpened transition range of 5. 0~5. 2 by forming one pair of complementary colors that are of the "opposite" hue in some color models (Mixed indicator bromocresol green-methyl red for carbonates in water. Stancil Cooper Ind Eng Chem Anal Ed, 1941, 13 (7): 466-470, DOI: 10. 1021/i560095a011). The mixed indicator solution is dark red at $pH < 5.0$, grayish green at pH 5. 1, and green at $pH > 5.2$. When sodium bicarbonate is titrated with HCl using this mixed indicator, the endpoint color change from green to dark red with a gray intermediate color is easy to observe. Indeterminate indicator error will be reduced when the transition range is narrowed.

Another approach to sharpen the endpoint observation is to mix an acid/base indicator with a dye that does not undergo a color transition itself, for example, a mixed indicator of methyl orange (transition range 3. 1~4. 4, red to yellow) and indigo carmine (blue). The indigo carmine is a non-pH-responsive dye, so that in the pH transition range the color will not change, but the complementary colors sharpen the color observation. This mixed indicator is purple (red+blue) at $pH < 3.1$ and green (yellow + blue) at $pH > 4.4$. Table 4. 1 gives a few examples of common indicators and their properties. Indicator information is available in the *Handbook of Acid-Base Indicators* (R. W. Sabnis. CRC Press, 2007).

Table 4.1 Some Commonly Used Acid/Base Indicators

Indicator	Color		pK_a	Transition Range, pH
	Acid Form	Base Form		
Methyl orange (MO)	Red	Yellow	3.4	3.1~4.4
Methyl red (MR)	Red	Yellow	5.0	4.4~6.2
Bromocresol purple	Yellow	Purple	6.1	5.2~6.8
Phenol red	Yellow	Red	7.8	6.4~8.2
Phenolphthalein (PP)	Colorless	Red	9.1	8.0~9.6
Thymolphthalein (TP)	Colorless	Blue	10.0	9.4~10.6
Bromocresol green(BCG)	Yellow	Blue	4.9	3.8~5.4

4.1.3 Titration Errors

There are two types of errors associated with acid/base indicators. The first type is the determinate error (systematic error) when the pH of endpoint detected by the indicator differs from the pH at the stoichiometric point. This type of indicator error can be minimized by carefully selecting an indicator.

Indeterminate error comes from the limited ability of the eye to distinguish reproducibly the intermediate color of the indicator. The visual uncertainty with an acid/base indicator is approximately ±0.3 pH units with variation from individual to individual.

4.1.4 Factors Influencing Performance

The factors that often affect the color transition include temperature, electrolyte concentration (ionic strength), solvents other than water, and colloidal particles. The first three factors affect the pK_a and thus the color transition range of the indicator. Colloidal particles disturb color observation by scattering the light.

An excess amount of indicator will cause a titration error, because the indicator which is an acid or base will consume the titrant. This error can be minimized by reducing the indicator concentration and by conducting a reagent blank titration (blank correction). In a blank titration, the solution being titrated should have everything but the analyte being titrated. The volume of titrant that the reagent blank uses is then subtracted form the total volume of titrant used in one single titration to correct for the reagent blank.

4.2 TITRATION CURVES AND SELECTION OF INDICATORS

4.2.1 Strong Acids (Bases)

The titration reaction of a strong acid/base with a strong base/acid is $OH^- + H^+ \rightleftharpoons H_2O$, and the reaction constant (titration constant) is $K_t = 1/K_w = 10^{14.00}$. Strong acid-base titration is the most complete titration among the acid-base titrations because the titration constant is the highest.

The titration of 20 mL, 0.1000 mol · L^{-1} HCl with 0.1000 mol · L^{-1} NaOH will be taken as an example to illustrate the construction of a titration curve and the selection of an indicator.

The pH of initial point, the pH of pre-stoichiometric point, the pH of stoichiometric point (SP), and the pH of post-stoichiometric point all need to be calculated to construct the hypothetical titration curve to get the necessary information for the selection of an appropriate indicator.

- Initial point. Before NaOH is added, $[H^+] = c(HCl) = 0.1000$ mol · L^{-1}, thus $pH = -\lg[H^+] = 1.00$.
- Pre-stoichiometric point. Before stoichiometric point is reached, $[H^+]$ of the titration product solution is determined by remaining amount of HCl. For example, after the addition of 19.98 mL NaOH, the concentration of unreacted HCl in the product solution is

$$[H^+] = c(HCl) = \frac{20.00 \times 0.1000 - 19.98 \times 0.1000}{20.00 + 19.98}$$

$$= \frac{0.02}{20.00 + 19.98} \times 0.1000 = 5.0 \times 10^{-5} (\text{mol} \cdot \text{L}^{-1})$$

$$pH = 4.30$$

- Stoichiometric point (SP). At the SP, all HCl is just neutralized by the added NaOH, thus the only acid-base equilibrium existing is the autoprotolysis of water.

$$[H^+] = [OH^-] = \sqrt{K_w} = 10^{-7.00} (\text{mol} \cdot \text{L}^{-1}), \quad pH = 7.00$$

- Post-stoichiometric point. Beyond the SP, the $[H^+]$ of the product solution is determined by the excess of NaOH.

For example, upon addition of 20.02 mL NaOH, the concentration of excess NaOH in the product solution is

$$[OH^-] = c(NaOH) = \frac{20.02 \times 0.1000 - 20.00 \times 0.1000}{20.00 + 20.02}$$

$$= \frac{0.02}{20.00 + 20.02} \times 0.1000 = 5.0 \times 10^{-5} (mol \cdot L^{-1})$$

Thus, $[H^+] = 2.0 \times 10^{-10}\ mol \cdot L^{-1}, \quad pH = 9.70$

Therefore, the pH of the titration product solution with any added volume of NaOH can be calculated. Additional data are provided in Table 4.2. The titration curve according to data in Table 4.2 is given in Figure 4.3 with pH as the y-axis and the percent titration ($T/\%$) as the x-axis.

Table 4.2 pH Change during the Titration of HCl with NaOH at the Same Concentration

Volume of NaOH/mL (20.00 mL HCl)	$T/\%$	pH		
		(0.1 mol · L^{-1})	(1.0 mol · L^{-1})	(0.01 mol · L^{-1})
0.00	0.0	2.00	1.00	0.00
10.00	50.0	2.48	1.48	0.48
15.00	75.0	2.85	1.85	0.85
19.98	99.9	5.30	4.30	3.30
20.00	100.0	7.00	7.00	7.00
20.02	100.1	8.70	9.70	10.70
25.00	125.0	11.05	12.05	13.05
30.00	150.0	11.30	12.30	13.30
40.00	200.0	11.52	12.52	13.52

It can be seen from Figure 4.3 that before 90% of the HCl is neutralized by NaOH, the remaining HCl is within its effective buffer range and the pH increases slowly. After the stoichiometric point when 10% excess NaOH is added, the pH of the product solution increases slowly, because the excess NaOH acts as a buffer at the very basic pH. When the percent titration changes from 99.9% to 100.1%, the 0.04 mL of NaOH is added (19.98 mL to 20.02 mL) and the pH of the product solution has a remarkable change from 4.3 to 9.7. This large pH change associated with a very small percentage change in the titration volume between post and pre-0.1% of the stoichiometric point (SP), is often called **titration break.** Note that the titration break is symmetric around the stoichiometric point. To ensure that the indicator error (titration error) is within ±0.1%, one needs to select an indicator with a transition range (or said better the **transition point**) falling within the titration break. Methyl orange (MO, 4.4), methyl red (MR, 6.0) and phenolphthalein (PP, 9.0) can be used for the endpoint detection in this titration.

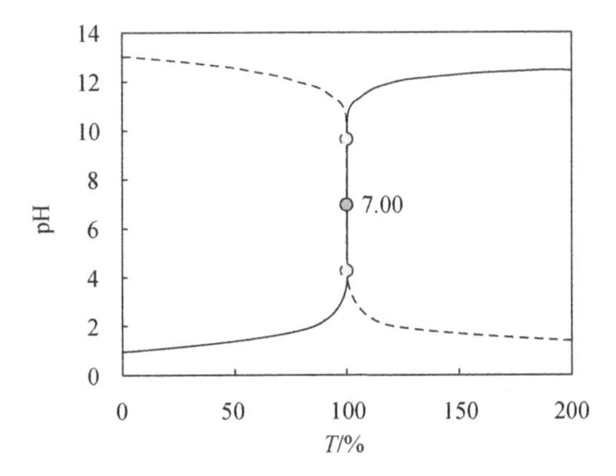

Figure 4.3 The titration curve for HCl with NaOH (solid line) and the titration curve for NaOH with HCl (dashed line). Volume of titrand: 20.00 mL; concentration for both titrand and titrant, 0.1000 mol · L^{-1}.

The titration curves of NaOH with HCl at the same concentration are similar to titrations of HCl with NaOH, but the pH changes occur from the other direction as the dashed line shown in Figure 4.3.

The effect of titrant and titrand concentration on the titration break and the selection of indicators for strong acid/base titration are summarized in Table 4.3. The change in titration concentration affects the titration break and the selection of the indicators but not the stoichiometric point. Increasing the concentration of titrant and titrand 10 fold, the titration break will be expanded 2 pH units with one pH unit being expanded at each end of the titration break (Figure 4.4).

Table 4.3 Summary of Titration Break and Indicator Selection for Strong Acid-base Titration with Titrand and Titrant at the Same Concentration

Titrand	Concentration /mol · L^{-1}	Stoichiometric Point (pH)	Titration Break (pH)	Indicators
HCl	1.0	7.0	3.3～10.7	Methyl orange, methyl red, phenolphthalein
	0.1	7.0	4.3～9.7	Methyl orange, methyl red, phenolphthalein
	0.01	7.0	5.3～8.7	Methyl red
NaOH	1.0	7.0	10.7～3.3	Methyl orange, methyl red, phenolphthalein
	0.1	7.0	9.7～4.3	Methyl red, phenolphthalein
	0.01	7.0	8.7～5.3	Phenolphthalein

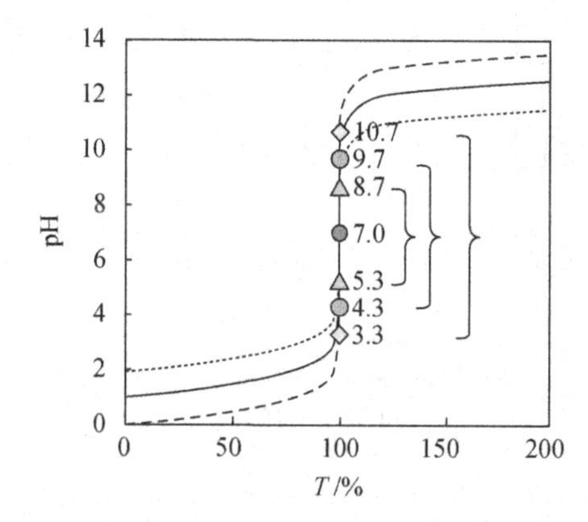

Figure 4.4 Titration curves for 20 mL of HCl with NaOH at the same concentration. Concentrations: 1 mol · L^{-1} (dashed), 0.1 mol · L^{-1} (solid), and 0.01 mol · L^{-1} (dotted).

4.2.2 Monoprotic Acids (Bases)

A weak acid is always titrated by a strong base to ensure the completeness of the reaction, thus enhancing the accuracy of the results. The reaction for titrating a monoprotic acid with a strong base and the titration constant expression are

$$HA + OH^- \rightleftharpoons A^- + H_2O \quad K_t = \frac{[A^-]}{[HA][OH^-]} = \frac{K_a}{K_w}$$

The titration constant for monoprotic acid titration (K_a/K_w) is smaller than that for strong acid titration because the completeness of the reaction is not as good as for a strong acid. One can tell from the titration constant expression that the strength of monoprotic acid determines the completeness of the reaction. The larger the K_a is, the larger the K_t value is, and the more complete the reaction is.

The titration of 20 mL, 0.1000 mol · L^{-1} HAc with 0.1000 mol · L^{-1} NaOH is taken as an example to construct the titration curve for the monoprotic acid. Again, the pH of initial point, the pH of pre-stoichiometric point, the pH of stoichiometric point, and the pH of post-stoichiometric point, each need to be calculated to construct the hypothetical titration curve in order to get the necessary information to select an appropriate indicator.

1. Initial Point

Before NaOH is added, the solution to be titrated is a 0.1000 mol · L^{-1} HAc solution.

$$[H^+] = \sqrt{K_a c} = \sqrt{10^{-4.76-1.00}} = 10^{-2.88}(\text{mol} \cdot L^{-1}), \quad pH = 2.88$$

2. Pre-stoichiometric Point

Before stoichiometric point is reached, a portion of the HAc is neutralized by NaOH, and the solution is a HAc-Ac^- buffer. For example, upon addition of 19.98 mL NaOH, the analytical concentrations of both HAc and NaAc are

$$c(\mathrm{HAc}) = \frac{20.00 \times 0.1000 - 19.98 \times 0.1000}{20.00 + 19.98} = \frac{0.02}{20.00 + 19.98} \times 0.1000$$

$$= 5.0 \times 10^{-5}\ (\mathrm{mol \cdot L^{-1}})$$

$$c(\mathrm{Ac^-}) = \frac{19.98 \times 0.1000}{20.00 + 19.98} = \frac{0.02}{20.00 + 19.98} \times 0.1000 = 5.0 \times 10^{-2}\ (\mathrm{mol \cdot L^{-1}})$$

According to Henderson-Hasselbalch equation,

$$\mathrm{pH} = \mathrm{p}K_a + \lg \frac{c_b}{c_a} = 4.76 + \lg \frac{5 \times 10^{-2}}{5 \times 10^{-5}} = 7.76$$

3. Stoichiometric Point

At the stoichiometric point, all the HAc is just neutralized by the added NaOH and converted to NaAc, thus the product solution is a NaAc aqueous solution.

$$[\mathrm{OH^-}] = \sqrt{K_b c} = \sqrt{\frac{10^{-14.00}}{10^{-4.76}} \times 10^{-1.30}} = 10^{-5.27}\ (\mathrm{mol \cdot L^{-1}}), \quad \mathrm{pH} = 8.73$$

4. Post-stoichiometric Point

Beyond the stoichiometric point, the product solution is composed of NaAc and NaOH, and the $[H^+]$ of the solution is determined by the excess NaOH. For example, upon addition of 20.02 mL NaOH, the concentration of excess NaOH in the product solution is

$$[\mathrm{OH^-}] = c(\mathrm{NaOH}) = \frac{20.02 \times 0.1000 - 20.00 \times 0.1000}{20.00 + 20.02}$$

$$= \frac{0.02}{20.00 + 20.02} \times 0.1000 = 5.0 \times 10^{-5}\ (\mathrm{mol \cdot L^{-1}})$$

Thus, $[\mathrm{H^+}] = 2.0 \times 10^{-10}\ \mathrm{mol \cdot L^{-1}}, \quad \mathrm{pH} = 9.70$

If a weak acid HA with $\mathrm{p}K_a = 7.00$ is titrated, the dissociation of water becomes significant when titration is before but close to the stoichiometric point, and dissociation of A^- becomes significant when titration is beyond but close to the SP. Calculation of $[H^+]$ in these two cases are shown in Examples 3.12 and 3.10, respectively.

Additional data is provided in Table 4.4 and the titration curve is given in Figure 4.5 with pH as the y-axis and the percent titration ($T/\%$) as the x-axis.

Table 4.4 pH Change during the Titration of 20.00 mL, 0.1 mol · L^{-1} HAc and HA, Respectively with NaOH at the Same Concentration

Volume of NaOH/mL	T/%	pH[HAc (pK_a=4.76)]	pH[HA (pK_a=7.00)]
0.00	0.0	2.88	4.00
10.00	50.0	4.76	7.00
18.00	90.0	5.71	7.95
19.80	99.0	6.76	9.00
19.96	99.8	7.46	9.56
19.98	99.9	7.70	9.70
20.00	100.0	8.70	9.85
20.02	100.1	9.70	10.00
20.04	100.2	10.00	10.14
20.20	101.0	10.70	10.70
22.00	110.0	11.68	11.70

The initial pH of a weak acid titration is higher than that of a strong acid titration. After the titration starts, the product solution is actually a $HAc\text{-}Ac^-$ buffer, thus pH changes slowly with each increase of titrant added. When [HAc] to [Ac^-] ratio is close to 1 : 1, the buffer capacity is the highest, thus the rate of pH change is the slowest. When percent titration is close to the stoichiometric point, HAc concentration in the product solution is so low that the $HAc\text{-}Ac^-$ has little buffer capacity, thus the pH increases rapidly when a small volume of NaOH is added. At the stoichiometric point, the product solution is simply a NaAc solution. After the stoichiometric point, the product solution is a mixture of NaAc and NaOH. The pH is largely determined by the excess NaOH, thus the titration is similar to the same stage for strong acid titration. Therefore, the titration curves beyond the stoichiometric point are almost identical for both a weak acid and a strong acid. The pH at the stoichiometric point is still approximately in the middle of the titration break.

Compared with the titration break for titration of 0.1000 mol · L^{-1} HCl, the titration break for 0.1000 mol · L^{-1} HAc is narrowed and lies within weak basic range, pH 7.76 ~ 9.70. Certainly, phenolphthalein can be used for endpoint indication, but methyl orange and methyl red cannot be used.

The strength of an acid is one of the major factors influencing the titration break (Figure 4.5). The titration curve before the stoichiometric point was governed by K_a of the acid titrated. The stronger the acid is, the lower the starting pH of titration break, and the larger the titration break will be. Consider the titration of 20 mL, 0.1000 mol • L^{-1} HA (pK_a=7.0) by NaOH at the same concentration, the titration break is pH 9.70~10.00. Even though an indicator is available with a transition point in the middle of the titration break (pH 9.85), it is not possible to have a titration error less than 0.1%, because the average human eye variation in color observation is ±0.3 pH unit and the required titration break is pH 9.55~10.15 with an anticipated titration error of ±0.2%. For a weak acid with K_a less than $10^{-7.0}$ ($pK_a \geq 7.0$), the titration break is much more difficult to identify and impossible to find an indicator to assure a titration error within ±0.2%.

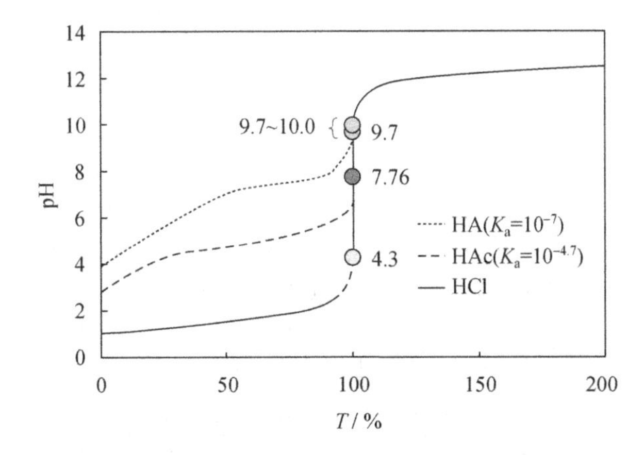

Figure 4.5 Titration curves for 20 mL of 0.1000 mol • L^{-1} acids with NaOH at the same concentration.

Concentration is the other major factor affecting the titration break for a weak acid (Figure 4.6). Before the stoichiometric point, the pH of solution is determined by the ratio of [HA] to [A] instead of the analytical concentration of HA. The titration curves before the stoichiometric point for different titrand concentrations are almost identical except the initial point and first few increments of NaOH added. Increasing the titrand concentration 10 fold will increase the pH of stoichiometric point 0.5 pH unit because $[H^+]=(K_a c)^{1/2}$. The change of the titration curve at the post-stoichiometric point is similar to a strong acid titration, that is, increasing titrand concentration 10 fold will expand titration break one pH unit at the post-stoichiometric side.

Because of the human eye uncertainty in color observation, at least 0.6 pH unit titration break is essential to ensure a titration error within ±0.2%. Considering the strength of a weak acid and the titration concentration, a $cK_a \geq 10^{-8}$ is required for the titration of weak acid with a titration error of less than ±0.2%.

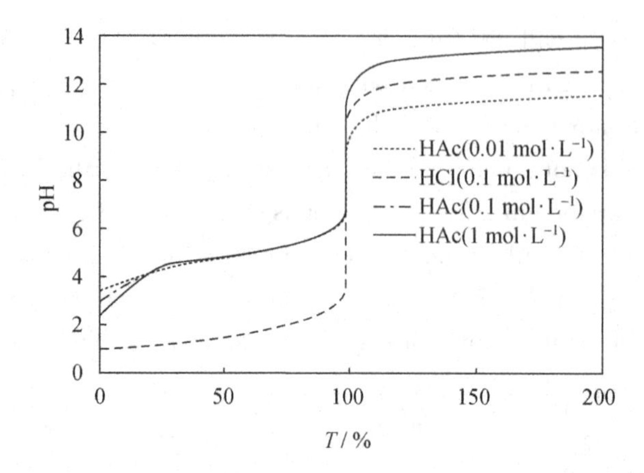

Figure 4.6 Titration curves for 20 mL of monoprotic acid with NaOH at the same concentration.

The titration curve of weak base with a strong acid can be derived in a similar way as the titration of a weak acid. For example, the titration reaction and titration constant expression of NH_3 (pK_b=4.75) with HCl is

$$NH_3 + H^+ \rightleftharpoons NH_4^+ \quad K_t = \frac{[NH_4^+]}{[NH_3][H^+]} = \frac{1}{K_a} = \frac{K_b}{K_w} = 10^{9.25}$$

The data derived for titration of 20 mL, 0.1000 mol · L^{-1} NH_3 with HCl at the same concentration is given in Table 4.5 and the titration curve is given in Figure 4.7. At the stoichiometric point (pH 5.28), the analyte is converted to its conjugate acid and the product solution is acidic. The titration break lies in the acidic pH range (pH 6.25~4.30); therefore, indicators in an acidic pH transition range should be the first choice. Methyl red is appropriate for this titration. If methyl orange is selected, the titration error will be +0.2% at the transition point of pH 4.00. Similar to the titration of a weak acid, the requirement for the titration of a weak base with a titration error less than ±0.2% is $cK_b \geqslant 10^{-8}$.

Table 4.5 pH Change during the Titration of 20 mL, 0.1000 mol · L^{-1} NH_3 with HCl at the Same Concentration

Volume of HCl/mL	T/%	pH	Volume of HCl/mL	T/%	pH
0.00	0.0	11.13	20.00	100.0	5.30
10.00	50.0	9.25	20.02	100.1	4.30
18.00	90.0	8.30	20.04	100.2	4.00
19.80	99.0	7.25	20.20	101.0	3.30
19.96	99.8	6.55	22.00	110.0	2.32
19.98	99.9	6.25			

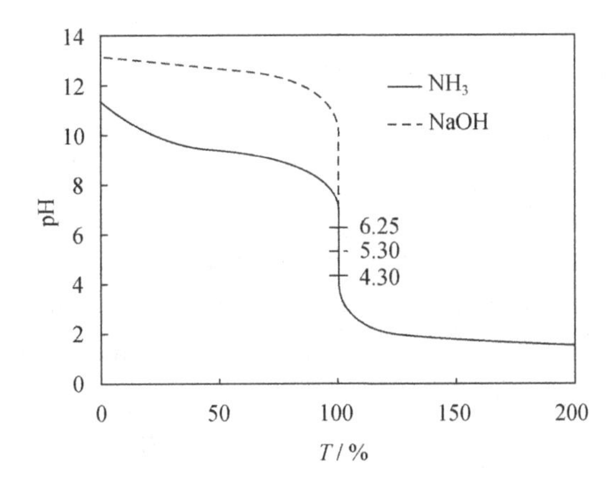

Figure 4.7 The titration curve for 20 mL, 0.1000 mol · L^{-1} NH_3 with HCl at the same concentration.

4.2.3 Strong and Weak Acids (Bases)

1. Titration Curves

Using a 0.1 mol · L^{-1} HCl and 0.2 mol · L^{-1} HA mixture as an example, let us discuss the feasibility to titrate selectively one or both components from the mixture with a titration error less than ±0.2%. The titration curves are given in Figure 4.8. Two titration breaks are expected for the titration of mixed acids.

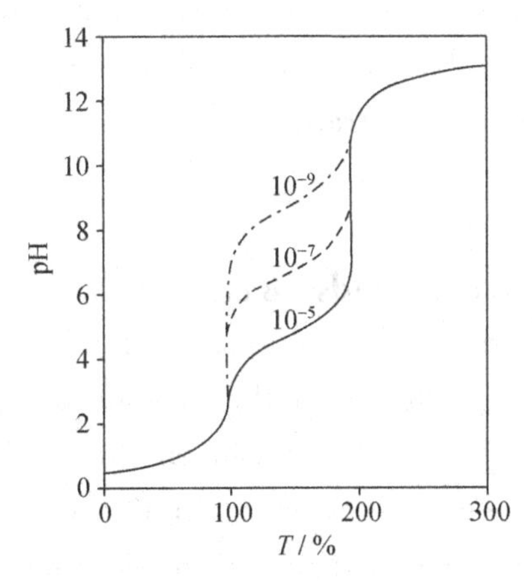

Figure 4.8 Titration curves for strong and weak acid mixture.

- If K_a of HA is 10^{-5}, the titration break for HCl is not large enough for endpoint detection, thus HCl cannot be selectively titrated. The second titration break is appreciably larger, thus facilitating the detection of the total acids in the solution.

- If K_a of HA is 10^{-7}, both the titration breaks are about 0.6 pH units, allowing selective titration of both constituents in the mixture with a titration error of 0.2%.
- If K_a of HA is 10^{-9}, a reasonable titration break of HCl can be obtained for endpoint detection, thus HCl can be selectively titrated in the mixed acids. The second titration break is not large enough to facilitate a selective titration of HA after titration of HCl, thus the determination of total acids is not realistic.

Conclusions for titration of strong and weak bases in a mixture can be reached by following similar discussions.

2. Can a Weak Acid/Base with $cK<10^{-8}$ be Titrated with Titration Error in 0.2%

The titration requirement for a weak acid/base with a titration error less than 0.2% is $cK \geqslant 10^{-8}$. A 0.1 mol · L^{-1} HAc ($pK_b=4.76$) can be accurately titrated by NaOH because $cK_a > 10^{-8}$, but the conjugated base, Ac^- ($pK_b=9.24$) cannot be titrated with titration error less than 0.2% because of the titrand $cK_b < 10^{-8}$.

Can Ac^- be determined by back-titration by adding excess HCl standard solution and then titrating with NaOH standard solution? After addition of HCl standard solution, the composition of the resulted solution is HCl + HAc. In this case, there is not a reasonable titration break for HCl, so the total acids will be titrated at the end. There is no way to determine the excess amount of HCl to calculate the amount of Ac^-. Back titration does not change the stoichiometric point and the titration break, thus back titration is not a feasible approach for the titration of a weak acid/base with $cK<10^{-8}$.

4.2.4 Polyfunctional Weak Acids (Bases)

A polyfunctional acid/base has more than one functional group to undergo stepwise dissociations. There is the possibility of the titration of total acid groups and the selective titration of one or more specific acid groups. In both cases, $cK \geqslant 10^{-8}$ is still the criteria for a feasible titration with titration error $\leqslant 0.2\%$. In stepwise titration, difference between the two adjacent dissociation constants should be big enough to assure a titration with the desired accuracy. For a diprotic acid (H_2A), the difference of the pK_a values, ΔpK, greater than 4 is essential to assure the titration error $\leqslant 1\%$ for the first titration, $\Delta pK > 5$ is necessary to assure a titration error $\leqslant 0.5\%$. Therefore, $cK_i \geqslant 10^{-8}$ and $K_i/K_{i+1} \geqslant 10^5$ are the requirements for the stepwise titration of a polyfunctional acid/base with a titration error $\leqslant 0.5\%$.

The ΔpK of many polyprotic acids is small (less than 4), thus a stepwise titration is not feasible. Examples are oxalic acid ($H_2C_2O_4$, $pK_{a1} = 1.25$, $pK_{a2} = 4.29$), tartaric acid ($HOOC(CHOH)_2COOH$, $pK_{a1} = 3.04$, $pK_{a2} = 4.37$), and citric acid ($HOOC(OH)C(CH_2COOH)_2$, $pK_{a1} = 3.13$, $pK_{a2} = 4.76$, and $pK_{a3} = 6.40$). Their last step dissociation constant is greater than 10^{-7}, thus the total titrable hydronium ion can be titrated with NaOH.

Phosphoric acid (H_3PO_4) is a typical polyfunctional acid with $pK_{a1} = 2.16$, $pK_{a2} = 7.21$, and $pK_{a3} = 12.32$. The difference between each dissociation constants is about 10^5, and the first two dissociation constants $\geqslant 10^{-7}$, making the first and the second stepwise titration feasible with the first stoichiometric point at $H_2PO_4^-$ and the second at HPO_4^{2-}. K_{a3} is much less than what is required for a feasible titration, thus the third stepwise titration is not possible.

PO_4^{3-} is a polyfunctional base with $pK_{b1} = 1.68$, $pK_{b2} = 6.79$, and $pK_{b3} = 11.84$. Similarly, PO_4^{3-} can be stepwise titrated to $H_2PO_4^-$, but the third stepwise titration is not feasible. The titration curve for $0.1\ mol \cdot L^{-1}\ H_3PO_4$ with $0.1\ mol \cdot L^{-1}$ NaOH is given in Figure 4.9.

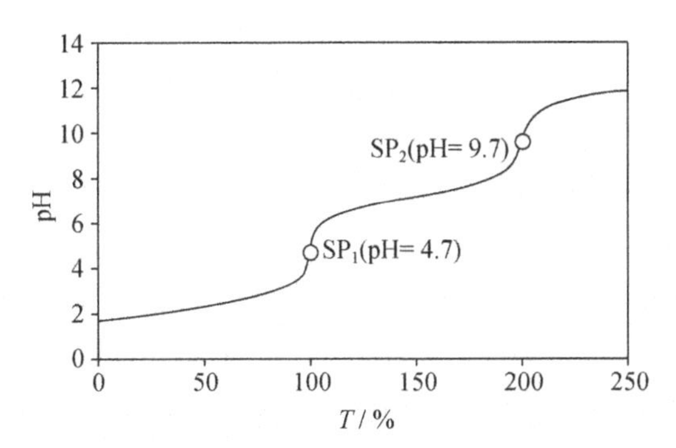

Figure 4.9 The titration curve for 20 mL, $0.05\ mol \cdot L^{-1}\ H_3PO_4$ with $0.1\ mol \cdot L^{-1}$ NaOH.

The composition of the product solution and the dissociation constants complicate the calculation of the titration of a polyfunctional acid. The following discussion is focused on the requirement for stepwise or continuous titration of polyfunctional acids and endpoint detection near the stoichiometric point.

For the titration of $0.1\ mol \cdot L^{-1}\ H_3PO_4$ with $0.1\ mol \cdot L^{-1}$ NaOH, the product for the first stoichiometric point is NaH_2PO_4,

$$[H^+] = \sqrt{\frac{cK_{a2}}{1 + c/K_{a1}}} = \sqrt{\frac{0.05 \times 10^{-7.21}}{1 + 0.05/10^{-2.16}}} = 10^{-4.71}\ (mol \cdot L^{-1}), \quad pH = 4.71$$

Methyl orange can be used for endpoint detection with a titration error $\leqslant -0.5\%$ when the color at the stoichiometric point is comparable to a methyl orange color reference solution containing the same concentration of $H_2PO_4^-$.

The reaction product is Na_2HPO_4 at the second stoichiometric point.

$$[H^+]=\sqrt{\frac{cK_{a3}+K_w}{c/K_{a2}}}=\sqrt{\frac{0.033\times 10^{-12.32}+10^{-14.00}}{0.033/10^{-7.21}}}=10^{-9.66}(\text{mol}\cdot\text{L}^{-1})$$

$$\text{pH}=9.66$$

Thymolphenolphthalein that has a transition point about 10 can be used for endpoint indication with a titration error of about $+0.5\%$. At the endpoint, color changes from colorless to light blue.

Amphiprotic species can be considered as the product of polyfunctional acid/base titration; however, both the acidic and basic properties need to be evaluated to determine the feasibility of the titration of an amphiprotic species.

Titration of sodium carbonate (Na_2CO_3) is a typical example for titration of a polyfunctional base, because primary standard grade sodium carbonate (Na_2CO_3) is often used to standardize HCl. Carbonate is the base form of CO_2, commonly denoted as carbonic acid (H_2CO_3) with $pK_{a1}=6.38$ and $pK_{a2}=10.25$. Two endpoints are observed in the titration of sodium carbonate (Na_2CO_3) (Figure 4.10).

$$CO_3^{2-}+H^+ \rightleftharpoons HCO_3^-$$

$$HCO_3^-+H^+ \rightleftharpoons CO_2+H_2O$$

The first stoichiometric point is pH 8.3 with a small titration break. Usually a cresol red and thymol blue mixed indicator is used for the endpoint detection. With a $\Delta pK <5$, a titration error about 0.5% can be achieved using a sodium bicarbonate ($NaHCO_3$) color reference solution with the same mixed indicator.

The second titration break is much larger than the first one; therefore, the second titration break is always used for standardization. The reaction products for the second titration break are carbonic acid (H_2CO_3) and carbon dioxide (CO_2). The pH at the stoichiometric point can be calculated from the product solution of a saturated carbon dioxide (CO_2) solution with a concentration at about 0.04 mol · L^{-1}.

$$[H^+]=\sqrt{K_{a1}\cdot c}=\sqrt{4.2\times 10^{-7}\times 0.04}=1.3\times 10^{-4}(\text{mol}\cdot\text{L}^{-1}),\quad \text{pH}=3.9$$

The endpoint observation can be sharpened by boiling the product solution to eliminate carbon dioxide (CO_2) when the indicator just starts to change color. The pH will increase to an alkaline range because the concentration of the dissolved CO_2 has decreased. The titration can be resumed after the solution has cooled. Further addition of acid titrant will show a sharp transition because the pH change is now large.

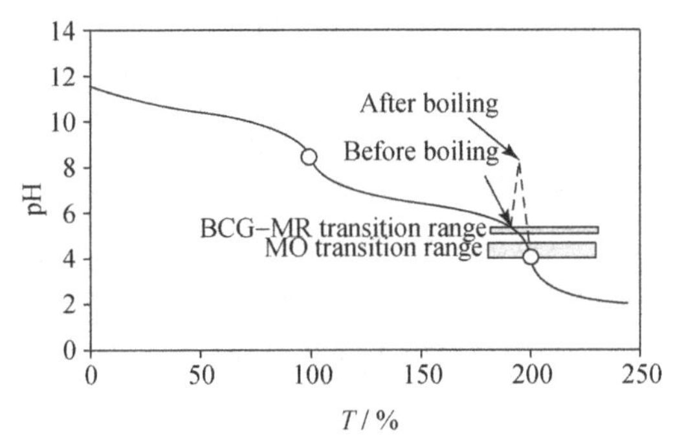

Figure 4.10 The titration curve for 25 mL, 0.05 mol · L^{-1} Na_2CO_3 with 0.1 mol · L^{-1} HCl.

4.2.5 Mixture of Weak Acids (Bases)

The mixture of weak acids, HA + HB (K_a(HA) > K_a(HB)) is much like a polyprotic acid. When $c(HA) \cdot K_a(HA) \geqslant 10^{-8}$ and $c(HA) \cdot K_a(HA)/c(HB) \cdot K_a(HB) \geqslant 10^5$, a stepwise titration is feasible with the composition of $A^- + HB$ at the first stoichiometric point. When $cK_a \geqslant 10^{-8}$ for both acids, and $c(HA) \cdot K_a(HA)/c(HB) \cdot K_a(HB) \geqslant 10^5$, a stepwise titration is feasible for both individual acids. The total titrable acid can be determined in an one-step titration, when $cK_a \geqslant 10^{-8}$ for both acids.

4.3 TITRATION ERROR CALCULATIONS

In most cases, the endpoint does not occur at the stoichiometric point, thus the amount of titrant between the endpoint and the stoichiometric point are different. This difference is called **titration error** (E_t), as defined by Equation (4-5).

$$E_t = \frac{n(\text{amount of titrant in excess or shortage})}{n(\text{amount of titrant at SP})} \tag{4-5}$$

4.3.1 Strong Acids (Bases)

For the titration of HCl with NaOH, the titration error (E_t) is

$$E_t = \frac{n(\text{NaOH in excess or shortage})}{n(\text{NaOH at SP})} = \frac{n_{ep}(\text{NaOH}) - n_{ep}(\text{HCl})}{n_{sp}(\text{NaOH})}$$

$$= \frac{c_{ep}(\text{NaOH})V_{ep} - c_{ep}(\text{HCl})V_{ep}}{c_{sp}(\text{HCl})V_{sp}}$$

As $V_{ep} \approx V_{sp}$,

$$E_t = \frac{c_{ep}(NaOH) - c_{ep}(HCl)}{c_{sp}(HCl)} \tag{4-6}$$

In practice, the pH at the endpoint is known, thus $c_{ep}(NaOH) - c_{ep}(HCl)$ can be calculated via their association with $[H^+]$ and $[OH^-]$ at the endpoint.

The proton-condition equation (PCE) at any point during titration is $c(NaOH) + [H^+] = c(HCl) + [OH^-]$, thus at the endpoint, $c_{ep}(NaOH) - c_{ep}(HCl) = [OH^-]_{ep} - [H^+]_{ep}$. Therefore, the titration error for HCl with NaOH is

$$E_t = \frac{[OH^-]_{ep} - [H^+]_{ep}}{c_{sp}(NaOH)} \tag{4-7}$$

At stoichiometric point, the amount of NaOH added equals the amount of HCl titrated. Water autoprotolysis is the only dissociation equilibrium at this point, therefore $[OH^-] = [H^+]$ and the pH=7.00.

If phenolphthalein is used for endpoint detection, the endpoint occurs beyond the stoichiometric point, then $[OH^-] > [H^+]$, a positive titration error occurs. If methyl orange or methyl red is used for endpoint detection, the endpoint occurs before the stoichiometric point, then $[H^+] > [OH^-]$, a negative titration error occurs.

【Example 4.1】 Calculate the titration error for 0.10 mol · L^{-1} HCl with 0.10 mol · L^{-1} NaOH when

(1) methyl orange is used for indicating the endpoint (pH 4.4);

(2) phenolphthalein is used for indicating the endpoint (pH 9.0).

Answer:

(1) $pH_{ep} = 4.4$: $E_t = \frac{[OH^-]_{ep} - [H^+]_{ep}}{c_{sp}(HCl)} = \frac{10^{-9.6} - 10^{-4.4}}{0.05} \times 100\% = -0.08\%$

(2) $pH_{ep} = 9.0$: $E_t = \frac{10^{-5.0} - 10^{-9.0}}{0.05} \times 100\% = +0.02\%$

4.3.2 Monoprotic Weak Acids (Bases)

For the titration of monoprotic acid (HA) with NaOH, the titration error E_t is

$$\begin{aligned} E_t &= \frac{n(\text{NaOH in excess or shortage})}{n(\text{NaOH at SP})} = \frac{n_{ep}(NaOH) - n_{ep}(HA)}{n_{sp}(NaOH)} \\ &= \frac{c_{ep}(NaOH)V_{ep} - c_{ep}(HA)V_{ep}}{c_{sp}(HA)V_{sp}} = \frac{c_{ep}(NaOH) - c_{ep}(HA)}{c_{sp}(HA)} \end{aligned} \tag{4-8}$$

The proton-condition equation (PCE) at any point during titration is $c(NaOH) +$

$[H^+]=[A^-]+[OH^-]$, thus at the endpoint, $c_{ep}(NaOH)=[OH^-]_{ep}-[H^+]_{ep}+[A^-]_{ep}$, and the mass-balance equation (MBE) is $c_{ep}(HA)=[HA]_{ep}+[A^-]_{ep}$, Therefore,

$$c_{ep}(NaOH)-c_{ep}(HA)=[OH^-]_{ep}-[H^+]_{ep}+[HA]_{ep}$$

Equation (4-8) becomes

$$E_t=\frac{c_{ep}(NaOH)-c_{ep}(HA)}{c_{sp}(HA)}=\frac{[OH^-]_{ep}-[H^+]_{ep}-[HA]_{ep}}{c_{sp}(HA)}$$

As $c_{sp}(HA)\approx c_{ep}(HA)$, $\frac{[HA]_{ep}}{c_{sp}(HA)}\approx x_{ep}(HA)$. Hence,

$$E_t=\frac{[OH^-]_{ep}-[H^+]_{ep}}{c_{sp}(HA)}-x_{ep}(HA) \tag{4-9}$$

[Example 4.2] Calculate the titration error of 0.10 mol · L^{-1} HAc with 0.10 mol · L^{-1} NaOH when endpoint pH is (1) 9.0 and (2) 7.0.

Answer:

(1) $pH_{ep}=9.0$: $E_t=\frac{[OH^-]_{ep}}{c_{sp}(HA)}-\frac{[H^+]_{ep}}{[H^+]_{ep}+K_a}$

$$=\left(\frac{10^{-5.0}}{10^{-1.3}}-\frac{10^{-9.0}}{10^{-9.0}+10^{-4.76}}\right)\times 100\%=+0.01\%$$

(2) $pH_{ep}=7.0$: $E_t=\frac{10^{-7.0}}{10^{-7.0}+10^{-4.76}}\times 100\%=-0.6\%$

4.3.3 Polyfunctional Acids (Bases)

A diprotic acid (H_2A) is taken as an example to discuss titration error.

1. A Stepwise Titration to the First Stoichiometric Point is Feasible

$$E_t=\frac{n_{ep}(\text{NaOH in excess or shortage})}{n_{sp}(H_2A)}$$

$$=\frac{n_{ep}(NaOH)-n_{ep}(H_2A)}{n_{sp}(H_2A)}=\frac{c_{ep}(NaOH)-c_{ep}(H_2A)}{c_{sp}(H_2A)} \tag{4-10}$$

The proton-condition equation (PCE) is $c(NaOH)+[H^+]=[HA^-]+2[A^{2-}]+[OH^-]$, and the mass-balance equation (MBE) is $c(H_2A)=[H_2A]+[HA^-]+[A^{2-}]$. Thus, $c(NaOH)-c(H_2A)=[A^{2-}]-[H_2A]+[OH^-]-[H^+]$. Therefore, at the endpoint

$$E_t=\frac{c_{ep}(NaOH)-c_{ep}(H_2A)}{c_{sp}(H_2A)}=\frac{[A^{2-}]-[H_2A]+[OH^-]-[H^+]}{c_{sp}(H_2A)}$$

$$=\frac{[OH^-]-[H^+]}{c_{sp}(H_2A)}+\frac{[A^{2-}]}{c_{sp}(H_2A)}-\frac{[H_2A]}{c_{sp}(H_2A)}$$

As $c_{sp}(HA) \approx c_{ep}(HA)$, $\frac{[H_2A]_{ep}}{c_{sp}(HA)} \approx x_{2(ep)}$, and $\frac{[A^{2-}]}{c_{sp}(H_2A)} \approx x_{0(ep)}$. Therefore,

$$E_t = \frac{[OH^-]-[H^+]}{c_{sp}(H_2A)} + x_{0(ep)} - x_{2(ep)} \tag{4-11}$$

2. Titration to A^{2-} is Feasible

$$E_t = \frac{n_{ep}(\text{NaOH in excess or shortage})}{n_{sp}\left(\frac{1}{2}H_2A\right)} = \frac{n_{ep}(NaOH) - n_{ep}\left(\frac{1}{2}H_2A\right)}{n_{sp}\left(\frac{1}{2}H_2A\right)}$$

$$= \frac{c_{ep}(NaOH) - 2c_{ep}(H_2A)}{2c_{sp}(H_2A)} \tag{4-12}$$

Based on the proton condition equation and mass balance equation,

$$c(NaOH) - c(H_2A) = [OH^-]-[H^+]-[HA^-]-2[H_2A]$$

Thus, at the endpoint

$$E_t = \frac{c_{ep}(NaOH) - c_{ep}(H_2A)}{2c_{sp}(H_2A)}$$

$$= \frac{[OH^-]-[H^+]-[HA^-]-2[H_2A]}{2c_{sp}(H_2A)}$$

$$= \frac{[OH^-]-[H^+]}{2c_{sp}(H_2A)} - \frac{[HA^-]}{2c_{sp}(H_2A)} - \frac{[H_2A]}{c_{sp}(H_2A)} \tag{4-13}$$

As $c_{sp}(HA) \approx c_{ep}(HA)$, $\frac{[H_2A]_{ep}}{c_{sp}(HA)} \approx x_{2(ep)}$, and $\frac{[HA^-]}{c_{sp}(H_2A)} \approx x_{1(ep)}$. Therefore,

$$E_t = \frac{[OH^-]-[H^+]}{2c_{sp}(H_2A)} - \frac{1}{2}x_{1(ep)} - x_{2(ep)} \tag{4-14}$$

4.4 PREPARATION OF STANDARD SOLUTIONS

4.4.1 Standard Acid Solutions

Hydrochloric acid is the usual titrant for the determination of bases. The standard solution of HCl is prepared by the standardization of diluted HCl prepared from commercially available concentrated HCl (about 12 mol · L^{-1}). Sodium carbonate (Na_2CO_3) is a low cost and easily available primary standard that is usually used to standardize HCl solution. Because sodium carbonate absorbs water easily at ambient conditions, it is usually dried to constant weight in an oven at 270～300℃ for 1 h and cooled in a desiccator before use. Sodium carbonate primary standard can also be prepared by converting analytical grade sodium bicarbonate ($NaHCO_3$) at 270～

300℃ for 1 h to the carbonate. The second titration break of sodium carbonate is always used for standardization of a HCl standard solution because the titration break is greater.

Methyl orange or the mixed indicator methyl orange-indigo carmine can be used for endpoint indication, but the color transition is not sharp and thus not easy to observe. Methyl red-bromocresol green mixed indicator is often used for standardizing a HCl standard solution with sodium carbonate. In this titration, when the product solution has just turned from green into red, the titration is stopped and the solution is boiled for 1 min to remove CO_2. After the removal of CO_2, the solution turns green, because the product solution at this point is composed of a small amount of $NaHCO_3$ (pH $\approx$ 8). After cooling the solution, further titration changes the green to red.

Sodium tetraborate decahydrate ($Na_2B_4O_7 \cdot 10H_2O$) sometimes called borax, can also be used for standardizing HCl. Borax becomes a H_3BO_3-$H_2BO_3^-$ buffer ($pK_a = 9.24$) when dissolved in water.

$$B_4O_7^{2-} + 5H_2O \rightleftharpoons 2H_3BO_3 + 2H_2BO_3^-$$

The primary standard crystalline borax is usually stored in a closed container with about 60% relative humidity to avoid losing the hydrated water. Borax is usually weighed directly to an Erlenmeyer flask or a beaker to standardize the acid solution, because borax has a large molecular mass. In the titration reaction, the stoichiometric ratio of borax to HCl is 2 : 1. To standardize 0.1 mol • L^{-1} HCl with 0.05 mol • L^{-1} borax, 20~25 mL water is added to the measured amount of borax to make a 0.05 mol • L^{-1} solution before titration. The product solution at stoichiometric point is a 0.10 mol • L^{-1} H_3BO_3,

$$[H^+] = \sqrt{K_a c} = \sqrt{10^{-9.24-1.0}} = 10^{-5.12}(\text{mol} \cdot L^{-1}), \quad pH = 5.12$$

Therefore, methyl red is an appropriate indicator for this titration.

4.4.2 Standard Base Solutions

Sodium hydroxide (NaOH) is usually used in the titration of acids. NaOH easily absorbs water and CO_2 from the atmosphere, thus sodium hydroxide solutions can contain significant amounts of sodium carbonate (Na_2CO_3).

Sodium carbonate is essentially insoluble in nearly saturated sodium hydroxide. The insoluble Na_2CO_3 will settle to the bottom of the container after the saturated NaOH has equilibrated for a couple of days. The supernatant can be withdrawn carefully to prepare diluted NaOH solution free of sodium carbonate. The water for preparing NaOH standard solution should be boiled to remove any dissolved CO_2, because dissolved CO_2 can cause a titration error. The titration error caused by CO_2 will be discussed in the subsequent section.

The primary standard, potassium hydrogen phthalate ($KHC_8H_4O_4$) used to standardize a NaOH standard solution is an amphiprotic compound ($pK_{a2}=5.41$) that acts as a weak acid in the titration reaction with NaOH. Potassium hydrogen phthalate is easy to store, because the phthalate does not absorb water from the ambient atmosphere.

The oxalic acid dihydrate ($H_2C_2O_4 \cdot 2H_2O$, $pK_{a1}=1.25$, $pK_{a2}=4.29$) can also be used to standardize NaOH. One mole oxalic acid reacts with two moles of NaOH with phenolphthalein as an indicator.

4.4.3 The Carbonate Error

Carbon dioxide can react quickly with sodium hydroxide in solution or in the solid state to produce the corresponding carbonate:

$$CO_2 + 2NaOH \rightleftharpoons Na_2CO_3 + H_2O$$

The production of one carbonate ion uses up two hydroxide ions; however, the carbonate ion produced in this manner does not always release the two hydroxide ions, thus the possibility of a systematic error, called **carbonate error**, can occur. Carbonate error is an error that often occurs on acid-base titration involving NaOH standard solution in the following situations:

(1) Carbon dioxide was not properly removed from NaOH saturated solution or water used for preparing the NaOH standard solution.

In this case, there are CO_3^{2-} ions existing in NaOH standard solution which is the reaction product of CO_2 with NaOH. In the standardization of NaOH using potassium hydrogen phthalate or oxalic acid, the transition range is in alkaline pH with phenolphthalein for endpoint indication. CO_3^{2-} reacting with acid is converted to HCO_3^-. If the NaOH standard solution prepared in this manner is used to determine the concentration of an acid, a systematic error will be produced when the titration break is in acidic pH with methyl orange or methyl red as an indicator. The carbonate ion (CO_3^{2-}) is converted to CO_2, making the "real" concentration of NaOH standard solution larger than that from the standardization of the sodium hydroxide. Less NaOH titrant is consumed, and a negative systematic error is produced for the concentration of an acid with sodium hydroxide as a standard solution.

(2) NaOH solution was properly prepared and standardized but carbon dioxide

was absorbed during storage.

One mole CO_2 reacts with 2 moles NaOH to produce one mole Na_2CO_3. If this NaOH standard solution is used as a titrant to determine the concentration of an acid with a titration break in alkaline pH with phenolphthalein to indicate the endpoint, a systematic error will occur. In this situation, Na_2CO_3 is neutralized and converted to HCO_3^-, which means one mole of Na_2CO_3 is equivalent to one mole NaOH, while the second mole NaOH is not used to neutralize the acid. Therefore, the "real" concentration of NaOH standard solution is smaller than that from standardization and more NaOH titrant is consumed, thus a positive systematic error is produced for determining the concentration of the acid.

(3) Effect of the convertion rate of CO_2 to H_2CO_3 on acid-base titration.

The converting equilibrium and constant expression of the dissolved CO_2 to H_2CO_3 is

$$CO_2 + H_2O \rightleftharpoons H_2CO_3 \quad K = \frac{[H_2CO_3]}{[CO_2]} = 2.2 \times 10^{-3}$$

About 0.2% of the total carbon dioxide species is in the form of carbonate (H_2CO_3) that can react with NaOH. At the phenolphthalein endpoint, the purple color in the solution may fade, because the dissolved CO_2 is slowly converted to H_2CO_3 to reduce the pH of the final solution; therefore, the phenolphthalein endpoint solution needs to be purple color for about 30 second to make sure that titration is complete.

4.5 EXAMPLES OF ACID-BASE TITRATIONS

4.5.1 Determination of Total Alkalinity

Total alkalinity is a measure of the ability of a solution to neutralize acids to the stoichiometric points of carbonate or bicarbonate. In the natural environment, carbonate alkalinity has a greater influence on the total alkalinity due to the common occurrence and dissolution of carbonate rocks and presence of carbon dioxide in the atmosphere. Total alkalinity is a useful measurement of natural water quality.

1. Total Alkalinity of Marine Water in Biochemical Processes (Wolf-Gladrow D A, Zeebe R E, Klaas C, et al. Total alkalinity: The explicit conservative expression and its application to biogeochemical processes. Marine Chemistry, 2007(106): 287-300)

Total alkalinity and dissolved inorganic carbon (DIC) are conservative quantities with respect to mixing and changes in temperature and pressure and are, therefore, used in oceanic carbon cycle models. The changes of total alkalinity will be helpful in understanding various biogeochemical processes such as the formation and remineralization of organic matter by the microalgae promotion of the precipitation and dissolution of calcium carbonate. Total alkalinity can be used in combination with other models to explore the correlation of changes in total alkalinity with uptake of nutrients, nitrogen fixation followed by remineralization of organic matter and subsequent nitrification of ammonia in tropical and subtropical regions.

2. Total Alkalinity of Soda Ash

Soda ash is a commercial product containing a high percentage of sodium carbonate (Na_2CO_3). The total alkalinity (total titrable bases) can be titrated with a HCl standard solution. The HCl standard solution volumetric titration for total alkalinity is the method used in the Chinese National Standard (GB210-1992) and American Society for Testing and Materials (ASTM) International Standard (ASTM E359-00(2005)).

Two titration breaks can be observed during titration of Na_2CO_3. The second acidic pH titration break is used to titrate the total alkalinity that is usually reported as percent Na_2CO_3 in the sample. When the amount of individual bases, such as NaOH and Na_2CO_3, are desired, volumetric information for both titration breaks can be used for calculations.

4.5.2 Determination of Nitrogen

Nitrogen determination is routinely performed for a variety of substances of interest in chemical, food and agricultural industries. Examples of samples for nitrogen determinations are amino acids, proteins, synthetic drugs, grains, meat, fertilizers, soil, and waste water. The Kjeldahl method, developed by Johan Kjeldahl in 1883, is the most commonly used method for determination of organic nitrogen. The Kjeldahl method and apparatus have been substantially modified over the years. Automated Kjeldahl systems are now available.

The Kjeldahl method includes three main steps: digestion, distillation, and titration. First, the sample is decomposed in hot, concentrated sulfuric acid with copper ion as a catalyst. The decomposition step is the most critical and the most time consuming step in the Kjeldahl method, because the sample needs to be completely digested and all the nitrogen compounds in the samples converted to NH_4^+ to ensure reasonable accuracy. To extract the converted NH_4^+ from the sample matrix, excess NaOH is added to the digesting solution to convert NH_4^+ into NH_3. The liberated ammonia is then distilled, collected in an acidic solution, and determined by acid-base titration.

- The liberated ammonia can be collected with an excess of HCl standard solution, and back titrated by a NaOH standard solution. The stoichiometric point is in an acidic pH range with pH≈5, because the product solution of nitrogen is as NH_4^+. Methyl red or methyl orange can be used for endpoint indication.
- The liberated ammonia (NH_3) can also be collected with an excess of boric acid (H_3BO_3) to produce $NH_4H_2BO_3$. The produced $H_2BO_3^-$ ($pK_b = 4.69$) is then titrated with an HCl standard solution. The product solution is composed of NH_4^+ and H_3BO_3 at pH≈5. Methyl red can be used for endpoint detection. In this situation, only a HCl standard solution is needed, because the stoichiometrically produced $H_2BO_3^-$ can be directly titrated by HCl standard solution while excess H_3BO_3 does not interfere with the determination.

The produced NH_4^+ from digestion can also be determined by reaction with formaldehyde. Four moles of NH_4^+ can react with 6 moles of formaldehyde to stoichiometrically produce 3 moles of hydronium ion and one mole of conjugated acid of hexamethylenetetramine($(CH_2)_6N_4$, urotropine),

$$4NH_4^+ + 6HCHO \rightleftharpoons (CH_2)_6N_4H^+ + 3H^+ + 6H_2O$$

The product from this reaction is a mixture of strong acid and weak acid ($(CH_2)_6N_4H^+$, $pK_a = 5.13$). The total acids can be titrated with NaOH standard solution to $(CH_2)_6N_4$ by using phenolphthalein for endpoint indication. One mole NaOH is equivalent to one mole NH_4^+. It is important that the small amount of formic acid (HCOOH) in formaldehyde must be pre-titrated with NaOH (phenolphthalein as the indicator) so that a systematic error can be avoided.

4.5.3 Determination of Boric Acid

Boric acid (H_3BO_3) cannot be directly titrated by NaOH because the dissociation constant is not large enough to assure the completeness of the titration reaction and thus the titration break is too small to ensure a sharp endpoint transition. The titration break of boric acid with sodium hydroxide can be enlarged with the addition of glycerol, mannitol, or similar polyols. Mannitol has been the most frequently used polyol for titration of boric acid. When the mannitol is present in a large excess over the boric acid, about 0.35 mole of mannitol per liter of solution at the equivalence point will yield both a small titration error and a sharp endpoint as measured potentiometrically (Max Hollander, and William Rieman, III. Titration of boric acid in presence of mannitol. Ind Eng Chem Anal Ed, 1945, 17 (9): 602-603).

4.6 ACID-BASE TITRATIONS IN NON-AQUEOUS SOLVENTS

Non-aqueous titration is the titration of substances dissolved in non-aqueous solvents. This titration is widely used in pharmaceutical analysis because many pharmaceuticals are weak acids or bases, which are nearly impossible to be titrated with reasonable accuracy in aqueous solutions. In addition, many pharmaceuticals are organic compounds which are insoluble in water but soluble in non-aqueous solvents.

4.6.1 Non-aqueous Solvents

Many organic solvents, like water, are amphiprotic solvents, which undergo autoprotolysis to form a pair of ionic species. Examples are methanol, ethanol, formic acid, acetic acid, liquid nitrogen, and ethylenediamine. For an amphiprotic solvent (SH), the autoprotolysis equation and the constant expression is

$$SH + SH \rightleftharpoons SH_2^+ + S^- \quad K_s = a(SH_2^+) \cdot a(S^-)$$

The ion-product (K_s) of amphiprotic organic solvent is associated with the acid-base property of the solvent itself. The ion-product of some amphiprotic solvents are listed in Table 4.6.

Table 4.6 Solvent Ion-product (pK_s) Constants at 25℃

Solvent	pK_s at 25℃	Solvent	pK_s at 25℃
Water (H_2O)	14.0	Formic acid (HCOOH)	6.2
Ethanol (C_2H_5OH)	19.1	Ethylenediamine	15.3
Methanol (CH_3OH)	16.7	Acetone (CH_3CN)	32.2
Acetic acid (HAc)	14.4	Methylisobutylketone (MIBK)	>30

Like K_w of H_2O, one can get information to judge the feasibility of a titration in SH solvent by evaluating K_s and the dissociation constant of an acid or base in this solvent. The conjugated acid and the conjugated base of SH are SH_2^+ and S^-, respectively. Therefore, strong acid/base titration reaction and the titration constant in an amphiprotic solvent (SH) would be

$$SH_2^+ + S^- \rightleftharpoons 2SH \qquad K_t = \frac{1}{a(SH_2^+) \cdot a(S^-)} = \frac{1}{K_s}$$

The smaller the K_s value is, the larger the K_t is, the more complete the titration is. For example, the titration break for titration of 0.1 mol · L^{-1} strong acid with strong base at the same concentration in water ($pK_{s(w)}=14.0$) is 5.4 pH units (pH 4.3~9.7). If the titration is carried out in ethanol ($pK_s=19.1$), the titration break will be 10.5 pH units ($p[C_2H_5OH_2^+]4.3 \sim p[C_2H_5O^-]4.3$), i.e. pH 4.3~14.8). The titration in ethanol has a larger titration constant, thus a larger titration break.

4.6.2 Examples of Non-aqueous Titrations

Many ketimines are instable in aqueous solution, thus their titration in aqueous solution is not feasible. Ketimines can be titrated in glacial acetic acid with perchloric acid as the titrant and crystal violet as the indicator. This titration is used to determine purity and yield in the study of the hydrogenation rates of ketimines (Pickard P L, Iddings. Titration of ketimines in glacial acetic acid. Anal Chem, 1959, 31(7): 1228-1230).

Amino acids can also be titrated in glacial acetic acid with perchloric acid as the titrant and crystal violet as the indicator (Nadeau G F, Branchen L E. The quantitative titration of amino acids in glacial acetic acid solution. J Am Chem Soc, 1935, 57(7): 1363-1365). The titration in glacial acetic acid gives a larger titration break.

Some compounds of pharmaceutical interests that are titrable in non-aqueous titrations are listed in Table 4.7 (Wollish E G, Pifer C W, Schmall M. Titration in nonaqueous solvent to pharmaceuticals. Anal Chem, 1954, 26(11): 1704-1706).

Table 4.7 Compounds Titrable with Non-aqueous Titrations

Category of Compounds	Examples
Acidic compounds	Carboxylic acids Amino acids Sulfonamides Barbiturates (enolic compounds) Phenolic compounds
Basic compounds	Amines (primary, secondary, tertiary) Quaternary ammonium compounds Amino acids Heterocyclic nitrogen compounds (narcotics, antihistaminics, antibiotics, oxazolines, pyrazolones)
Reducible compounds	Aldehydes, ketones Esters Alcohols Amides Aromatic nitro compounds

Chapter 4 Questions and Problems

4.1 A 0.1000 mol · L^{-1} formic acid (HCOOH, $pK_a = 3.77$) is titrated with 0.1000 mol · L^{-1} NaOH.

(1) Calculate the pH at stoichiometric point (SP);

(2) Calculate the titration error using PP as the indicator (pT=9.0).

4.2 A 20.00 mL aliquot of 0.0020 mol · L^{-1} barium hydroxide ($Ba(OH)_2$) is titrated with 0.0020 mol · L^{-1} HCl.

(1) Calculate the pH at the following points in the titration:

i. Pre-SP 0.1%; ii. SP; iii. Post-SP 0.1%.

(2) Calculate the titration error using PP as the indicator (color transition point pH≈8.0).

4.3 A 0.5000 mol · L^{-1} monobase ($pK_b = 6.00$) is titrated with 0.5000 mol · L^{-1} HCl.

(1) Calculate the pH at the following points in the titration:

i. Pre-SP 0.1%; ii. SP; iii. Post-SP 0.1%.

(2) When both concentrations change to 0.02000 mol · L^{-1}, calculate the pH at the following points in the titration:

i. Pre-SP 0.1%; ii. SP; iii. Post-SP 0.1%.

4.4 A 0.1000 mol · L^{-1} H_3PO_4 ($pK_{a1} \sim pK_{a3}$: 2.16, 7.21, and 12.32) solution is titrated with 0.1000 mol · L^{-1} NaOH. Calculate the titration error when titration stops at

(1) pH=5.0; (2) pH=10.0.

4.5 A solution consists of 0.1000 mol · L^{-1} NaOH and 0.2000 mol · L^{-1} sodium acetate (NaAc, $pK_a = 4.76$). A 20.00 mL aliquot of such solution is titrated with 0.1000 mol · L^{-1} HCl.

(1) Calculate the pH at the following points in the titration:

i. Pre-SP 0.1%; ii. SP; iii. Post-SP 0.1%.

(2) Calculate the titration error when titration stops at pH 7.0.

4.6 A sample is analyzed by the Kjeldahl method. In the process, the sample is decomposed in hot, concentrated sulfuric acid to convert the all the nitrogen compounds to ammonium ion. The resulting solution is then cooled, diluted and made basic. The liberated ammonia (NH_3, $pK_b = 4.75$) is distilled, collected in 100 mL of 0.3 mol · L^{-1} HCl, judged to be an excess amount of acid. The acidic ammonium solution is then back titrated with 0.2 mol · L^{-1} NaOH. Assuming the analytical concentration of ammonium in the solution is 0.2 mol · L^{-1}, calculate

(1) pH at SP; (2) titration error at pH 4.0; (3) titration error at pH 7.0.

4.7 Can the analytes in the following list be directly titrated using an acid or base standard solution? If so, name the standard solution of the titrant and the indicator; If not, what can be done to make acid-base titration feasible for this specific analyte?

(1) $CH_3CH_2NH_2$ (ethylamine); (2) NH_4Cl; (3) HF (hydrogen fluoride);

(4) NaAc; (5) H_3BO_3 (boric acid); (6) $NaHCO_3$ (sodium bicarbonate);

(7) $Na_2B_4O_7$ $10H_2O$; (8) C_6H_7N (aniline or phenylamine); (9) Glycine;

(10) NaHS (sodium hydrosulfide); (11) Na_2HPO_4.

4.8 Write the product at stoichiometric point, the indicator that can be used for endpoint judgement around the stoichiometric point, and also the standard solution used for titration.

(1) $C_6H_8O_7$ (citric acid); (2) $C_4H_4O_4$ (maleic acid);

(3) $NaOH+(CH_2)_6N_4$ (hexamethylene tetramine); (4) $H_2SO_4+H_3PO_4$;

(5) $HCl+H_3BO_3$; (6) HF+HAc.

4.9 Write the experimental design for the determination of each component in the mixture.

(1) $HCl+NH_4Cl$; (2) $Na_2B_4O_7 \cdot 10H_2O + H_3BO_3$; (3) $HCl+H_3PO_4$.

CHAPTER 5

COMPLEXATION REACTION AND COMPLEXOMETRIC TITRATION

5.1 Complexes and Formation Constants

5.1.1 Formation Constants

5.1.2 Concentration of ML_n in Complexation Equilibria

5.1.3 Ethylenediaminetetraacetic Acid (EDTA) and Metal-EDTA Complexes

5.1.4 Side Reaction Coefficients and Conditional Formation Constants in Complexation Reactions

5.2 Metallochromic Indicators

5.2.1 How a Metallochromic Indicator Works

5.2.2 Color Transition Point pM ($(pM)_t$) for Metallochromic Indicators

5.2.3 Frequently Used Metallochromic Indicators

5.3 Titration Curves and Titration Errors

5.3.1 Titration Curves

5.3.2 Titration Errors

5.3.3 pH Control in Complexometric Titrations

5.4 Selective Titrations of Metal Ions in the Presence of Multiple Metal Ions

5.4.1 Selective Titration by Regulating pH

5.4.2 Selective Titration Using Masking Reagents

5.5 Applications of Complexometric Titrations

5.5.1 Buffer Selection in Complexometric Titrations

5.5.2 Titration Methods and Applications

5.5.3 Preparation of Standard Solutions

Many metal ions form slightly dissociated complexes. The formation of these complexes can serve as the basis for the accurate and convenient titrations of metal ions. Complexometric titrations are especially useful for the quantification of a large number of metal ions. Indirect titration and back titration can also be employed for determination of anions and organic compounds. Complexometry is widely applied to drug, clinical, food and biological product analyses as well as environmental analysis.

A complexometric titration usually involves one or more complexation equilibria as well as acid-base equilibria. In this chapter, the side reaction coefficients are introduced to give an idea how to evaluate quantitatively the effect of many contributing factors on complexation equilibria. This approach simplifies the way to deal with complicated complexation equilibria and is applicable to other equilibria.

5.1 COMPLEXES AND FORMATION CONSTANTS

5.1.1 Formation Constants

Most metal ions react with electron-pair donors to form coordination compounds and complexes. The donor species, or ligand, must have at least one pair of unshared electrons available for bond formation. Water, ammonia and halide ions are common inorganic ligands. In fact, most metal ions in aqueous solution actually exist as aqua complexes. For example, copper (Ⅱ) in aqueous solution is readily complexed with water molecules to form species such as $Cu(H_2O)_4^{2+}$. We often simplify such complexes in chemical equations by writing the metal ion as if it were uncomplexed Cu^{2+}; however, such ions are actually aqua complexes in aqueous solutions.

The number of covalent bonds that a cation tends to form with electron donors is the cation coordination number. Typical values for coordination numbers are 2, 4, and 6. The species formed as a result of coordination can be electrically positive, neutral, or negative. For example, copper (Ⅱ), which has a coordination number of 4, forms a cationic amine complex ($Cu(NH_3)_4^{2+}$), a neutral complex with glycine ($Cu(NH_2CH_2COO)_2$), and an anionic complex with chloride ion ($CuCl_4^{2-}$).

Titrimetric methods based on complex formation, also called complexometric titrations, have been used for more than a century. The truly remarkable growth in their analytical application began in the 1940s based on a particular class of coordination compounds called chelates. A chelate is produced when a metal ion coordinates with two or more donor groups of a single ligand to form a five- or six-member heterocyclic ring. The copper complex of glycine mentioned in the previous paragraph is one example. Here, copper bonds to both the oxygen of the carboxyl group and the nitrogen of the amine group.

Complexation reactions involve a metal ion (M) reacting with a ligand (L) to form a complex (ML), as shown in the equation

$$M + L \rightleftharpoons ML$$

The charges on the ions have been omitted so as to be general in the description.

Complexation reactions occur in a stepwise manner; the reaction in above equation is often followed by additional reactions:

$$\begin{aligned}
&M + L \rightleftharpoons ML && K_1 = \frac{[ML]}{[M][L]} \\
&ML + L \rightleftharpoons ML_2 && K_2 = \frac{[ML_2]}{[ML][L]} \\
&\vdots && \vdots \\
&ML_{n-1} + L \rightleftharpoons ML_n && K_n = \frac{[ML_n]}{[ML_{n-1}][L]}
\end{aligned} \tag{5-1}$$

Each of the reactions is associated with a **stepwise formation constant** (also called stepwise stability constant, consecutive stability constant), K_1 through K_4 as written on right side of each equation above. We can also write the equilibria as the sum of individual steps. These equilibria have **overall formation constants** (cumulative constants, gross constants) designated by the symbol, β_n. Thus,

$$\begin{aligned}
&M + L \rightleftharpoons ML && \beta_1 = \frac{[ML]}{[M][L]} = K_1 \\
&M + 2L \rightleftharpoons ML_2 && \beta_2 = \frac{[ML_2]}{[M][L]^2} = K_1 \cdot K_2 \\
&\vdots && \vdots \\
&M + nL \rightleftharpoons ML_n && \beta_n = \frac{[ML_n]}{[M][L]^n} = K_1 \cdot K_2 \cdot \cdots \cdot K_n
\end{aligned} \tag{5-2}$$

Except for the first step, the overall formation constants are products of the stepwise formation constants for the individual steps leading to the products.

The stepwise dissociation constants (instability constants) are the reciprocal of each stepwise formation constant in reverse order.

$$\begin{aligned}
&ML_n \rightleftharpoons ML_{n-1} + L && K_{diss1} = \frac{1}{K_n} \\
&ML_{n-1} \rightleftharpoons ML_{n-2} + L && K_{diss2} = \frac{1}{K_{n-1}}
\end{aligned} \tag{5-3}$$

$$\vdots \qquad\qquad \vdots$$

$$ML \rightleftharpoons M + L \qquad K_n = \frac{1}{K_1}$$

The bigger the K is, the smaller the K_{diss} is, the more stable the complex will be.

5.1.2 Concentration of ML_n in Complexation Equilibria

We can obtain $[ML_n]$ from β_n expressions above,

$$[ML] = \beta_1[M][L]$$
$$[ML_2] = \beta_2[M][L]^2$$
$$\vdots \qquad \vdots$$
$$[ML_n] = \beta_n[M][L]^n$$

The total concentration $c(M)$ can be written as

$$c(M) = [M] + [ML] + [ML_2] + \cdots + [ML_n]$$

From the overall formation constants, the concentration of the complexes can be expressed in terms of the free metal concentration $[M]$, to give

$$c(M) = [M](1 + \beta_1[L] + \beta_2[L]^2 + \cdots + \beta_n[L]^n)$$

For a given species like ML, we can calculate the fraction of the total metal concentration existing in that form, x. Thus, x_0, x_1, x_2, $\cdots$, x_n is the fraction of M, ML, ML_2, $\cdots$, ML_n.

$$x_0 = \frac{[M]}{c(M)} = \frac{[M]}{[M] + [ML] + [ML_2] + \cdots + [ML_n]}$$
$$= \frac{[M]}{[M] + \beta_1[M][L] + \beta_2[M][L]^2 + \cdots + \beta_n[M][L]^n}$$
$$= \frac{1}{1 + \beta_1[L] + \beta_2[L]^2 + \cdots + \beta_n[L]^n} \tag{5-4}$$
$$x_1 = \frac{[ML]}{c(M)} = \frac{\beta_1[L]}{1 + \beta_1[L] + \beta_2[L]^2 + \cdots + \beta_n[L]^n}$$
$$\vdots \qquad \vdots$$
$$x_n = \frac{[ML_n]}{c(M)} = \frac{\beta_n[L]^n}{1 + \beta_1[L] + \beta_2[L]^2 + \cdots + \beta_n[L]^n}$$

If $[L]$ is known, the concentration of M, ML, ML_2, $\cdots$, ML_n can be calculated by $[ML_n] = c(M) \cdot x_n$. Generally, the initial concentration for ligand, $c(L)$, is much more than $c(M)$; thus, the amount of L forming complexes with metal ions can be neglected.

【Example 5.1】 A 10 mL of 0.20 mol · L^{-1} $CuSO_4$ solution is mixed with a 10 mL of 2.00 mol · L^{-1} $NH_3 \cdot H_2O$ solution. Calculate $[Cu^{2+}]$, $[Cu(NH_3)^{2+}]$, $[Cu(NH_3)_2^{2+}]$,

$[Cu(NH_3)_3^{2+}]$ and $[Cu(NH_3)_4^{2+}]$ in solution.

Answer: $c(Cu^{2+})=0.10\ mol\cdot L^{-1}$, $c(NH_3)=1.00\ mol\cdot L^{-1}$, $[NH_3]\approx 1.00-4\times 0.10 = 0.60(mol\cdot L^{-1})$.

$\beta_1=K_1=1.3\times 10^4$, $\beta_2=K_1\cdot K_2=4.1\times 10^7$

$\beta_3=K_1\cdot K_2\cdot K_3=3.0\times 10^{10}$, $\beta_4=K_1\cdot K_2\cdot K_3\cdot K_4=3.9\times 10^{12}$

$$x_0=\frac{1}{1+0.60\beta_1+(0.60)^2\beta_2+(0.60)^3\beta_3+(0.60)^4\beta_4}=\frac{1}{5.1\times 10^{11}}$$

$$[Cu^{2+}]=c(Cu^{2+})\cdot x_0=0.10\times\frac{1}{5.1\times 10^{11}}=2.0\times 10^{-13}(mol\cdot L^{-1})$$

$$[Cu(NH_3)^{2+}]=c(Cu^{2+})\cdot x_1=0.10\times\frac{0.60\beta_1}{5.1\times 10^{11}}=1.5\times 10^{-9}(mol\cdot L^{-1})$$

$$[Cu(NH_3)_2^{2+}]=c(Cu^{2+})\cdot x_2=0.10\times\frac{(0.60)^2\beta_2}{5.1\times 10^{11}}=2.9\times 10^{-6}(mol\cdot L^{-1})$$

$$[Cu(NH_3)_3^{2+}]=c(Cu^{2+})\cdot x_3=0.10\times\frac{(0.60)^3\beta_3}{5.1\times 10^{11}}=1.3\times 10^{-3}(mol\cdot L^{-1})$$

$$[Cu(NH_3)_4^{2+}]=c(Cu^{2+})\cdot x_4=0.10\times\frac{(0.60)^4\beta_4}{5.1\times 10^{11}}=0.099(mol\cdot L^{-1})$$

The concentration ratio is:

$$[Cu^{2+}]:[Cu(NH_3)^{2+}]:[Cu(NH_3)_2^{2+}]:[Cu(NH_3)_3^{2+}]:[Cu(NH_3)_4^{2+}]$$
$$=1:7.5\times 10^3:1.5\times 10^7:6.5\times 10^9:5.0\times 10^{11}$$

It can be seen from the calculated results that the complexes tend to adopt a higher coordination number when an excess amount of ligands exists. Plots of x_n versus pNH_3 gives the distribution diagrams of copper ammonia complexes at various ammonia concentrations (Figure 5.1).

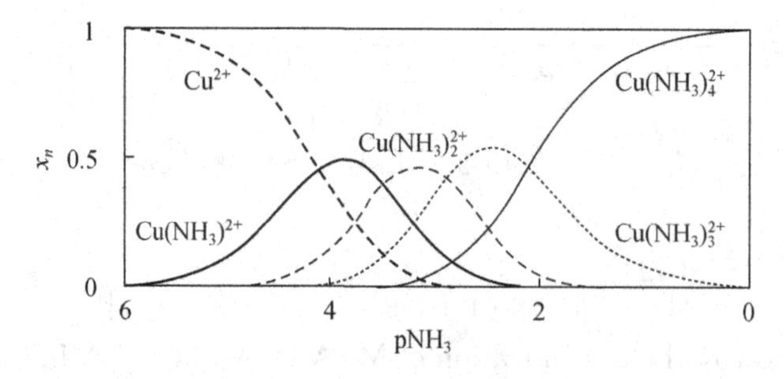

Figure 5.1 Composition of copper-ammonia complex solution as a function of pNH_3.

This figure shows that the distribution diagrams of copper ammonia complexes are similar to multiprotic acids. Again, the complexes tend to adopt a higher coordination number as pNH_3 decrease ($[NH_3]$ increases). The correspondent pNH_3 value at the point of intersection of adjacent curves ($x_i = x_{i-1}$) equals the formation constant, $\lg K_i$ related to ML_{i-1} and ML_i. The stepwise formation constants for copper ammonia complexes are close to each other, because several forms of complexes exist in a wide NH_3 concentration range. Similar to the acid-base distribution diagram, a predominance-area diagram (ladder diagram) can be obtained based on the distribution diagram.

5.1.3 Ethylenediaminetetraacetic Acid (EDTA) and Metal-EDTA Complexes

Some inorganic complexing agents, such as NH_3, Cl^-, F^-, and CN^-, are unidentate which means the molecule has a single donor group. These agents react with metal ions in steps as mentioned above with relatively small stepwise formation constants, so the overall formation constant is not large enough for titration purpose, so few unidentate entities have been applied to titrimetry. Titration of Ag^+ using CN^- and of Hg^{2+} using Cl^- are two successful examples.

Organic complexing agents, such as glycine and ethylenediaminetetraacetic acid (EDTA), have more than two groups available for covalent binding with metal ions to form stable five- or six-member heterocyclic rings. The complexes formed are called chelates. Amongst the multidentate chelating agents, complexones with aminocarboxylic groups is widely used for complexometric titrations. These complexone molecules contain both oxygen and nitrogen as donor atoms, which easily form complexes with almost all metal ions.

Ethylenediaminetetraacetic acid, also called (ethylenedinitrilo) tetraacetic acid, which is commonly shortened to EDTA, is a widely used complexometric titrant. EDTA itself has four acidic groups in the molecule to dissociate four protons, thus denoted as H_4Y. In aqueous solution, two protons of —COOH groups will move to nitrogen to form double zwitterion in the molecule with the structure given.

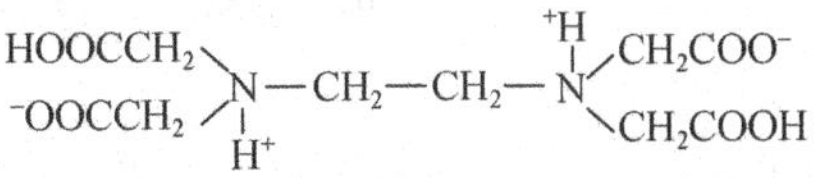

$$\begin{matrix} HOOCCH_2 \diagdown & & & & {}^{+}H \quad \diagup CH_2COO^- \\ & N—CH_2—CH_2—N & & & \\ {}^{-}OOCCH_2 \diagup & H^+ & & & \diagdown CH_2COOH \end{matrix}$$

The net charge on this species is zero. The two nitrogen atoms in EDTA (2 —COO^- groups in aqueous solution) can accept two more protons to form H_6Y^{2+}; thus, EDTA becomes a hexaprotic acid with six stepwise dissociation constants with $pK_{a1} \sim pK_{a6}$ of 0.9, 1.6, 2.07, 2.75, 6.24, and 10.34. In aqueous solution, there are seven species for EDTA, H_6Y^{2+}, H_5Y^+, H_4Y, H_3Y^-, H_2Y^{2-}, HY^{3-} and Y^{4-}. The distribution diagram is presented in Figure 5.2.

To relate acid-base dissociation equilibrium with complexation equilibrium,

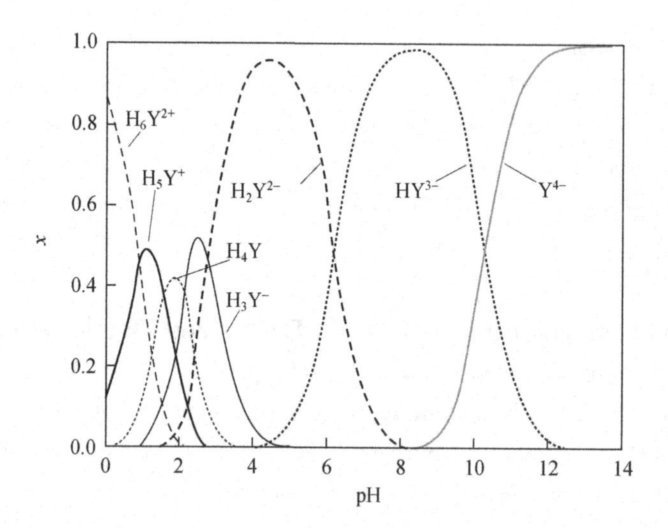

Figure 5.2 Composition of EDTA solution as a function of pH.

EDTA can be considered as the stepwise complexation products of Y^{4-} with protons. The stepwise reactions, formation constants (K^H, also called protonation constants), and cumulative formation constants (β^H) are

$$Y^{4-} + H^+ \rightleftharpoons HY^{3-} \qquad K_1^H = \frac{[HY^{3-}]}{[H^+][Y^{4-}]} = \frac{1}{K_{a6}}, \quad \beta_1^H = K_1^H = \frac{[HY^{3-}]}{[H^+][Y^{4-}]} = 10^{10.34}$$

$$HY^{3-} + H^+ \rightleftharpoons H_2Y^{2-} \qquad K_2^H = \frac{[H_2Y^{2-}]}{[H^+][HY^{3-}]} = \frac{1}{K_{a5}}, \quad \beta_2^H = K_1^H K_2^H = \frac{[H_2Y^{2-}]}{[H^+]^2[Y^{4-}]} = 10^{16.58}$$

$$\vdots \qquad\qquad \vdots \qquad\qquad \vdots$$

$$H_5Y^+ + H^+ \rightleftharpoons H_6Y^{2+} \qquad K_6^H = \frac{[H_6Y^{2+}]}{[H^+][H_5Y^+]} = \frac{1}{K_{a1}}, \quad \beta_6^H = K_1^H K_2^H \cdots K_6^H = \frac{[H_6Y^{2+}]}{[H^+]^6[Y^{4-}]} = 10^{23.9}$$

EDTA forms chelates with all cations except alkali metals, and most of these chelates are sufficiently stable for titration. This great stability undoubtedly results from the several complexing sites within the molecule that give rise to a cage-like structure, in which the common structure for metal-EDTA complexes is shown in Figure 5.3. EDTA behaves as a hexadentate ligand that six donor atoms are involved in bonding the divalent metal cation.

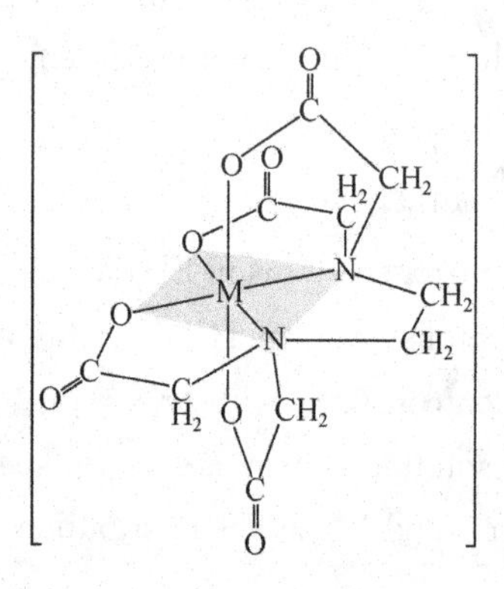

Figure 5.3 Structure of a metal-EDTA complex.

Most metal-EDTA chelates are colorless which makes endpoint observation less complicated. Colored metal ions form darker colored chelates which can make the endpoint difficult to observe when the concentration is high.

5.1.4 Side Reaction Coefficients and Conditional Formation Constants in Complexation Reactions

EDTA complexes with most metal ions in a 1 : 1 ratio:

$$M + Y \rightleftharpoons MY$$

Formation constant refers to the equilibrium involving the fully unprotonated species Y^{4-} with the metal ion

$$K(MY) = \frac{[MY]}{[M][Y]} \tag{5-5}$$

In complexometric titration, common side reactions can occur with hydrogen ions, hydroxide ions, buffering substances, masking reagents, and interfering metal ions. These reactions schematically shown below, produce compounds not included in the schemes upon which the analytical procedures are based.

M	+	Y	$\rightleftharpoons$	MY	Main reaction
A/ \OH		H/\N		H/ \OH	
MA M(OH)		HY NY		MHY M(OH)Y	
⋮ ⋮		⋮			Side reaction
MA_n $M(OH)_n$		H_6Y			
M′		Y′		(MY)′	

The completeness of the main reaction is altered by side reactants of M and Y, as well as product MY. Occurrence of side reaction of MY will be favorable for titration, and occurrence of side reactions of M and Y is the opposite. For M, there are other species like $MA, MA_2, \cdots$, $M(OH)$, $M(OH)_2, \cdots$, which do not take part in the main reaction. A sum of concentration of all metal species, $[M']$, denotes the concentration not only of the free metal ion but also all the other forms of the metal in solution that have not reacted with the complexing agent. In a corresponding manner, $[Y']$ represents not only the concentration of the free ligand but also the concentration of all the other forms of the complexing agent not bound to the metal. $[(MY)']$ represents not only MY but also other forms of MY. Thus, $K'(MY)$ factors in the side reactions of the metal, the ligand, and MY.

$$K'(MY) = \frac{[(MY)']}{[M'][Y']} \tag{5-6}$$

K' is a measure of completeness of the main reaction when side reactions take place, and is called conditional formation constant (or simplified as conditional constant). Side reaction coefficients were introduced by Schwarzenbach to calculate the conditional constant, and were defined by the following equations

$$\alpha_M = [M']/[M], \quad \alpha_Y = [Y']/[Y], \quad \text{and} \quad \alpha_{MY} = [(MY)']/[MY]$$

The side reaction coefficients are measures of the extent of the side reactions. Side reaction of Y, M and MY will be described in the following paragraphs.

1. Side Reaction Coefficient of Y

When the titration is carried out in a solution with pH<12, EDTA is in the protonated form H_iY. The protonation reactions can be considered as side reaction of Y with metal ion M. The side reaction coefficient of Y with H is

$$\begin{aligned}\alpha_{Y(H)} &= \frac{[Y']}{[Y]} = \frac{[Y]+[HY]+\cdots+[H_6Y]}{[Y]} \\ &= \frac{[Y]+[H][Y]\beta_1^H+[H]^2[Y]\beta_2^H+\cdots+[H]^6[Y]\beta_6^H}{[Y]} \qquad (5\text{-}7) \\ &= 1+[H]\beta_1^H+[H]^2\beta_2^H+\cdots+[H]^6\beta_6^H\end{aligned}$$

The $\alpha_{Y(H)}$ is actually the reciprocal value of Y fraction, x_0, as defined in acid-base equilibrium chapter for the multiprotic acid, H_6Y. $\alpha_{Y(H)}$ is a function of concentration of hydronium ion. When acidity of the solution increases, $\alpha_{Y(H)}$ increases; thus, $\alpha_{Y(H)}$ is also called the **acidic effective coefficient.**

【Example 5.2】 Calculate $\alpha_{Y(H)}$ of EDTA at pH 5.00.

Answer: According to Appendix D2, $K_{a1} \sim K_{a6}$ for EDTA are $10^{-0.9}$, $10^{-1.6}$, $10^{-2.07}$, $10^{-2.75}$, $10^{-6.24}$, and $10^{-10.34}$, respectively. Stepwise formation constants for EDTA, $K_1^H \sim K_6^H$, are $10^{10.34}$, $10^{6.24}$, $10^{2.75}$, $10^{2.07}$, $10^{1.6}$, and $10^{0.9}$; thus, the accumulative constants, $\beta_1^H \sim \beta_6^H$, are $10^{10.34}$, $10^{16.58}$, $10^{19.33}$, $10^{21.40}$, $10^{23.0}$, and $10^{23.9}$.

According to Equation (5-7), we will find

$$\begin{aligned}\alpha_{Y(H)} &= 1+[H]\beta_1^H+[H]^2\beta_2^H+\cdots+[H]^6\beta_6^H \\ &= 1+10^{-5.00+10.34}+10^{-10.00+16.58}+10^{-15.00+19.33}+10^{-20.00+21.40} \\ &\quad +10^{-25.00+23.0}+10^{-30.00+23.9} \\ &= 1+\underline{10^{5.34}}+\underline{10^{6.58}}+10^{4.33}+10^{1.40}+10^{-2.0}+10^{-6.1} \\ &= 10^{6.60}\end{aligned}$$

There are a few terms in the equation for calculating $\alpha_{Y(H)}$. In a solution with a specific pH, the equation can be simplified to a composition of 2~3 terms with the others neglected. It can be seen from the above calculation that H_2Y is the predominant form for EDTA at pH 5.00, and HY is second in predominance.

$\alpha_{Y(H)}$ is a very important information in complexometric titration, So the $\alpha_{Y(H)}$

values for the whole pH range are given in Appendix D4 for convenience.

The plot of $\alpha_{Y(H)}$ versus pH illustrates clearly the influence of acidity. The pH of solution dramatically affects the $\alpha_{Y(H)}$ value. At pH 1.0, $\alpha_{Y(H)} = 10^{18.3}$, which means side reaction of Y with H is considerable. [Y] is only $10^{-18.3}$ of [Y′]. $\alpha_{Y(H)}$ decreases as pH increases. At pH 12, $\alpha_{Y(H)} = 1$; thus, at pH>12, there is no side reaction of Y with H.

2. Side Reaction Coefficient of M

When there is a side reaction between M and A with M and Y as the main reaction, we have

$$\alpha_{M(A)} = \frac{[M]+[MA]+[MA_2]+\cdots+[MA_n]}{[M]}$$

$$= 1+[A]\beta_1+[A]^2\beta_2+\cdots+[A]^n\beta_n \qquad (5\text{-}8)$$

$\alpha_{M(A)}$ changes as [A] changes. A can be a buffer agent, other complexation agent to prevent formation of hydro complex, or masking agent to mask co-existing metal ions. As an example, if titration is carried out in a solution with high pH, A can be OH^- which forms hydroxo complexes with M. To illustrate the side reactions of M with A, plots of $\lg\alpha_{M(OH)}$ versus pH and $\lg\alpha_{M(NH_3)}$ versus $\lg[NH_3]$ for some metal ions are presented in Figures 5.4 and 5.5.

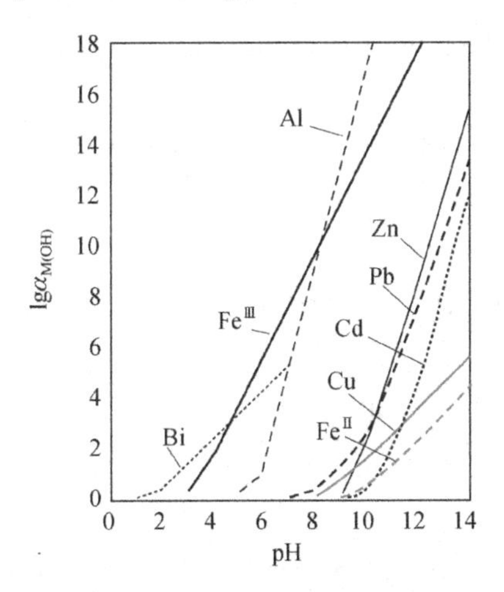

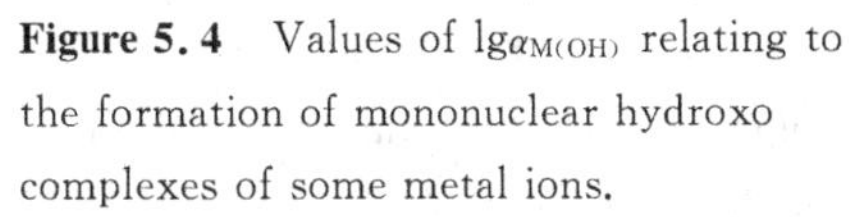
Figure 5.4 Values of $\lg\alpha_{M(OH)}$ relating to the formation of mononuclear hydroxo complexes of some metal ions.

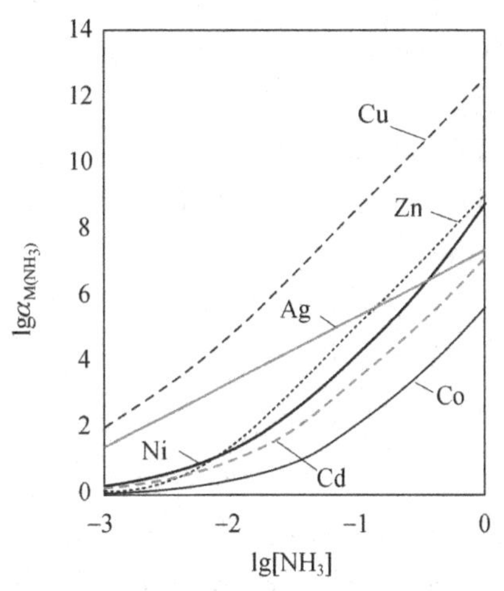

Figure 5.5 Plots of $\lg\alpha_{M(NH_3)}$ versus $\lg[NH_3]$ for selected metal ions.

The analyst is often confronted with the problem of calculating conditional constants for a system containing not one, but several side reactions of M. For example, if there are side reactions of M with both A and B, an overall side reaction coefficient, α_M, can be calculated from the individual α values of the various side reactions using the equation

$$\alpha_M = \frac{[M']}{[M]} = \frac{[M]+[MA]+[MA_2]+\cdots+[MB]+[MB_2]+\cdots}{[M]}$$

$$= \frac{[M]+[MA]+[MA_2]+\cdots}{[M]} + \frac{[M]+[MB]+[MB_2]+\cdots}{[M]} - \frac{[M]}{[M]}$$

$$= \alpha_{M(A)} + \alpha_{M(B)} - 1$$

If the number of side reactions is P, then

$$\alpha_M = \alpha_{M(A_1)} + \alpha_{M(A_2)} + \cdots + (1-P) \qquad (5\text{-}9)$$

The last term in parentheses can be understood by noting that every individual α_M equals $[M']/[M]$ by definition. In this expression, $[M']$ contains $[M]$ in addition to the concentrations of all the species formed by the reactions between M and the interfering ligand. Every individual α_M thus contains the term 1 accounting for $[M]$, and the last term in the parenthesis is to subtract the number of times when $[M]$ is included in the calculation while it should not be. As a rule, the equations are used when the extent of side reactions is appreciable, i. e. , when one or a few of the terms are large (several power of ten). Then $(1-P)$ is small in comparison and can be neglected. Usually, one of the individual values will predominate, and the other coefficients can be disregarded.

【Example 5.3】 Calculate $\lg\alpha_{Zn}$ at pH 11.0 with $[NH_3]=0.10\ mol \cdot L^{-1}$. $\lg\beta_1 \sim \lg\beta_4$ for $Zn(NH_3)_4^{2+}$ are 2.27, 4.61, 7.01, 9.06, respectively.

Answer:

$$\alpha_{Zn(NH_3)} = 1 + [NH_3]\beta_1 + [NH_3]^2\beta_2 + [NH_3]^3\beta_3 + [NH_3]^4\beta_4$$

$$= 1 + 10^{-1.00+2.27} + 10^{-2.00+4.61} + 10^{-3.00+7.01} + 10^{-4.00+9.06} = 10^{5.10}$$

From Appendix D5, $\lg\alpha_{Zn(OH)}=5.4$ at pH 11.0. Thus,

$$\alpha_{Zn} = \alpha_{Zn(NH_3)} + \alpha_{Zn(OH)} - 1 = 10^{5.1} + 10^{5.4} - 1 = 10^{5.6}$$

$$\lg\alpha_{Zn} = 5.6$$

One should be aware that $[A]$ is the equilibrium concentration for A, i. e. , concentration of A not complexed with M. In most cases, A is a weak base and easily protonated; therefore, $[A]$ will change as pH changes. If the protonation reaction of A with H^+ is considered to be the side reaction of A with M, then

$$\alpha_{A(H)} = \frac{[A']}{[A]} = \frac{[A]+[HA]+\cdots+[H_nA]}{[A]}$$

$$= 1 + [H]\beta_1 + \cdots + [H]^n\beta_n \qquad (5\text{-}10)$$

When the titration reaction is close to the stoichiometric point, [M] will be extremely small. The amount of A reacted with M can be neglected. [A′] is approximately the analytical concentration, $c(A)$, thus [A] can be obtained from the expression:

$$[A] = \frac{[A']}{\alpha_{A(H)}} \approx \frac{c(A)}{\alpha_{A(H)}}$$

【Example 5.4】 Calculate $\lg\alpha_{Zn(NH_3)}$ in solution with pH = 9.0 and $c(NH_3) = 0.1\ mol \cdot L^{-1}$, assuming the amount of NH_3 reacting with zinc can be neglected. $pK_b(NH_3) = 4.6$.

Answer:

$\lg K^H(NH_4^+) = 9.4$, $\alpha_{NH_3(H)} = 1 + [H] \cdot K^H(NH_4^+) = 1 + 10^{-9.0+9.4} = 10^{0.5}$

Thus,

$$[NH_3] = \frac{[NH_3']}{\alpha_{NH_3(H)}} \approx \frac{c(NH_3)}{\alpha_{NH_3(H)}} = \frac{10^{-1.0}}{10^{0.5}} = 10^{-1.5}(mol \cdot L^{-1})$$

$$\alpha_{Zn(NH_3)} = 10^{3.2}, \quad \lg\alpha_{Zn(NH_3)} = 3.2$$

3. Side Reaction of Complex, α_{MY}

MY will be protonated in very acidic solution to form acid complex MHY. This reaction can be considered as the side reaction of MY with proton. Thus,

$$\alpha_{MY(H)} = 1 + [H] \cdot K^H(MHY) \tag{5-11}$$

MY will form basic complexes at very basic pH. This reaction can be considered as the side reaction of MY with OH^-; thus,

$$\alpha_{MY(OH)} = 1 + [OH] \cdot K^{OH}(MOHY) \tag{5-12}$$

The $K^H(MHY)$ and $K^{OH}(MOHY)$ of some MY complexes are given in Appendix D3. Usually, $\alpha_{MY(H)}$ and $\alpha_{MY(OH)}$ can be neglected when titration is carried out in a solution with pH between 3～11.

4. The Conditional Stability Constant

If consideration is given to all the side reactions in which the components, MY, M, and Y, are involved, the conditional constant equals

$$K'(MY) = \frac{[(MY)']}{[M'][Y']}$$

According to the definition of side reaction, $[M'] = \alpha_M \cdot [M]$, $[Y'] = \alpha_Y \cdot [Y]$, and $[(MY)'] = \alpha_{MY} \cdot [MY]$. By introducing the above equations, we obtain

$$K'(MY) = \frac{\alpha_{MY} \cdot [MY]}{\alpha_M \cdot [M] \cdot \alpha_Y \cdot [Y]} = \frac{\alpha_{MY}}{\alpha_M \cdot \alpha_Y} \cdot K(MY) \tag{5-13}$$

When the solution parameters are known for a specific titration, α_M, α_Y and α_{MY} can be considered as constants in this situation; therefore, $K'(MY)$ can be considered as a constant.

$$\lg K'(\mathrm{MY}) = \lg K(\mathrm{MY}) - \lg\alpha_{\mathrm{M}} - \lg\alpha_{\mathrm{Y}} + \lg\alpha_{\mathrm{MY}} \tag{5-14}$$

In most cases when titration is carried out in a media with pH between 3～11. The above equation can be simplified as

$$\lg K'(\mathrm{MY}) = \lg K(\mathrm{MY}) - \lg\alpha_{\mathrm{M}} - \lg\alpha_{\mathrm{Y}} \tag{5-15}$$

【Example 5.5】 Calculate $\lg K'(\mathrm{ZnY})$ at pH 2.0 and 5.0.

Answer: $\lg K(\mathrm{ZnY})=16.5$, $\lg K^{\mathrm{H}}(\mathrm{ZnHY})=3.0$.

At pH=2.0, $\lg\alpha_{\mathrm{Y(H)}}=13.8$ and $\lg\alpha_{\mathrm{Zn(OH)}}=0$.

$$\alpha_{\mathrm{ZnY(H)}} = 1+[\mathrm{H}]\cdot K^{\mathrm{H}}(\mathrm{ZnHY}) = 1+10^{-2.0+3.0} = 10^{1.0}$$

$$\lg K'(\mathrm{ZnY}) = \lg K(\mathrm{ZnY}) - \lg\alpha_{\mathrm{Zn(OH)}} - \lg\alpha_{\mathrm{Y(H)}} + \lg\alpha_{\mathrm{ZnY(H)}}$$
$$= 16.5 - 0 - 13.8 + 1.0 = 3.7$$

At pH=5.0, $\lg\alpha_{\mathrm{Y(H)}}=6.6$, $\lg\alpha_{\mathrm{Zn(OH)}}=0$, $\lg\alpha_{\mathrm{ZnY(H)}}=0$.

$$\lg K'(\mathrm{ZnY}) = \lg K(\mathrm{ZnY}) - \lg\alpha_{\mathrm{Y(H)}} = 16.5 - 6.6 = 9.9$$

It is clear from the above calculation that the side reaction of Y with H is highly significant at pH 2 ($\lg\alpha_{\mathrm{Y(H)}} = 13.8$). In this situation, the logarithmic value of conditional constant is only 3.7 even though $\lg K(\mathrm{ZnY})$ is as high as 16.5, which makes the complexes very instable for titration purpose. At pH 5.0, $\lg\alpha_{\mathrm{Y(H)}}=6.6$, then $\lg K'(\mathrm{ZnY})= 9.9$, the completeness of the complexation reaction is reasonably high. In a less acidic solution, $\lg\alpha_{\mathrm{Y(H)}}$ is smaller, thus the conditional constant increases and consequently the completeness of the complexation reaction increases. At the same time, as the pH increases, $\lg\alpha_{\mathrm{M(OH)}}$ increases, which is not favorable for complexation reaction (main reaction). The influence of pH on conditional constant is presented in Figure 5.6. When there is no other ligand existing in the solution, conditional constant becomes much less than the thermodynamic constant of the reaction as pH increases. For HgY, $K(\mathrm{HgY}) = 10^{21.8}$, the maximum of $K'(\mathrm{HgY})$ is less than $10^{12.0}$; for Fe(Ⅲ) and Cu(Ⅱ), $K(\mathrm{Fe^{Ⅲ}Y}) \gg K(\mathrm{CuY})$;

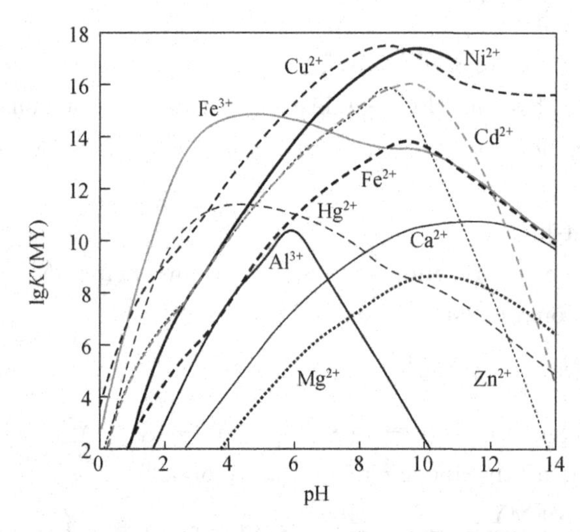

Figure 5.6 Conditional stability constants ($\lg K'(\mathrm{MY})$) of various metal-EDTA complexes as functions of pH.

$K'(Fe^{III}Y)$ is much less than $K'(CuY)$ at pH>8 because Fe^{III} has a higher $\alpha_{M(OH)}$ value.

【Example 5.6】 Calculate $\lg K'(ZnY)$ at pH $=9.0$ and $c(NH_3)=0.10\ mol\cdot L^{-1}$.

Answer: The following equilibria exist in the solution.

$$\begin{array}{ccccc} & Zn & + & Y & \rightleftharpoons ZnY \\ NH_4^+ \xleftarrow{H^+} NH_3 & \swarrow \quad \searrow OH^- & & \downarrow H^+ & \\ & Zn(NH_3) \quad Zn(OH) & & HY & \\ & \vdots \qquad\quad \vdots & & \vdots & \end{array}$$

$\alpha_{Zn(NH_3)} = 10^{3.2}$ (Example 5.4), $\alpha_{Zn(OH)} = 10^{0.2}$ (Appendix D5)

$\alpha_{Zn} = 10^{3.2} + 10^{0.2} - 1 = 10^{3.2}$

$\lg\alpha_{Y(H)} = 1.4$ (Appendix D4)

$\lg K'(ZnY) = \lg K(ZnY) - \lg\alpha_{Zn} - \lg\alpha_{Y(H)} = 16.5 - 3.2 - 1.4 = 11.9$

The plots of $\lg K'(ZnY)$-pH at various concentrations of ammonia are presented in Figure 5.7. In a more acidic solution, the side reaction of Y with H is the predominant factor affecting the conditional constant; therefore, $\lg K'(ZnY)$ increases as the pH increases. When the side reaction of Y with H is the major factor, curves for different concentrations of ammonia are essentially the same. As pH continues to increase, side reaction of Zn^{2+} with NH_3 becomes predominant, which makes $\lg K'(ZnY)$ decrease; the bigger $c(NH_3)$, the smaller $\lg K'(ZnY)$. When pH>12, side reaction of Zn^{2+} with OH^- becomes predominant regardless how large $c(NH_3)$ is. The curves become identical in this range, and $\lg K'(ZnY)$ decreases as pH increases.

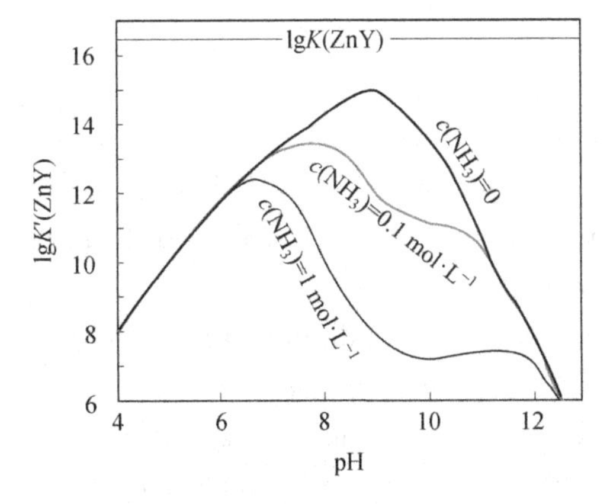

Figure 5.7 Conditional stability constant ($\lg K'(ZnY)$) of the zinc-EDTA complex as a function of pH at different $c(NH_3)$.

5.2 METALLOCHROMIC INDICATORS

Metallochromic indicators are substances that undergo a marked color change when the concentration of free metal ion in a solution changes.

5.2.1 How a Metallochromic Indicator Works

As an organic dye, the metallochromic indicator can form colored complexes with metal ions. The color of complex is different from the organic dye itself and thus becomes an indication of metal ion concentration in solution. For example, the complex of eriochrome black T (EBT) is red for magnesium complex but the dye itself is blue.

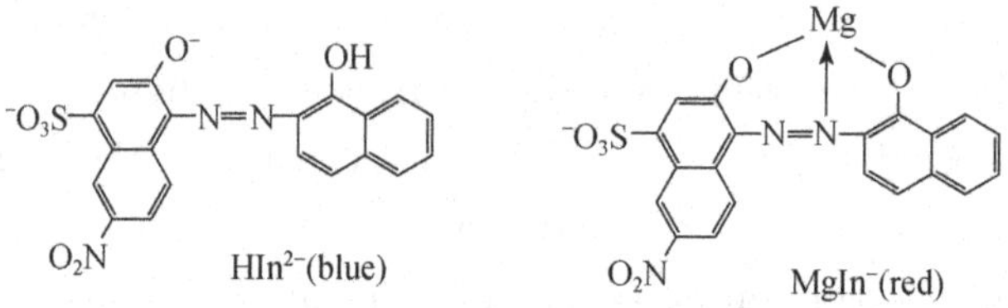

HIn^{2-}(blue) $MgIn^{-}$(red)

In the titration of Mg^{2+} with EDTA, initially there is excess Mg^{2+} in the solution compared to EBT. Red Mg-EBT complex forms, thus the solution is red. As the titration proceeds, more EDTA is added to the solution and more MgY^{2-} forms. When the titration approaches the stoichiometric point, the concentration of Mg^{2+} is very low. An additional increment of EDTA will exchange with EBT in the complex to release the indicator ligand, i. e. ,

$$\underset{\text{(red)}}{MgIn^{-}} + HY^{3-} \rightleftharpoons MgY^{2-} + \underset{\text{(blue)}}{HIn^{2-}}$$

The solution changes from red to blue, which indicates the end of titration.

Metallochromic indicators must meet the following criteria:

(1) The color of indicator must be different from the metal-indicator complex to facilitate endpoint detection.

In most cases, metallochromic indicators are weak organic acids and change color as pH changes. Therefore, the titration needs to be carried out in a controlled pH buffer to ensure that the color of indicator is different from the color of metal-indicator complex. For EBT, the following equilibria exist in solution:

$$\underset{\text{(red-violet)}}{H_2In^{-}} \xrightleftharpoons{pK_{a2} = 6.3} \underset{\text{(blue)}}{HIn^{2-}} \xrightleftharpoons{pK_{a3} = 11.6} \underset{\text{(orange)}}{In^{3-}}$$

It is easy to know that pH 7~11 will be appropriate for indicating endpoint because dye itself is blue and complex is red in this pH range.

(2) Conditional constant for MIn must be in an appropriate range.

The metal-indicator complex (MIn) must be less stable than that of the metal-EDTA complex, or the EDTA will not displace the indicator complexed with the metal at the endpoint. For example, in the titration of the total amount of Ca^{2+} and Mg^{2+} at pH 10, coexisting Al^{3+}, Fe^{3+}, Cu^{2+}, Co^{2+}, and Ni^{2+} will block the indicator from reacting with Ca^{2+} and Mg^{2+}; thus, the endpoint will be impossible to see. One approach to prevent interfering ions from blocking the indicator is to add masking reagents to the test solution such as triethanolamine to prevent Al^{3+} and Fe^{3+} from blocking the indicator. Addition of KCN can prevent Cu^{2+}, Co^{2+}, and Ni^{2+} from blocking the indicator. On the other hand, the metal-indicator complex must have reasonable stability, or EDTA will start replacing the indicator at the beginning of the titration, and a diffused endpoint will result. In general, the formation constant for metal-indicator complex should be less than that of the metal-titrant complex by 10～100 times.

(3) The reaction between indicator and metal ion should be fast and easily reversed close to the endpoint by addition of one or half drop of EDTA standard solution close to the endpoint. If the solubility of an indicator or metal-indicator complex is low, the exchange rate of titrant with metal-indicator complex will be small; thus, the color transition is spread and the endpoint decision is difficult to make. This slow change of color through endpoint is called ossification of the indicator. The color change can be accelerated by adding organic solvent or carrying out the titration at an elevated temperature.

Most metallochromic indicators have double bonds, so the indicator solutions are less stable than the dry powder for EBT and calcon. The indicator compounds are mixed with NaCl or KCl to make a more stable solution.

5.2.2 Color Transition Point pM ($(pM)_t$) for Metallochromic Indicators

The use of conditional constant is a convenient way to simplify the mathematics. As an example, when an acid indicator ($H_j In$) is used in the titration to form a metal-indicator complex (MIn), the main reaction for metal with indicator and the side reaction of the indicator are given below provided that M does not participate in any side reactions.

$$\begin{array}{ccccc} M & + & In & \longrightarrow & MIn \\ & & \downarrow H & & \\ & & HIn & & \\ & & \vdots & & \end{array}$$

Conditional constant $$K(\text{MIn}') = \frac{[\text{MIn}]}{[\text{M}][\text{In}']} = \frac{K(\text{MIn})}{\alpha_{\text{In(H)}}}$$

$$\text{pM} + \lg\frac{[\text{MIn}]}{[\text{In}']} = \lg K(\text{MIn}) - \lg\alpha_{\text{In(H)}}$$

The color change occurs when $[\text{MIn}]=[\text{In}']$, the corresponding pM value at the color transition point is denoted by $(\text{pM})_t$.

$$(\text{pM})_t = \lg K(\text{MIn}') = \lg K(\text{MIn}) - \lg\alpha_{\text{In(H)}} \tag{5-16}$$

【Example 5.7】 Calculate $(\text{pMg})_t$ for titration of Mg^{2+} with EDTA at pH 10.0, using EBT ($\lg\beta_1=11.6$ and $\lg\beta_2=17.9$) as the indicator ($\lg K(\text{MgIn})=7.0$).

Answer:

$$\alpha_{\text{In(H)}} = 1+[\text{H}]\beta_1+[\text{H}]^2\beta_2 = 1+10^{-10.0+11.6}+10^{-20.0+17.9} = 10^{1.6}$$

Thus, $(\text{pMg})_t = \lg K(\text{MgIn}') = \lg K(\text{MgIn}) - \lg\alpha_{\text{In(H)}} = 7.0-1.6 = 5.4$

For MIn complex of 1 : 2 and 1 : 3 as well as acidic complex, calculation of $(\text{pM})_t$ becomes complicated. In reality, endpoint is not taken exactly at $(\text{pM})_t$ as calculated. For titration of M with EDTA, the endpoint is taken when the color of indicator is more obvious. For titration of EDTA with M, the endpoint is taken when the color of MIn is more obvious. $(\text{pM})_t$ for titration of most metal ions are experimentally determined. $(\text{pM})_t$ as a function of pH for a couple of general indicators are given in Table 5.1.

Table 5.1 $(\text{pM})_t$ for XO and EBT at Different pH

pH		1.0	2.0	3.0	4.0	4.5	5.0	5.5	pH		6.0	7.0	8.0	9.0	10.0	11.0	12.0
XO	Bi^{3+}	4.0	5.4	6.8					EBT	Ca^{2+}			1.8	2.8	3.8	4.7	5.3
	Cd^{2+}					4.0	4.5	5.0		Mg^{2+}	1.0	2.4	3.4	4.4	5.4	6.3	
	Pb^{2+}			4.2	4.8	6.2	7.0	7.6		Zn^{2+}	6.9	8.3	9.3	10.5	12.2	13.9	
	Zn^{2+}					4.1	4.8	5.7		Mn^{2+}	3.6	5.0	6.2	7.8	9.7	11.5	

When a metal ion does not participate in the side reaction, then $(\text{pM})_{ep} = (\text{pM})_t$. If metal ion is involved in the side reaction, concentration of uncomplexed metal ion can be denoted as $[\text{M}']$ at the endpoint.

$$[\text{M}'] = \alpha_\text{M}[\text{M}]$$

$$(pM')_{ep} = (pM)_t - \lg\alpha_M \tag{5-17}$$

Titration error is less than 0.1% only when pM′ falls in the break range of the titration curve.

5.2.3 Frequently Used Metallochromic Indicators

The frequently used metallochromic indicators can be found in Appendix B2. The following are the most frequently used ones.

1. Eriochrome Black T (EBT)

EBT can be applied to titration of Mg^{2+}, Zn^{2+}, and Pb^{2+} ions in weakly basic solution. The formation constant for EBT with Ca^{2+} is relatively small compared with the metal ions mentioned above. As a consequence, the significant conversion of $CaIn^-$ to HIn^{2-} color change during the transition is not easy to observe; thus, EBT is not a good indicator for titration of calcium.

2. Xylenol Orange (XO)

XO is yellow colored in acidic solution($pH<6.0$), and forms red complexes with metal ions. XO has been used as the indicator for direct complexometric titration of ZrO, Hf, Th, Sc, and In, rare earths, as well as Yb, Bi, Pb, Zn, Cd, and Hg. Some metal ions, such as ions of Al, Ni, Co, and Cu, form extremely stable complexes with XO so that the direct titration is not feasible, thus a back titration is used. The titration can be conducted by addition of excess EDTA standard solution in pH 5.0～5.5, and back titration is carried out using Zn^{2+} or Pb^{2+} standard solution. For determination of Fe^{3+}, back titration is carried out using $Bi(NO_3)_3$ standard solutions at pH 2～3.

3. 4-(2-Pyridylazo)-1-Naphthol (*p*-PAN)

The reaction of PAN with Cu^{2+} is quick enough to indicate color change. For many other metal ions, such as Ni^{2+}, Co^{2+}, Zn^{2+}, Pb^{2+}, Bi^{3+}, and Ca^{2+}, either the reaction with PAN is slow or complexes are not intensively colored at transition range (low sensitivity). This situation can be improved by using Cu-PAN as the indirect indicator for determination of several multiple metal ions in one continuous titration because Cu-PAN can be used in a wide pH range. At pH 1.9～12.2, PAN is yellow and the Cu-PAN complex is red. The wide pH application range of Cu-PAN is very useful for continuous titration of more than two metal ions. because it is hard to observe color change close to the endpoint when more than one indicator is used.

4. Other Indicators

At pH 2, colorless salicylic acid forms mauve complexes with Fe^{3+} and can be used as the indicator for titration of Fe^{3+}. At pH 12.5, blue calconcarboxylic acid forms mauve complexes with Ca^{2+} and can be used as the indicator for titration of calcium.

5.3 TITRATION CURVES AND TITRATION ERRORS

5.3.1 Titration Curves

During a titration, the concentrations of unreacted metal ions ($[M']$) is reduced with the addition of titrant. The titration curve can be constructed by plotting pM′ versus $T(\%)$.

- Pre-SP, $[M'] = c(M) - [MY]$
- Post-SP, excess titrant is added to the solution

$$[M'] = \frac{[MY]}{[Y'] \cdot K'(MY)}$$

- At SP, $[M'] = [Y']$ (At this point, the assumption that $[M] = [Y]$ may not hold). If complexes are stable enough, then $[MY]_{sp} = c_{sp}(M) - [M']_{sp} \approx c_{sp}(M)$. Substitute into conditional constant $K'(MY) = \frac{[MY]}{[M'][Y']}$, we get

$$[M']_{sp} = \sqrt{\frac{c_{sp}(M)}{K'(MY)}} \tag{5-18}$$

That is,

$$(pM')_{sp} = \frac{1}{2}(\lg K'(MY) + pc_{sp}(M)) \tag{5-19}$$

$c_{sp}(M)$ denotes analytical concentration of the metal ion at SP and is half of the initial concentration of metal ions, if the same concentration level of EDTA is used in the titration.

【Example 5.8】 In the titration of 0.020 mol · L^{-1} Zn^{2+} using 0.020 mol · L^{-1} EDTA at pH=9.0 and $c(NH_3)=0.20$ mol · L^{-1}, calculate pZn′, pZn, pY′, and pY at SP, as well as pZn′ and pY′ at pre 0.1% SP and post 0.1% SP.

Answer: At SP, pH=9.0, $\lg\alpha_{Y(H)} = 1.4$, $c(NH_3) = 0.20/2 = 0.10$ (mol · L^{-1}); from Example 5.6: $\lg\alpha_{Zn} = \lg\alpha_{Zn(NH_3)} = 3.2$, $\lg K'(ZnY) = 11.9$ (At ±0.1%, $K'(ZnY)$ remains constant); $c_{sp}(Zn) = 0.010$ mol · L^{-1}.

According to Equation (5-19),

$$(pZn')_{sp} = \frac{1}{2}(\lg K'(ZnY) + pc_{sp}(Zn)) = \frac{1}{2}(11.9 + 2.0) = 7.0$$

$$[Zn] = \frac{[Zn']}{\alpha_{Zn}}$$

$$(pZn)_{sp} = (pZn')_{sp} + \lg\alpha_{Zn} = 7.0 + 3.2 = 10.2$$

and $(\text{pY}')_{sp} = (\text{pZn}')_{sp} = 7.0$

$$(\text{pY})_{sp} = (\text{pY}')_{sp} + \lg\alpha_{Y(H)} = 7.0 + 1.4 = 8.4$$

At pre 0.1% SP,

$$[\text{Zn}'] = 0.010 \times 0.1\% = 1.0 \times 10^{-5}(\text{mol} \cdot \text{L}^{-1}), \quad \text{pZn}' = 5.0$$

$$[\text{Y}'] = \frac{[\text{ZnY}]}{[\text{Zn}'] \cdot K'(\text{ZnY})} = \frac{0.010}{1.0 \times 10^{-5} \times 10^{11.9}} = 10^{-8.9}(\text{mol} \cdot \text{L}^{-1}), \quad \text{pY}' = 8.9$$

At post 0.1% SP,

$$[\text{Y}'] = 0.010 \times 0.1\% = 1.0 \times 10^{-5}(\text{mol} \cdot \text{L}^{-1}), \quad \text{pY}' = 5.0$$

$$[\text{Zn}'] = \frac{[\text{ZnY}]}{[\text{Y}'] \cdot K'(\text{ZnY})} = \frac{0.010}{1.0 \times 10^{-5} \times 10^{11.9}} = 10^{-8.9}(\text{mol} \cdot \text{L}^{-1}), \quad \text{pZn}' = 8.9$$

Accuracy for complexometric titration is dictated by transition range of the titration curve, which is affected by:

(1) Conditional constant ($K'(\text{MY})$)

When titration is carried out at a given concentration, transition range is expanded as $K'(\text{MY})$ increases (Figure 5.8). At pre-SP, pM′ is calculated according to the amount of the unreacted M′. The titration curves of different $\lg K'(\text{MY})$ at the same concentration coincide with each other before SP, thus the titration curve before SP does not change with $\lg K'(\text{MY})$. For post-SP, $\text{pM}' = \lg K'(\text{MY}) - \lg \frac{[\text{MY}]}{[\text{Y}']}$. At post 0.1% SP, $\text{pM}' = \lg K'(\text{MY}) - 3.0$. This indicates that pM′ is dictated only by $\lg K'(\text{MY})$ and will increase as $\lg K'(\text{MY})$ increases. If $K'(\text{MY})$ increases 10 fold, pM′ will increase one pH unit. By altering pH of the solution and concentration of the titrant, $\lg K'(\text{MY})$ can be increased so that the transition range can be expanded.

(2) Concentration of metal ions ($c(\text{M})$)

For a given $K'(\text{MY})$, transition range increases as concentration increases (Figure 5.9). Before SP, pM′ decreases as $c(\text{M})$ increases. At pre 0.1% SP, $\text{pM}' = \text{p}c_{sp}(\text{M}) + 3.0$. pM′ decreases one pM unit as concentration increases 10 fold. After SP, titration curves with different concentration levels coincide with each other, indicating that pM′ is not responsive to concentration change. Pre-SP side of titration curve changes with concentration.

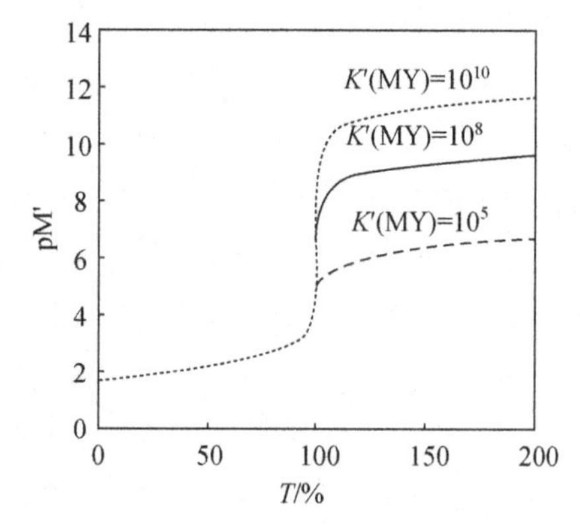

Figure 5.8 Titration curves with different conditional formation constants($c=10^{-2}$ mol · L^{-1}).

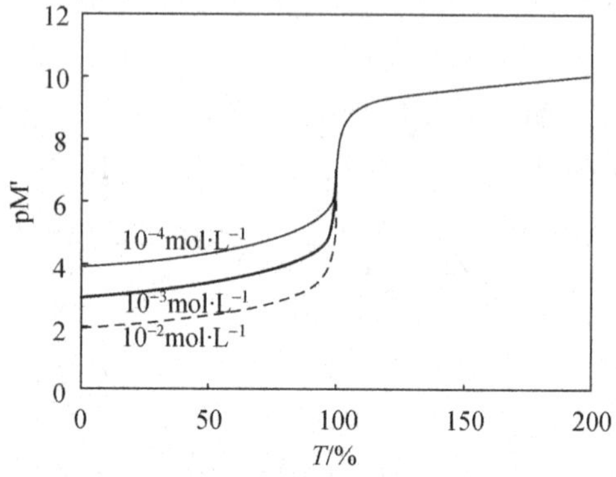

Figure 5.9 Titration curves with different concentrations ($K'=10^{10}$).

5.3.2 Titration Errors

According to definition of titration error, the E_t for titration of M is expressed as

$$E_t = \frac{[Y']_{ep} - [M']_{ep}}{c_{sp}(M)} \tag{5-20}$$

According to conditional constant

$$[Y'] = \frac{[MY]_{ep}}{[M']_{ep} \cdot K'(MY)} \approx \frac{c_{sp}(M)}{[M']_{ep} \cdot K'(MY)}$$

Substitute into Equation (5-20), we obtain

$$E_t = \frac{1}{[M']_{ep} \cdot K'(MY)} - \frac{[M']_{ep}}{c_{sp}(M)} \tag{5-21}$$

By referring to E_t equation for neutralization titration, Ringbom E_t equation can be derived.

From $\quad \Delta pM' = (pM')_{ep} - (pM')_{sp}, \quad \Delta pY' = (pY')_{ep} - (pY')_{sp}$

We have $\quad [M']_{ep} = [M']_{sp} \cdot 10^{-\Delta pM'}, \quad [Y']_{ep} = [Y']_{sp} \cdot 10^{-\Delta pY'} = [M']_{sp} \cdot 10^{\Delta pM'}$

Substitute into Equation (5-20), we have

$$E_t = \frac{[M']_{sp}(10^{\Delta pM'} - 10^{-\Delta pM'})}{c_{sp}(M)}$$

According to Equation (5-18), we have

$$[M']_{sp} = (c_{sp}(M)/K'(MY))^{1/2}$$

Because $\Delta pM' = \Delta pM$, we obtain E_t equation for complexometric titration

$$E_t = \frac{10^{\Delta pM} - 10^{-\Delta pM}}{(K'(MY) \cdot c_{sp}(M))^{1/2}} \tag{5-22}$$

Criteria for complexometric titration are also dictated by E_t and the indicator.

For visual observation of endpoint, the precision is $\pm(0.2 \sim 0.5)\Delta pM$ as varied with indicators and observers. For $\Delta pM = \pm 0.2$, if $E_t = \pm 0.1\%$, we have $\lg(c_{sp}(M) \cdot K'(MY)) \geqslant 6$; thus, $\lg(cK') \geqslant 6$ is required for complexometric titration with $E_t < 0.1\%$.

【Example 5.9】 Calculate transition range and select an appropriate indicator for titration of $0.020\ mol \cdot L^{-1}\ Pb^{2+}$ in pH=5.0 solution buffered by $(CH_2)_6N_4$, using same concentration of EDTA as the titrant.

Answer: At pH=5.0, $\lg\alpha_{Pb(OH)} = 0$, $\lg\alpha_{Y(H)} = 6.6$. Therefore,

$$\lg K(PbY') = 18.0 - 6.6 = 11.4$$

According to Equation (5-22), when

$$\lg(cK') = 11.4 - 2.0 = 9.4, \quad E_t = \pm 0.1\%$$

then $\quad \Delta pM = \pm 1.7$

$$(pPb)_{sp} = \frac{1}{2}(\lg K(PbY') + pc_{sp}(Pb)) = \frac{1}{2}(11.4 + 2.0) = 6.7$$

Transition range for titration of Pb^{2+} is 6.7 ± 1.7, or expressed as $5.0 \sim 8.4$. At pH=5.0, $(pPb)_t = 7.0$ (Table 5.1). The color transition point falls in the transition range and can be used for this titration.

【Example 5.10】 Calculate E_t for titration of $0.020\ mol \cdot L^{-1}\ Mg^{2+}$ in NH_4^+-NH_3 buffer at pH 10.0, using same concentration of EDTA as the titrant and EBT as the indicator.

Answer: At pH=10.0, $\lg\alpha_{Y(H)} = 0.5$, $(pMg)_{ep} = (pMg)_t = 5.4$ (Example 5.7).

$$\lg K(MgY') = \lg K(MgY) - \lg\alpha_{Y(H)} = 8.7 - 0.5 = 8.2$$

$$(pMg)_{sp} = \frac{1}{2}(\lg K(MgY') + pc_{sp}(Mg)) = \frac{1}{2}(8.2 + 2.0) = 5.1$$

$$\Delta pM = (pMg)_{ep} - (pMg)_{sp} = 5.4 - 5.1 = +0.3$$

$$E_t = \frac{10^{0.3} - 10^{-0.3}}{10^{(8.2-2.0)/2}} \times 100\% = +0.1\%$$

5.3.3 pH Control in Complexometric Titrations

In complexometric titration, pH should be kept at least above a value (pH_{min}, also called the highest acidity), at which $\alpha_{Y(H)}$ is not too high to ensure that $K'(MY) \geqslant 10^8$. If side reaction does not happen to M, and side reaction of MY can be neglected, the conditional constant is denoted as $K(MY')$. When $c_{sp}(M) = 0.01 mol \cdot L^{-1}$, $\lg K(MY')$ must be greater than 8 to make sure that titration error is less than $\pm 0.1\%$.

【Example 5.11】 Calculate pH_{min} for titration of $0.020\ mol \cdot L^{-1}\ Zn^{2+}$ with

EDTA at the same concentration, based on the following criteria: $\Delta pM = \pm 0.2$ and $E_t \leqslant \pm 0.1\%$.

Answer: As $c_{sp}(Zn) = 0.01\ mol \cdot L^{-1}$, $\lg K'(MY)$ must be greater than 8 to meet the criteria. If no side reaction occurs to Zn^{2+}, then

$$\lg K(ZnY') = \lg K(ZnY) - \lg \alpha_{Y(H)}$$

$$\lg \alpha_{Y(H)} = \lg K(ZnY) - \lg K(ZnY') = 16.5 - 8.0 = 8.5$$

With this $\alpha_{Y(H)}$ value, the pH obtained from Appendix D4 is 4.0. This means the titration must be carried out at a pH value greater than 4.0.

$\lg K(MY)$ changes as metal ion varies, which makes the permitted $\alpha_{Y(H)}$ value differ based on $\lg K(MY') \geqslant 8.0$, resulting in altered pH_{min}. For stable complexes, such as BiY ($\lg K(BiY) = 27.9$), titration can be carried out in a very acidic solution ($pH \approx 1$). For less stable complexes, such as MgY ($\lg K(MgY) = 8.7$), titration must be carried out at pH=10.

On the other hand, metal ions tend to form metal hydroxide precipitate at a suitable pH. The formation of metal hydroxide precipitate reduces the reaction rate and makes it difficult to observe color change at the endpoint. The pH right before hydroxide precipitate occurs is called the lowest acidity (pH_{max}) and can be calculated from solubility product of $M(OH)_n$. For example, in the titration of Zn^{2+},

$$[OH^-] \leqslant \sqrt{\frac{K_{sp}(Zn(OH)_2)}{[Zn^{2+}]}} = \sqrt{\frac{10^{-15.3}}{0.020}} = 10^{-6.8}(mol \cdot L^{-1})$$

thus $pH_{max} = 7.2$. pH must be kept lower than 7.2 so that $Zn(OH)_2$ precipitate does not occur. The addition of auxiliary complex reagents can prevent the metal ions from forming hydroxide precipitate. For example, tartaric acid is often added for titration of Pb^{2+} and ammonia is added for the titration of Zn^{2+} and Cd^{2+}, so that titration can be conducted in a much less acidic pH. One should remember that $K'(MY)$ will be reduced in the presence of auxiliary reagent, therefore the amount of auxiliary reagent should be adequate to prevent metal hydroxide formation. If the concentration of auxiliary reagent is too high in the solution, the $\lg K'(MY)$ may be reduced to a value of less than 8. Then the titration with $E_t \leqslant 0.1\%$ is not feasible.

5.4 SELECTIVE TITRATIONS OF METAL IONS IN THE PRESENCE OF MULTIPLE METAL IONS

As a complexone, EDTA finds a wide application in determination of metal ions.

Practical samples often have more than one metal ion, which makes selective titration impossible due to interference from metal ions present. Selective titration remains a key problem to be solved for the titration of such samples.

5.4.1 Selective Titration by Regulating pH

Take the analysis of a specimen with two metal ions, M and N, as an example. If both form complexes with EDTA and $K(MY) > K(NY)$, M will react with EDTA first in the titration. If the difference between $K(MY)$ and $K(NY)$ is large enough, EDTA will react with N only after M is titrated quantitatively, that is, M can be titrated quantitatively in the presence of N. Can N be titrated quantitatively after M is titrated? The judgment can be based on the criteria for titration of single metal ion. The questions are, "How large should the difference between $K(MY)$ and $K(NY)$ be to titrate M in the presence of N?" and "If titration is feasible, what pH should be used for the titration?"

If we consider the reaction of N with EDTA and the reaction of H^+ with EDTA as side reactions of M with EDTA, a conditional constant $K(MY')$ can be derived, then the feasibility of quantitative titration of M can be easily answered.

1. pH Dependency of $K(MY')$

If there is no side reaction concerning M, the following equilibria are existing:

$$\begin{array}{ccccc} M & + & Y & \rightleftharpoons & MY \\ & & H \swarrow \searrow N & & \\ & & HY \quad NY & & \\ & & \vdots & & \end{array}$$

$\alpha_{Y(H)}$ decreases as pH increases. Whereas,

$$\alpha_{Y(N)} = \frac{[Y]+[NY]}{[Y]} = 1 + [N] \cdot K(NY)$$

[NY] at the stoichiometric point must be very small so that quantitative titration of M can be carried out with a reasonable accuracy. If there is no other reagent to complex N, then

$$[N] = c(N) - [NY] \approx c(N)$$

thus,
$$\alpha_{Y(N)} \approx 1 + c(N) \cdot K(NY) \approx c(N) \cdot K(NY) \tag{5-23}$$

It is clear that $\alpha_{Y(N)}$ is dictated by $c(N)$ and $K(NY)$. When the pH is not so high that hydroxide for N does not form, $\alpha_{Y(N)}$ is a constant. For Y, the total side reaction coefficient is

$$\alpha_Y = \alpha_{Y(H)} + \alpha_{Y(N)} - 1$$

- If titration is carried out in a very acidic solution, $\alpha_{Y(H)} > \alpha_{Y(N)}$, thus $\alpha_Y \approx \alpha_{Y(H)}$. then,

$$K(MY') = K(MY)/\alpha_{Y(H)} \tag{5-24a}$$

The effect of N can be neglected for. $K(MY')$ increases as pH increases.

- If titration is carried out in a solution with low acidity, $\alpha_{Y(N)} > \alpha_{Y(H)}$, thus $\alpha_Y \approx \alpha_{Y(N)}$,

$$K(MY') = K(MY)/\alpha_{Y(N)} \tag{5-24b}$$

In this situation, the side reaction of Y with H can be neglected as long as M and N neither form hydroxo complexes nor undergo other side reactions to reduce $\alpha_{Y(N)}$. Therefore, $K(MY')$ will remain at the maximum without changing with pH.

2. Criteria for Selective Titration and pH Control

For a pH range to make $\alpha_{Y(N)} > \alpha_{Y(H)}$, $\alpha_Y \approx \alpha_{Y(N)} \approx c(N) \cdot K(NY)$, conditional constant $K(MY')$ is affected by only $c(N)$ and $K(NY)$.

$$K(MY') = K(MY)/(c(N) \cdot K(NY))$$

Logarithm is taken after both sides are multiplied by $c(M)$. We get

$$\begin{aligned} \lg(c(M) \cdot K(MY')) &= \lg(c(M) \cdot K(MY)) - \lg(c(N) \cdot K(NY)) \\ &= \Delta\lg(cK) \end{aligned} \tag{5-25}$$

Criteria for selective titration is dictated by the value of $\Delta\lg(cK)$, which can be derived from E_t and ΔpM. If $E_t = \pm 0.1\%$ and $\Delta\text{pM} = \pm 0.2$, to make sure $\lg(c(M) \cdot K(MY')) \geqslant 6$, we have $\Delta\lg(cK) \geqslant 6$. If $c(M) = c(N)$, then $\Delta\lg K \geqslant 6$; if $c(M) \geqslant 10c(N)$, then $\Delta\lg K \geqslant 5$.

In most cases, it is desirable that selective titration is carried out when $\lg K(MY')$ reaches maximum. At this point where $\alpha_{Y(H)} = \alpha_{Y(N)}$, the pH is considered to be pH_{min}. pH_{max} is calculated in the same way as that in single metal titration, namely, the pH value at which there is no formation of $M(OH)_n$ precipitate.

Some multivalent metal ions are more apt to form hydroxo complexes even in a very acidic solution; however, their EDTA complexes are relatively stable so that the titration can still be carried out. To titrate Bi^{3+} in a solution containing Pb^{2+} is one example.

If $c_{sp}(Pb) = 10^{-2.0}$ mol · L^{-1}, then $\alpha_{Y(H)} = \alpha_{Y(Pb)} = 10^{-2.0+18.0} = 10^{16.0}$, thus pH=1.4. To ensure that $\lg K(BiY)' > 8$, titration should be carried out in a solution with pH>1; however, Bi^{3+} will form hydroxo complexes at pH 1.4, making the color change at the endpoint difficult to detect. Generally, a quantitative titration at pH = 1.0 is feasible ($\lg K(BiY') = 9.6$). Afterwards, Pb^{2+} can be titrated by adjusting pH to 4~6 using $(CH_2)_6N_4$. XO can form red complexes with both Bi^{3+}

and Pb^{2+}, thus is used for continuous selective titration of Bi^{3+} at pH 1 and Pb^{2+} at pH 4~6.

5.4.2 Selective Titration Using Masking Reagents

In some cases, $\lg K(MY)$ is not big enough to be selectively titrated by controlling the solution pH. Sometimes $\lg K(MY)$ is even smaller than $\lg K(NY)$, therefore titration of M by simply controlling pH may be impossible. A masking reagent can be added to react with N to reduce or even eliminate the interference of N to the titration of M. Masking, another approach to achieve selective titration, includes complexation, precipitation, and redox reaction. Masking by complexation is used most frequently as it does not involve phase separation or removal of excess masking reagents.

1. Masking by Complexation

The following equilibria exist when complexation masking reagent A is added to the solution:

M + Y ⇌ MY

H / \ N —A→ NA

HY NY

⋮

If masking is sufficient to reduce [N] to make $\alpha_{Y(N)} < \alpha_{Y(H)}$, then $\alpha_Y \approx \alpha_{Y(H)}$, thus N will no longer interfere with the titration. Like the titration of single metal ion, $\lg K(MY')$ changes only as pH changes.

After addition of masking reagent, if $\alpha_{Y(N)} > \alpha_{Y(H)}$, then $\alpha_Y \approx \alpha_{Y(N)}$.

$$\alpha_{Y(N)} = 1 + [N] \cdot K(NY) \approx \frac{c(N)}{\alpha_{N(A)}} \cdot K(NY)$$

Therefore, $\lg K(MY') = \lg K(MY) - \lg\alpha_{Y(N)} = \Delta\lg K + pc(N) + \lg\alpha_{N(A)}$ (5-26)

【Example 5.12】 Calculate E_t for titration of 0.020 mol · L^{-1} Zn^{2+} using EDTA when Al^{3+} is present and KF is used for masking Al^{3+}. At the endpoint, $[F'] = 0.01$ mol · L^{-1}; pH=5.5; and XO is used as the indicator.

Answer: For AlF_6^{3-}, $\lg\beta_1 \sim \lg\beta_6$ are 6.1, 11.2, 15.0, 17.7, 19.4, 19.7; $pK_a(HF) =$ 3.1. At pH=5.5, $[F^-] = [F'] = 0.01$ mol · L^{-1}.

$$\begin{aligned}\alpha_{Al(F)} &= 1 + [F]\beta_1 + [F]^2\beta_2 + \cdots + [F]^6\beta_6 \\ &= 1 + 10^{-2.0+6.1} + 10^{-4.0+11.2} + 10^{-6.0+15.0} \\ &\quad + 10^{-8.0+17.7} + 10^{-10.0+19.4} + 10^{-12.0+19.7} \\ &= 10^{10.0}\end{aligned}$$

$$[\mathrm{Al}] = \frac{[\mathrm{Al}']}{\alpha_{\mathrm{Al(F)}}} \approx \frac{c(\mathrm{Al})}{\alpha_{\mathrm{Al(F)}}} = \frac{10^{-2.0}}{10^{10.0}} = 10^{-12.0}(\mathrm{mol\cdot L^{-1}})$$

$$\alpha_{\mathrm{Y(Al)}} = 1+[\mathrm{Al}]\cdot K(\mathrm{AlY}) = 1+10^{-12.0+16.1} = 10^{4.1} < \alpha_{\mathrm{Y(H)}} = 10^{5.7}$$

This result shows masking of Al^{3+} by F^- is satisfactory; therefore, $\alpha_Y \approx \alpha_{Y(H)}$.

$$\lg K(\mathrm{ZnY'}) = \lg K(\mathrm{ZnY}) - \lg\alpha_{\mathrm{Y(H)}} = 16.5 - 5.7 = 10.8$$

At pH=5.5, $(\mathrm{pZn})_{ep}=5.7$(XO), hence

$$E_t = \frac{1}{[\mathrm{Zn}]_{ep}\cdot K(\mathrm{ZnY'})} - \frac{[\mathrm{Zn}]_{ep}}{c_{sp}(\mathrm{Zn})} = \left(\frac{1}{10^{-5.7+10.8}} - \frac{10^{-5.7}}{10^{-2.0}}\right)\times 100\%$$
$$= -0.02\%$$

【Example 5.13】 How to titrate Zn^{2+} in the presence of Cd^{2+} with both concentrations at 0.020 mol · L^{-1}?

Titration of Zn^{2+} using EDTA is carried out at pH=5 with XO as the indicator, and KI is used to mask Cd^{2+}. At the endpoint, $[I^-]=0.5$ mol · L^{-1}. Calculate: (1) E_t for this titration; (2) E_t when EDTA is replaced by HEDTA.

Answer: For CdI_4^{2-}, $\lg\beta_1 \sim \lg\beta_4$ are 2.4, 3.4, 5.0, 6.2; $[I^-]=0.5$ mol · L^{-1}.

$$\alpha_{\mathrm{Cd(I)}} = 1+10^{-0.3+2.4}+10^{-0.6+3.4}+10^{-0.9+5.0}+10^{-1.2+6.2} = 10^{5.1}$$

Concentration of free Cd^{2+} is

$$[\mathrm{Cd}] = \frac{[\mathrm{Cd}']}{\alpha_{\mathrm{Cd(I)}}} \approx \frac{c(\mathrm{Cd})}{\alpha_{\mathrm{Cd(I)}}} = \frac{10^{-2.0}}{10^{5.1}} = 10^{-7.1}(\mathrm{mol\cdot L^{-1}})$$

(1) EDTA as the titrant, $\lg K(\mathrm{ZnY})=\lg K(\mathrm{CdY})=16.5$; at pH=5.5, $\alpha_{Y(H)}=10^{5.5}$.

$$\alpha_{\mathrm{Y(Cd)}} = 1+[\mathrm{Cd}]\cdot K(\mathrm{CdY}) = 10^{-7.1+16.5} = 10^{9.4} \gg \alpha_{\mathrm{Y(H)}}$$

$$\lg K(\mathrm{ZnY'}) = \lg K(\mathrm{ZnY}) - \lg\alpha_{\mathrm{Y(Cd)}} = 16.5 - 9.4 = 7.1$$

At pH = 5.5, $(\mathrm{pZn})_{ep}$ = 5.7 (XO),

$$E_t = \left(\frac{1}{10^{-5.7+7.1}} - \frac{10^{-5.7}}{10^{-2.0}}\right)\times 100\% = +4\%$$

Iodide ion (I^-) cannot effectively mask Cd^{2+} for selective titration of Zn^{2+}.

(2) HEDTA(X) as the titrant, $\lg K(\mathrm{ZnX})=14.5$, $\lg K(\mathrm{CdX})=13.0$; At pH=5.5, $\alpha_{X(H)}=10^{4.6}$.

$$\alpha_{\mathrm{X(Cd)}} = 1+[\mathrm{Cd}]\cdot K(\mathrm{CdX}) \approx 10^{-7.1+13.0} = 10^{5.9} \gg \alpha_{\mathrm{X(H)}}$$

$$\lg K(\mathrm{ZnX'}) = \lg K(\mathrm{ZnX}) - \lg\alpha_{\mathrm{X(Cd)}} = 14.5 - 5.9 = 8.6$$

$$E_t = \left(\frac{1}{10^{-5.7+8.6}} - \frac{10^{-5.7}}{10^{-2.0}}\right)\times 100\% = +0.1\%$$

Changing titrant from EDTA to HEDTA reduces titration error from 4% to 0.1%. This is a good example about how to selectively titrate one metal ion in the presence of other metal ions. Choosing suitable titrant and masking reagent makes this selective titration successful.

The most commonly used masking reagents via complexation formation are given in Table 5.2.

Table 5.2 The Commonly Used Complexation Agents for Masking Purpose

Masking Reagent	Ion Masked	pH
Triethanolamine	Al^{3+} Fe^{3+} Sn^{4+} TiO_2^{2+}	10*
Fluoride	Al^{3+} Sn^{4+} TiO_2^{2+} Zr^{4+}	>4
Acetylacetone	Al^{3+} Fe^{3+}	5~6
Phenanthroline	Zn^{2+} Cu^{2+} Co^{2+} Ni^{2+} Cd^{2+} Hg^{2+}	5~6
Cyanide	Zn^{2+} Cu^{2+} Co^{2+} Ni^{2+} Cd^{2+} Hg^{2+} Fe^{2+}	10**
2,3-Dithioglycerine	Zn^{2+} Pb^{2+} Bi^{3+} Sb^{3+} Sn^{4+} Cd^{2+} Cu^{2+}	10
Thiourea	Hg^{2+} Cu^{2+}	Weakly acidic media
Iodide	Hg^{2+}	

* When the tested solution is acidic, triethanolamine must be added prior to adjusting the pH to 10, otherwise, the hydroxo complex will form before the addition of masking reagent.

** KCN must be used only in basic media, or toxic HCN gas will form. After titration, $FeSO_4$ should be added in excess to the solution to change CN^- into $Fe(CN)_6^{4-}$.

When titration of N is necessary in the above situation, a demasking reagent is added to liberate N so that N can also be titrated after titration of M. For the determination of both Pb^{2+} and Zn^{2+} in a solution, a selective titration is not feasible because the EDTA formation constants with these two metal ions are very close to each other. In this case, The selective titration of Pb^{2+} becomes feasible by masking Zn^{2+} using KCN in a ammonia-tartrate solution. After Pb^{2+} is titrated by EDTA with EBT as the indicator, formaldehyde is added to react with cyanide to liberate Zn^{2+} by the following reaction:

$$4\,HCHO + Zn(CN)_4^{2-} + 4\,H_2O \rightleftharpoons Zn^{2+} + 4\,H_2C\begin{matrix}\diagup CN \\ \diagdown OH\end{matrix} + 4\,OH^-$$

The released Zn^{2+} can then be titrated by EDTA. Cd^{2+} can be liberated in a similar manner from the Cd-cyanide complex. The cyanide complexes of Cu^{2+}, Co^{2+}, Ni^{2+}, and Hg^{2+} are too stable to be demasked by formaldehyde. Increasing the concentration of formaldehyde can only partially demask the above metal ions. Other masking reagents or approaches have to be adopted for this purpose. For example, Cu^{2+} can be masked by being reduced to Cu^+ with thiourea and then can be demasked by oxidization with H_2O_2.

2. Masking by Redox Reaction

For the determination of Zr in the presence of iron, a direct titration by simply controlling pH is not feasible because $\Delta \lg K$ is not large enough to facilitate this titration ($\lg K(ZrOY^{2-}) = 29.9$, $\lg K(FeY^{-}) = 25.1$). However, Fe^{3+} can be masked by reducing reagents such as ascorbic acid or hydroxylamine hydrochloride that reduce Fe^{3+} to Fe^{2+}. With a much smaller formation constant than FeY^{-}, FeY^{2-} will not interfere with titration of Zr^{4+} ($\lg K(FeY^{2-}) = 14.3$). The same approach can be adopted for titration of Th^{4+}, Bi^{3+}, In^{3+}, and Hg^{2+} in the presence of Fe^{3+}.

3. Masking by Precipitation

The formation constants for Ca^{2+} and Mg^{2+} with EDTA are close to each other, thus the selective titration by controlling pH is impossible. We can still find a way to mask one to titrate the other: The solubility of magnesium hydroxide is 5 orders of magnitude less than that of calcium hydroxide, thus Mg^{2+} forms a $Mg(OH)_2$ precipitate at pH 12 and Ca^{2+} can be titrated after Mg^{2+} is precipitated. The commonly used precipitating reagents for masking purpose and examples of application are listed in Table 5.3.

Table 5.3 The Commonly Used Precipitating Reagents for Masking Purpose

Masking Reagent	Ion to be Masked	Ion to be Titrated	pH	Indicator
Hydroxide	Mg^{2+}	Ca^{2+}	12	Calconcarboxylic acid
Potassium iodide (KI)	Cu^{2+}	Zn^{2+}	5~6	PAN
Fluoride	Ba^{2+} Sr^{2+} Ca^{2+} Mg^{2+}	Zn^{2+} Cd^{2+} Mn^{2+}	10	EBT
Sulfate	Ba^{2+} Sr^{2+}	Ca^{2+} Mg^{2+}	10	EBT
Sodium sulfide	Hg^{2+} Pb^{2+} Bi^{3+} Cu^{2+} Cd^{2+}	Ca^{2+} Mg^{2+}	10	EBT

In practical applications, masking by precipitation is not widely used because: (1) Completeness of some precipitation reactions is not satisfactory. Sometimes, the precipitate is hard to collect because of supersaturation. (2) The absorption of the analyte on the precipitate will cause error. (3) Some precipitates have dark color and some have a large volume so that the color change at the endpoint is difficult to observe.

Other than EDTA, there are other complexones, which can form complexes with alternated formation constants. Those complexones can also be used for titration or selective titration of metal ions. EGTA forms a more stable complex with

Ca^{2+} than with Mg^{2+} ($\lg K(MgY)=5.2$, $\lg K(CaY)=11.0$). Selective titration of Ca^{2+} in the presence of Mg^{2+} would be easier by using EGTA as the titrant. In bioassays, EGTA is often used for masking Ca^{2+}.

5.5 APPLICATIONS OF COMPLEXOMETRIC TITRATIONS

5.5.1 Buffer Selection in Complexometric Titrations

Hydronium ions are continuously released during EDTA titrations. In the titration of Pb^{2+} at about pH 5, the complexation reaction is written as

$$Pb + H_2Y \rightleftharpoons PbY + 2H^+$$

As a result, the increase in acidity of the solution will reduce $K(MY')$, thus reducing the completeness of the reaction. Meanwhile, $K(MIn')$ will also reduce the sensitivity for detecting the endpoint. Consequently, pH control in complexometric titration is necessary and often a pH buffer is added to the solution being analyzed.

Acetate buffer or $(CH_2)_6N_4$ buffer is often used for titration in weakly acidic pH (pH 5~6); ammonia buffer is used for titration in weakly basic pH (pH 8~10). For titrations in strongly acidic (e.g., titration of Bi^{3+} at pH 1) or basic pH (e.g., titration of Ca^{2+} at pH >12), it is unnecessary to add buffer because strong acid/base itself can buffer the solution to maintain pH at a desired value. The following factors need to be included for selection of buffer in complexometric titration: (1) The pK_a of the buffer system should be close to the pH of the titration. (2) The buffer system should not cause side reactions with metal ion thus affecting the completeness of titration. For example, acetate buffer is not used for titration of Pb^{2+} at pH≈5, because acetate will form complexes with Pb^{2+} at this pH and reduce $\lg K(PbY)'$. (3) Buffer capacity of the selected buffer system should be large enough to keep pH in the desired range during the titration.

【Example 5.14】 How many grams of $(CH_2)_6N_4$ and how many milliliters of HNO_3 are needed to ensure that pH≥4.8 in titration of 25 mL of 0.02 mol · L^{-1} Pb^{2+} at pH 5.0 using the same concentration of EDTA.

Answer: The reaction for titration of Pb^{2+} is

$$Pb^{2+} + H_2Y^{2-} \rightleftharpoons PbY^{2-} + 2H^+$$

Two fold H^+ ions are produced in the titration reaction. At stoichiometric point, $[H^+]$ is increased by 0.02 mol · L^{-1}.

Hexamethylene-tetramine (urotropine) and its conjugate acid is an appropriate buffer for titration of Pb^{2+}, because it does not react with Pb^{2+} and has a buffer range greater than pH 5. The produced H^+ in the titration will increase the fraction of conjugate acid.

$$\Delta c(H^+) = c \cdot (x_2(HA) - x_1(HA)) = 0.02(mol \cdot L^{-1})$$

At pH = 5.0, $$x_1(HA) = \frac{10^{-5.0}}{10^{-5.3} + 10^{-5.0}} = 0.67$$

At pH = 4.8, $$x_2(HA) = \frac{10^{-4.8}}{10^{-5.3} + 10^{-4.8}} = 0.76$$

Then $$c = \frac{0.02}{0.76 - 0.67} = 0.22(mol \cdot L^{-1})$$

$$m((CH_2)_6N_4) = 0.22 \times 0.025 \times 2 \times 140 = 1.5(g)$$

$$n(HNO_3) = 0.22 \times 0.67 \times 0.025 \times 2 = 7.4(mmol)$$

Therefore, 1.5 g urotropine and 7.4 mmol HNO_3 are needed to add to Pb^{2+} solution before titration.

5.5.2 Titration Methods and Applications

1. Direct Titration

Direct complexometric titration is based on indicators that respond to the analyte itself or an added metal ion. The advantages of direct titration are obvious: simple, fast, and less likely to introduce errors; therefore, a direct titration is always the first choice for a selected method. When the reaction of a metal ion with EDTA meets the requirements for complexometric titration, a direct titration can be used to determine this metal ion. Examples using direct complexometric titrations with EDTA are given in Table 5.4.

Table 5.4 Examples for Direct Titration

Metal Ions	pH	Indictor	Remarks
Bi^{3+}	1	XO	HNO_3
Fe^{3+}	2	Sulfosalicylic acid(SSal)	50～60℃
Th^{4+}	2.5～3.5	XO	
Cu^{2+}	2.5～10	PAN	Ethanol or heating
	8	Murexide	
Zn^{2+}, Cd^{2+}, Pb^{2+}, rare earths	≈5.5	XO	Tartaric acid as an auxiliary complexing reagent for Pb^{2+}
	9～10	EBT	
Ni^{2+}	9～10	Murexide	Ammonia buffer, 50～60℃
Mg^{2+}	10	EBT	
Ca^{2+}	12～13	Calconcarboxylic acid	

The determination of water hardness by complexometric titration is a good example to show application of complexometric titration. Historically, water "hardness" was defined in terms of the capability of cations in the water to replace the sodium or potassium ions in soaps and to form sparingly soluble products that cause "scum" in the sink or bathtub. Most multiply charged cations share this undesirable property; however in natural waters the concentration of calcium and magnesium ions generally far exceed those of any other metal ion. Consequently, hardness is now expressed in terms of the concentration of calcium carbonate that is equivalent to the total concentration of all the multivalent cations in the sample.

Water hardness is determined by an EDTA titration with EBT as the indicator after the sample has been buffered to pH 10. Magnesium, which forms the least stable EDTA complex of all of the common multivalent cations in typical water samples, will not be titrated until enough reagents have been added to complex calcium. The formation constants of these two metal ions with EBT are in reverse order as with EDTA ($\lg K(\mathrm{CaIn}) = 5.4$, $\lg K(\mathrm{MgIn}) = 7.0$). When the solution changes color from mauve to blue, Mg^{2+} is completely chelated with EDTA and Ca^{2+} has completely reacted with EDTA before Mg^{2+}. As a result, the titration measures the total concentration of Ca^{2+} and Mg^{2+}. Another aliquot of testing sample is adjusted to pH > 12 to precipitate Mg^{2+} as $Mg(OH)_2$. The Ca^{2+} is titrated when using calcon as the indicator. The amount of Mg^{2+} in solution can be obtained by subtracting Ca^{2+} from the total concentration.

2. Back Titration

Back titration is useful for the following situations: (1) the reaction rate between metal ion and EDTA is slow; (2) metal ions are apt to undergo hydrolysis to form hydroxo complex at the titration pH and a proper auxiliary complexing reagent is not available; (3) the complex of the metal ion with the indicator is more stable than the metal-EDTA complex so that the indicator is blocked. Titration of Al^{3+} using EDTA is an example: the reaction rate of Al^{3+} with EDTA is slow. Al^{3+} tends to form hydroxo complex in weakly acidic solution, which makes the complexation reaction with EDTA even slower. The indicator (e.g. XO) is blocked. All the characters mentioned above make direct titration of Al^{3+} impossible.

The problem presented in the titration of Al^{3+} can be solved by back titration combined with pH control during the titration. An excess amount of EDTA is added to the test solution followed by adjusting pH to about 3.5. Heating is necessary to accelerate the complexation reaction. After cooling the solution to room temperature and adjusting the pH to 5～6, XO indicator is added to the solution. All Al^{3+} have reacted to form an AlY complex, so Al^{3+} will not block the indicator and the excess EDTA can then be back titrated using a Zn^{2+} standard solution. The complex of EDTA with metal ion used for back titration purpose should not be more stable than the metal ion being analyzed to avoid displacement of tested metal ion from EDTA complex. The frequently used metal ions for back titration purpose are given in Table 5. 5.

Table 5. 5 Frequently Used Metal Ions for Back Titration Purpose

pH	Metal Ion	Indicator	Analyte
1～2	Bi^{3+}	XO	ZrO^{2+} Sn^{4+}
5～6	Zn^{2+} Pb^{2+}	XO	Al^{3+} Cu^{2+} Co^{2+} Ni^{2+}
5～6	Cu^{2+}	PAN	Al^{3+}
10	Mg^{2+} Zn^{2+}	EBT	Ni^{2+} rare earths
12～13	Ca^{2+}	Calconcarboxylic acid	Co^{2+} Ni^{2+}

3. Demasking

Demasking approach is a simple way to cope with a sample containing more than one metal ions. For the determination of Al^{3+} in the presence of Pb^{2+}, Zn^{2+}, and Cd^{2+}, the metal ions can be analyzed by back titration. Masking the other metal ions one by one is nearly impossible; however, Al^{3+} can be liberated after back titration: when back titration reaches the endpoint, NaF is added to the solution to liberate Al^{3+} from AlF_6. Heating is necessary for the demasking reaction as below,

$$AlY^- + 6F^- + 2H^+ \rightleftharpoons AlF_6^{3-} + H_2Y^{2-}$$

After cooling to room temperature, the liberated EDTA can be titrated with Zn^{2+} standard solution. According to the reaction stoichiometry, concentration of Al^{3+} can be obtained. This method has very high selectivity and only Zr^{4+}, Ti^{4+}, and Sn^{4+} interfere with the determination. This method can also be used for determination of Sn^{4+} in a gunmetal sample (containing Sn^{4+}, Cu^{2+}, Pb^{2+}, and Zn^{2+}). Hg^{2+} can be liberated by KI; Cu^{2+} can be liberated by thiourea; Zn^{2+}, Cd^{2+}, Cu^{2+}, Co^{2+}, Ni^{2+}, and Hg^{2+}, etc., can be liberated by KCN or phenanthroline.

4. Indirect Titration

Indirect titration is applied to the following cases: (1) metal ions which form EDTA complexes are not stable enough to facilitate titration; (2) non-metal ions which react stoichiometrically with metal ions. For example, K^+ can be precipitated to form stoichiometrically a precipitate, $K_2NaCo(NO_2)_6 \cdot 6H_2O$. After separation and dissolution of the precipitate, Co^{2+} is titrated by EDTA and the concentration of K^+ can be calculated. This approach has been applied to the determination of K^+ in human serum and urine.

PO_4^{3-} can be precipitated stoichiometrically as $MgNH_4PO_4 \cdot 6H_2O$. After separation and dissolution in HCl, a back titration is carried out to determine Mg^{2+} in the solution. The excess EDTA is added to the solution following by adjustment of pH to 10 with ammonia, and standard Mg^{2+} is used to back titrate the excess EDTA. The amount of P in the sample can be calculated from the titration. To determine SO_4^{2-}, excess standard Ba^{2+} is added to form a $BaSO_4$ precipitate. The remaining Ba^{2+} is back titrated using EDTA with MgY-EBT as the indicator. CO_3^{2-}, CrO_4^{2-}, and S^{2-} can also be determined in the same way. Medicines containing metal ions, such as Pepto-Bismol and calcium lactate, as well as alkaloids which can react with metal ions (such as caffeine), can be determined using complexometric titration.

The Ag-EDTA complex does not meet the requirement for direct titration, because the formation constant is not large enough($\lg K(AgY)=7.8$). When excess $Ni(CN)_4^{2-}$ is added to a sample containing Ag^+, the following reaction will take place:

$$2Ag^+ + Ni(CN)_4^{2-} \rightleftharpoons 2Ag(CN)_2^- + Ni^{2+}$$

This reaction has a relatively large formation constant, which enables stoichiometric liberation of Ni^{2+}.

$$K = \frac{(K(Ag(CN)_2^-))^2}{K(Ni(CN)_4^{2-})} = \frac{(10^{21.1})^2}{10^{31.3}} = 10^{10.9}$$

EDTA is used to titrate the released Ni^{2+}. For determination of Ag and Cu in a silver coin, the sample is dissolved in nitric acid and the pH adjusted to 8 with ammonia. Then the solution is titrated by EDTA using murexide as the indicator. Excess $Ni(CN)_4^{2-}$ is added to the solution after titration, and liberated Ni^{2+} is then titrated by EDTA. The amount of silver can be calculated from the released Ni^{2+}. Murexide is a classical indicator for complexometric titration of Ca^{2+}, Ni^{2+}, Co^{2+}, and Cu^{2+}.

Sometimes indirect metal indicators are used for complexometric titration. The best known example is the use of MgY in the titration of Ca^{2+} using EBT to indicate endpoint. The reaction of Ca^{2+} and EBT does not generate sharp endpoint color change, but the reaction of Mg^{2+} and EDTA does. Upon the addition of a small amount of MgY to a Ca^{2+} solution at pH 10, the following displacement reaction will occur:

$$Ca^{2+} + MgY \rightleftharpoons CaY + Mg^{2+}$$

The liberated Mg^{2+} forms a mauve complex with EBT. After Ca^{2+} has completely reacted with EDTA, a small amount of EDTA will release Mg^{2+} in Mg-EBT complex. The color at the endpoint changes from mauve to blue. In this application, the amount of MgY formed during titration equals the amount of MgY added from the beginning. The endpoint is decided through reaction of EBT with Mg^{2+}. The mechanism for Cu-PAN to indicate endpoint is the same.

5.5.3 Preparation of Standard Solutions

EDTA is slightly soluble in water (0.02 g/100 mL at 22℃), insoluble in acids and organic solvents, and soluble in NaOH and NH_3. The sodium salt commercially available as the dihydrate, $Na_2H_2Y \cdot 2H_2O$ (EDTA), has higher solubility (11.1 g/100 mL at 22℃, about 0.3 $mol \cdot L^{-1}$). Usually the concentration for titration is 0.01～0.05 $mol \cdot L^{-1}$. This sodium salt of EDTA can be used as a primary standard. As metal ions often exist in water and other reagents to be used for titration purpose, EDTA standard solution is usually prepared by standardization.

The primary standards for EDTA include zinc, copper, bismuth, ZnO, $CaCO_3$, and $MgSO_4 \cdot 7H_2O$. Zinc which can be obtained in high purity (99.99%) is often used because it is stable in the air; can be used at pH 5～6 (XO as indicator) or pH 9～10 (EBT as indicator); and has a sharp color transition at the endpoint.

The quality of distilled water is the key to success of complexometric titration because: (1) Al^{3+} and Cu^{2+} in water can block indicators, which makes judging the endpoint difficult; (2) Ca^{2+}, Mg^{2+}, Pb^{2+}, and Sn^{2+} in water consume EDTA. Consequently, the analytical results will be affected. When there are a small amount of Ca^{2+} or Pb^{2+} in water, the following situation may be considered: (1) these two metal ions can both react with EDTA in basic solution; (2) Pb^{2+} reacts with EDTA in a weakly acidic solution, but Ca^{2+} does not; (3) both ions do not react with EDTA in a strongly acidic solution. The influence from the two metal ions can be avoided by conducting standardization and determination at the same pH. It is always recommended that standardization and determination be carried out at the same conditions and it is better to use the standard solution of the same metal ion as the analyte for standardization.

Chapter 5 Questions and Problems

5.1 For phosphoric acid, fill the blank in the table:

Protonation Formation Constant	Equation	pH
$\lg K_1 = 11.7$	$[H_3PO_4] = [H_2PO_4^-]$	
$\lg K_2 = 6.9$	$[H_2PO_4^-] = [HPO_4^{2-}]$	
$\lg K_3 = 2.0$	$[HPO_4^{2-}] = [PO_4^{3-}]$	

5.2 For EDTA, Fill the blank in the table:

Dissociation Constant	pK_{a1}	pK_{a2}	pK_{a3}	pK_{a4}	pK_{a5}	pK_{a6}
	0.9	1.6	2.07	2.75	6.24	10.34
Stepwise Formation Constant	$\lg K_1$	$\lg K_2$	$\lg K_3$	$\lg K_4$	$\lg K_5$	$\lg K_6$
Cumulative Constant	$\lg\beta_1$	$\lg\beta_2$	$\lg\beta_3$	$\lg\beta_4$	$\lg\beta_5$	$\lg\beta_6$

5.3 For copper-ammonia complexes,

$K_{diss1} = 7.8 \times 10^{-3}$, $K_{diss2} = 1.4 \times 10^{-3}$, $K_{diss3} = 3.3 \times 10^{-4}$, $K_{diss4} = 7.4 \times 10^{-5}$

(1) Calculate stepwise formation constants $K_1 \sim K_4$ and cumulative constants $\beta_1 \sim \beta_4$;

Dissociation Constant	pK_{diss1}	pK_{diss2}	pK_{diss3}	pK_{diss4}
	2.11	2.85	3.48	4.13
Stepwise Formation Constant	$\lg K_1$	$\lg K_2$	$\lg K_3$	$\lg K_4$
Cumulative Constant	$\lg\beta_1$	$\lg\beta_2$	$\lg\beta_3$	$\lg\beta_4$

(2) Calculate $[NH_3]$ in a solution where $[Cu(NH_3)_4^{2+}] = 10[Cu(NH_3)_3^{2+}]$;

(3) Calculate the equilibrium concentration of cupper-ammonia complexes in each form in a solution where $c(NH_3) = 1.0 \times 10^{-2}$ mol · L^{-1}, and $c(Cu) = 1.0 \times 10^{-4}$ mol · L^{-1}. which is the predominant form for Cu(II) in such a solution (the side reactions of Cu^{2+} and NH_3 are neglected)?

5.4 Fill the blank in the table for Fe-acetylacetone (L) complexes. $\lg\beta_1 \sim \lg\beta_3$: 11.4, 22.1, and 26.7.

pL	22.1	11.4	7.7	3.0
Predominant Form				

5.5 A Pb^{2+} solution of 0.02 mol · L^{-1} is titrated with the same concentration of EDTA at pH 10.0 of ammonia buffer. At the initial point, analytical concentration of tartarate is 0.2 mol · L^{-1}. Calculate $\lg K'(PbY)$, $[Pb']$ and $[PbL]$ (concentration of lead-tartarate complex) at SP ($\lg K(PbL)=3.8$, $\lg K(PbY)=18.0$).

5.6 A 15 mL of 0.020 mol · L^{-1} EDTA is mixed with 10 mL of 0.020 mol · L^{-1} Zn^{2+}.

(1) Calculate $[Zn^{2+}]$ at pH 4.0;

(2) What's the pH value of the solution so that $[Zn^{2+}]=10^{-7}$ mol · L^{-1}?

Constants: $\lg K(ZnY)=16.5$, $\lg K^{H}(ZnHY)=3.0$; pH=4.0, $\lg\alpha_{Y(H)}=8.6$, $\lg\alpha_{Zn(OH)}=0.0$.

5.7 A 20.00 mL of 0.010 mol · L^{-1} metal ion is titrated using EDTA at the same concentration under a controlled condition. The titration reaction is complete at the condition with pM spread by 1 pM unit with addition of 19.98~20.02 mL EDTA. Calculate $K'(MY)$.

5.8 A 0.020 mol · L^{-1} EDTA is used to titrate 0.020 mol · L^{-1} Cu^{2+} in the presence of Ca^{2+} at the same concentration. Providing that pH=5.0 and $(pCu)_t=8.8$, calculate

(1) E_t;

(2) $[(CaY)_{sp}]$ and $[(CaY)_{ep}]$.

5.9 For titration of Ca^{2+} using EDTA at pH 13, fill the blank based on the information provided:

c	pCa		
	Pre 0.1% SP	SP	Post 0.1% SP
0.01 mol · L^{-1}	5.3	6.5	
0.1 mol · L^{-1}			

5.10 The dissociation constants for erio blue black R(EBB R) are $K_{a1}=10^{-7.3}$, $K_{a2}=10^{-13.5}$, and the formation constant for Mg-EBB R is $K(MgIn)=10^{7.6}$.

(1) Calculate $(pMg)_t$ at pH 10.0;

(2) Calculate the titration error for titration of 2×10^{-2} mol · L^{-1} Mg^{2+} using EDTA at the same concentration when EBB R is used as the indicator.

5.11 In a solution with thorium ion (Th^{4+}) and lanthanum ion (La^{3+}) both at 2×10^{-2} mol · L^{-1}, Th^{4+} is stepwise titrated first using EDTA at the same concentration by controlling pH of the solution.

(1) What is the appropriate pH range for titration of Th^{4+} (That is the pH range between the value where $\lg K'(ThY)$ reaches maximum and precipitation of $Th(OH)_4$ does not occur);

(2) Calculate the titration error for continuous stepwise titration of La^{3+} at pH 5.5 using XO as the indicator.

$\lg K(ThY)=23.2$, $\lg K(LaY)=15.4$.

5.12 In a solution with Pb^{2+} and Al^{3+} both at 2×10^{-2} mol · L^{-1}, Pb^{2+} is stepwise titrated using EDTA at the same concentration by masking Al^{3+} with acetylacetone (E, $pK_a(HE)=8.8$). At the endpoint, $c(HE)=0.1$ mol · L^{-1}. If titration is carried out at pH=5.0 with XO as the indicator, calculate titration error (The complexation of E with Pb^{2+} can be neglected). AlE_j: $\lg\beta_1 \sim \lg\beta_3$ are 8.1, 15.7, 21.2; $\lg K(AlY)=16.1$ and $\lg K(PbY)=18.0$.

5.13 Masking and pH adjustment are both used for titration of 0.020 mol·L^{-1} Zn^{2+} in the presence of the same concentration of Cd^{2+}. The titration is carried out at pH 5.5 using HEDTA(X) at the same concentration with XO as the indicator. KI is used for masking Cd^{2+}. The following information are available: $\lg K(ZnX)=14.5$, $\lg K(CdX)=13.0$, $\lg\alpha_{X(H)}=4.6$, $(pZn)_t(XO)=5.7$; The following constants are calculated: $\lg\alpha_{Cd(I)}=5.1$, $\lg\alpha_{X(Cd)}=5.9$, $\lg K(ZnX')=8.6$, $(pZn)_{sp}=5.3$, $E_t=+0.1\%$.

Fill the blank in the table according to what is provided above (concentration unit: mol·L^{-1}):

	[X']	[X]	$\sum_{i=1\sim3}[H_iX]$	$[Cd^{2+}]$
SP				
EP				

5.14 The Bi^{3+} in a 25.00 mL Bi^{3+} and Pb^{2+} solution (pH 1.0) is titrated with 15.00 mL of 0.02000 mol·L^{-1} EDTA. If it is desirable to titrate Pb^{2+} after titration of Bi^{3+}, calculate the mass of hexamethylene tetramine ($N_4(CH_2)_6$) in grams needed for adjusting pH to 5.0 after titration of Bi^{3+}.

5.15 Write the experimental design for the following analyses by complexometric titrations. State clear the experimental steps, standard solutions, buffers, indicators, masking and demasking reagents wherever necessary.

(1) Determination of Bi^{3+} in a solution containing Fe^{3+};

(2) Determination of Pb^{2+} and Zn^{2+} in a copper alloy;

(3) Determination of Fe^{3+}, Al^{3+} Ca^{2+}, Mg^{2+} in a concrete sample;

(4) Determination of Zn^{2+} in a solution containing Al^{3+}, Zn^{2+}, and Mg^{2+};

(5) Determination of concentration of each metal ion in a Bi^{3+}, Al^{3+}, and Pb^{2+} solution.

CHAPTER 6

REDOX EQUILIBRIUM AND TITRATION

6.1 Standard Electrode Potentials, Formal Potentials and Redox Equilibria

6.1.1 Standard Electrode Potentials

6.1.2 The Nernst Equation and Formal Potentials

6.1.3 Factors Affecting the Formal Potential

6.1.4 The Equilibrium Constant of Redox Reaction

6.2 Factors Affecting the Reaction Rate

6.2.1 Concentrations

6.2.2 Temperature

6.2.3 Catalysts and Reaction Rate

6.2.4 Induced Reaction

6.3 Redox Titrations

6.3.1 Constructing Redox Titration Curves

6.3.2 Indicators

6.3.3 Auxiliary Oxidizing and Reducing Agents

6.4 Examples of Redox Titrations

6.4.1 Potassium Permanganate ($KMnO_4$)

6.4.2 Potassium Dichromate ($K_2Cr_2O_7$)

6.4.3 Iodine: Iodimetry and Iodometry

6.4.4 Potassium Bromate ($KBrO_3$)

6.4.5 Ceric Sulfate ($Ce(SO_4)_2$)

Redox titration is based on oxidation-reduction reaction with one or more electrons as the reactive species, which is transferred from the reductant to the oxidant. Redox titrations can be used for quantitation of many types of inorganic and organic compounds based on oxidation and reduction reaction. Redox titration was introduced shortly after the development of acid-base titrimetry. The earliest methods took advantage of the oxidizing power of chlorine. The number of redox titrimetric methods increased in the mid-1800s with the introduction of permanganate (MnO_4^-), dichromate ($Cr_2O_7^{2-}$) and iodine (I_2) as oxidizing titrants, and ferrous ion (Fe^{2+}) and thiosulfate ($S_2O_3^{2-}$) as reducing titrants. The reaction rate of redox reactions can be very slow, and sometimes a specific stoichiometry is difficult to obtain because side reactions occur; therefore, it is very important to maintain reaction conditions that assure a reaction with a reasonable rate and stoichiometry.

6.1 STANDARD ELECTRODE POTENTIALS, FORMAL POTENTIALS AND REDOX EQUILIBRIA

A redox reaction is one that occurs between a reducing and an oxidizing agents:

$$Ox_1 + Red_2 \rightleftharpoons Red_1 + Ox_2 \qquad (6\text{-}1)$$

where Ox_1 is reduced to Red_1 and Red_2 is oxidized to Ox_2. Ox_1 is the **oxidizing agent**, and Red_2 is the **reducing agent**. The reducing or oxidizing tendency of a substance will depend on its reducing potential, which will be obtained from an understanding of the **electrochemical cells** and **electrode potentials**.

6.1.1 Standard Electrode Potentials

Considering the following redox reaction:

$$Fe^{2+} + Ce^{4+} \rightleftharpoons Fe^{3+} + Ce^{3+} \qquad (6\text{-}2)$$

If Fe^{2+} is mixed with Ce^{4+}, there is a certain tendency for the ions to transfer electrons. The reactants can be separated and allowed to react under controlled conditions by placing the reactants in separate electrode compartments as shown in Figure 6.1A. The salt bridge allows charge transfer through the solutions but prevents the physical mixing of the solutions. Two inert platinum wires are placed in each solution to establish a galvanic cell. Closing the switch S allows electrons to flow from the left electrode compartment to the right electrode compartment. The reactions are:

Left side (oxidation occurs): $Fe^{2+} \rightleftharpoons Fe^{3+} + e^-$ (6-3)

Right side (reduction occurs): $Ce^{4+} + e^- \rightleftharpoons Ce^{3+}$ (6-4)

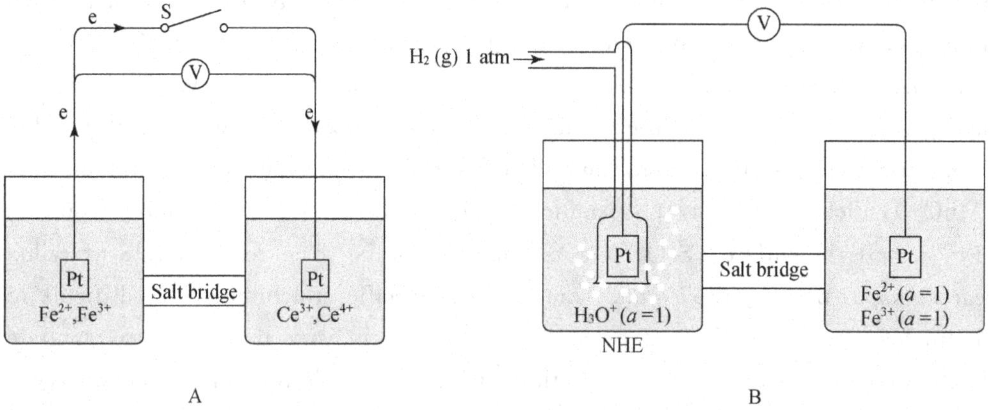

A

B

Figure 6.1 A, Electrochemical cell for carrying out a redox reaction; B, measurement of the Fe^{3+}/Fe^{2+} electrode potential, using the normal hydrogen electrode (NHE) as the reference.

The platinum wires can be considered as the electrodes. Each wire will adopt an electrical potential, called **electrode potential** determined by the tendency of the ions to lose or to accept electrons. A voltmeter placed between the electrodes will indicate the difference in the potential (cell potential) between the two electrodes.

Equations (6-3) and (6-4) are half-reactions. There must be an electron donor (a reducing agent, e.g., Fe^{2+}) and an electron acceptor (an oxidizing agent, e.g., Ce^{4+}) to compose an electrochemical cell. Each half reaction will generate a definite potential that would be adopted by an inert electrode dipped in the solution.

Unfortunately, there is no way to measure individual electrode potentials, but the difference between two electrode potentials can be measured. The electrode potential of the half-reaction

$$2H^+ + 2e^- \rightleftharpoons H_2 \quad (6\text{-}5)$$

has arbitrarily been assigned a value of 0.000 V at $p_{H_2} = 1.00$ atm and $a_{H^+} = 1.00$ mol · L^{-1} with temperature at any value. This is called the **normal hydrogen electrode** (**NHE**) or the **standard hydrogen electrode** (**SHE**). The potential differences between NHE half-reaction and other half-reactions can be measured (Figure 6.1B, Appendix D7). Electrode potentials are dependent on the activities of species. All standard potentials refer to conditions of unit activity for all species (For the hydrogen in NHE, the partial pressure is 1).

According to the Gibbs-Stockholm electrode potential convention, adopted at the 17th Conference of the International Union of Pure and Applied Chemistry in Stockholm, 1953, the half-reaction is written as a reduction. Some general conclusions can be drawn from the electrode potentials: (1) the more positive the electrode potential, the greater the tendency of the oxidized form to be reduced; (2) the more negative the electrode potential, the greater the tendency of the reduced form to be oxidized.

6.1.2 The Nernst Equation and Formal Potentials

The electrode potentials of the reversible half-reactions are dependent on the activity of the species. An electrode reaction is called reversible when the same reaction occurs, but in the opposite direction when the flow of current is reversed. An electrode reaction is called irreversible when a different reaction occurs when the flow of current is reversed. The dependence of the potential on activity is described by the **Nernst equation**:

$$\mathrm{Ox} + n\mathrm{e}^- \rightleftharpoons \mathrm{Red} \qquad (6\text{-}6)$$

$$\varphi = \varphi^\ominus + \frac{0.059}{n}\lg\frac{a(\mathrm{Ox})}{a(\mathrm{Red})} \quad (\text{at } 25℃) \qquad (6\text{-}7)$$

where $\varphi^\ominus$ is the standard potential, n is the number of electrons involved in the half-reaction, and $a(\mathrm{Ox})$ and $a(\mathrm{Red})$ are the activities of the oxidized and the reduced forms, respectively.

The analytical concentrations of the species are used rather than species activity. Deviation may occur when high ionic strength concentrations are used for calculations rather than using species activity. The potential is also altered significantly when the oxidizing form or reducing form undergoes side reactions, e. g. , pH change, precipitation or complex formation.

Considering the effects of ionic strength, as well as side reactions, activity can be expressed as follows:

$$a(\mathrm{Ox}) = [\mathrm{Ox}] \cdot \gamma(\mathrm{Ox}) = c(\mathrm{Ox}) \cdot \gamma(\mathrm{Ox})/\alpha_{\mathrm{Ox}}$$

$$a(\mathrm{Red}) = [\mathrm{Red}] \cdot \gamma(\mathrm{Red}) = c(\mathrm{Red}) \cdot \gamma(\mathrm{Red})/\alpha_{\mathrm{Red}}$$

where $\gamma(\mathrm{Ox})$ and $\gamma(\mathrm{Red})$ are the activity coefficients, $[\mathrm{Ox}]$ and $[\mathrm{Red}]$ are the equilibrium concentrations, $c(\mathrm{Ox})$ and $c(\mathrm{Red})$ are the analytical concentrations, and α_{Ox} and α_{Red} are the side reaction coefficients, respectively.

Substitute the above expressions for the activities equation (6-7), we get

$$\varphi = \varphi^{\ominus} + \frac{0.059}{n}\lg\frac{\gamma(\mathrm{Ox}) \cdot \alpha_{\mathrm{Red}}}{\gamma(\mathrm{Red}) \cdot \alpha_{\mathrm{Ox}}} + \frac{0.059}{n}\lg\frac{c(\mathrm{Ox})}{c(\mathrm{Red})} \tag{6-8}$$

When $c(\mathrm{Ox}) = c(\mathrm{Red}) = 1\ \mathrm{mol \cdot L^{-1}}$, the calculated potential, noted as $\varphi^{\ominus\prime}$ is the conditional potential or **formal potential.**

$$\varphi^{\ominus\prime} = \varphi^{\ominus} + \frac{0.059}{n}\lg\frac{\gamma(\mathrm{Ox}) \cdot \alpha_{\mathrm{Red}}}{\gamma(\mathrm{Red}) \cdot \alpha_{\mathrm{Ox}}} \tag{6-9}$$

where $\varphi^{\ominus\prime}$ denotes the potential when $c(\mathrm{Ox}) = c(\mathrm{Red}) = 1\ \mathrm{mol \cdot L^{-1}}$ and the concentrations of other species are all carefully specified. $\varphi^{\ominus\prime}$ is influenced by ionic strength and side reactions. $\varphi^{\ominus\prime}$ is a constant when solution parameters are specified, and the Nernst equation can be expressed as

$$\varphi = \varphi^{\ominus\prime} + \frac{0.059}{n}\lg\frac{c(\mathrm{Ox})}{c(\mathrm{Red})} \tag{6-10}$$

Formal potentials for many half-reactions are listed in Appendix D7, which are useful data for judging the possibility of a redox reaction occurring or whether a redox titration is possible.

6.1.3 Factors Affecting the Formal Potential

1. Dependence of Potential on Ionic Strength

Most analytical redox reactions are carried out in solutions of high ionic strength so the activity coefficients will be well below 1, which results in a big difference between the standard electrode potential and the formal potential. Taking $Fe(CN)_6^{3-}$ / $Fe(CN)_6^{4-}$ electrode potential as an example, the formal potential increases as ionic strength of the solution increases ($\varphi^{\ominus} = 0.355$ V).

Ionic Strength/mol • kg^{-1}	0.00064	0.0128	0.112	1.6
Formal Potential/V	0.3619	0.3814	0.4094	0.4584

Only in an extremely diluted solution, $\varphi^{\ominus\prime} = \varphi^{\ominus}$.

To simplify the calculation when side reactions occur, activity is substituted by equilibrium concentration in the Nernst equation as

$$\varphi = \varphi^{\ominus} + \frac{0.059}{n}\lg\frac{[\mathrm{Ox}]}{[\mathrm{Red}]} \tag{6-11}$$

All calculations in redox titration are based on constants at $I = 0.1$.

2. Dependence of Potential on Precipitation

In redox reactions, precipitation of the oxidizing or reducing forms will alter the electrode potentials. The precipitation of the oxidizing form will reduce the potential;

however, the precipitation of the reducing form will increase the potential. For example, when iodometry is chosen to determine Cu^{2+} ($\varphi^{\ominus}(Cu^{2+}/Cu^{+}) = 0.17$ V, $\varphi^{\ominus}(I_2/I^-) = 0.54$ V), one would expect I_2 to oxidize Cu^+ by just comparing the standard potentials. However, $[Cu^+]$ is remarkably reduced by the formation of a CuI precipitate according to the following reaction: $2Cu^{2+} + 4I^- \rightleftharpoons 2CuI\downarrow + I_2$. Therefore, $\varphi^{\ominus\prime}(Cu^{2+}/Cu^{+})$ is increased remarkably, which makes Cu^{2+} a strong oxidizing agent as compared to iodide in this solution and assures complete oxidation of I^- by Cu^{2+}.

【Example 6.1】 Calculate $\varphi^{\ominus\prime}(Cu^{2+}/Cu^{+})$ (25℃) in a 1 mol · L^{-1} KI solution. $\varphi^{\ominus}(Cu^{2+}/Cu^{+}) = 0.17$ V; $K_{sp}(CuI) = 2\times10^{-12}$.

Answer:

$$\begin{aligned}\varphi &\approx \varphi^{\ominus}(Cu^{2+}/Cu^{+}) + 0.059\lg\frac{[Cu^{2+}]}{[Cu^{+}]}\\ &= \varphi^{\ominus}(Cu^{2+}/Cu^{+}) + 0.059\lg\frac{[Cu^{2+}]}{K_{sp}/[I^-]}\\ &= \varphi^{\ominus}(Cu^{2+}/Cu^{+}) + 0.059\lg\frac{[I^-]}{K_{sp}} + 0.059\lg[Cu^{2+}]\end{aligned}$$

Because there is no side reaction occurring for Cu^{2+}, $[Cu^{2+}] = c(Cu^{2+})$; thus,

$$\begin{aligned}\varphi^{\ominus\prime} &= \varphi^{\ominus}(Cu^{2+}/Cu^{+}) + 0.059\lg\frac{[I^-]}{K_{sp}}\\ &= 0.17 - 0.059\lg(2\times10^{-12}) = 0.86(V)\end{aligned}$$

Therefore, Cu^{2+} can oxidize I^- to I_2, and this serves as the basis for the determination of copper by iodometry.

3. Dependence of Potential on Complexation

If complexation occurs to one or both of the ions in a redox couple, the concentration of the free ion is reduced, thus potential of the redox couple will change. This complexation reaction can be used for masking purpose in developing an analytical method. For example, in the iodometry titration of Cu^{2+}, Fe^{3+} as a coexisting substance can also oxidize I^-. By adding sodium fluoride (F^-), Fe^{3+} forms complexes with F^-. As a result, $\varphi^{\ominus\prime}(Fe^{3+}/Fe^{2+})$ is reduced to a potential which is lower than $\varphi^{\ominus}(I_3^-/I^-)$, therefore Fe^{3+} will not interfere with the determination of Cu^{2+}.

【Example 6.2】 Calculate $\varphi^{\ominus\prime}(Fe^{3+}/Fe^{2+})$ (25℃) in a pH 3.0 medium with $[F'] = 0.1$ mol · L^{-1} (The effect of ionic strength is neglected in this situation). $\lg\beta_1 \sim \lg\beta_3$ for Fe(Ⅲ)-F complexes are 5.2, 9.2, and 11.9; $\lg K^{H}(HF) = 3.1$; $\varphi^{\ominus}(Fe^{3+}/Fe^{2+}) = 0.77$ V.

Answer: According to Equation (6-11),

$$\varphi = \varphi^{\ominus}(Fe^{3+}/Fe^{2+}) + 0.059\lg\frac{[Fe^{3+}]}{[Fe^{2+}]}$$
$$= \varphi^{\ominus}(Fe^{3+}/Fe^{2+}) + 0.059\lg\frac{c(Fe^{3+})/\alpha_{Fe^{3+}(F)}}{c(Fe^{2+})/\alpha_{Fe^{2+}(F)}}$$
$$= \varphi^{\ominus}(Fe^{3+}/Fe^{2+}) + 0.059\lg\frac{\alpha_{Fe^{2+}(F)}}{\alpha_{Fe^{3+}(F)}} + 0.059\lg\frac{c(Fe^{3+})}{c(Fe^{2+})}$$

That is $$\varphi^{\ominus\prime} = \varphi^{\ominus}(Fe^{3+}/Fe^{2+}) + 0.059\lg\frac{\alpha_{Fe^{2+}(F)}}{\alpha_{Fe^{3+}(F)}}$$

At pH=3.0, $\alpha_{F(H)} = 1 + [H^+]\cdot K^H(HF) = 1 + 10^{-3.0+3.1} = 10^{0.4}$

Then $[F^-] = [F']/\alpha_{F(H)} = 10^{-1.0}/10^{0.4} = 10^{-1.4}(mol\cdot L^{-1})$

Hence $$\alpha_{Fe^{3+}(F)} = 1 + [F^-]\beta_1 + [F^-]^2\beta_2 + [F^-]^3\beta_3$$
$$= 1 + 10^{-1.4+5.2} + 10^{-2.8+9.2} + 10^{-4.2+11.9} = 10^{7.7}$$

As $\alpha_{Fe^{2+}(F)} = 1$, then

$$\varphi^{\ominus\prime} = 0.77 + 0.059\lg(1/10^{7.7}) = 0.32(V)$$

In most cases, complexation of the oxidized form is more stable than that of the reduced form, which reduces the electrode potential; however, the reverse effect can occur when the complexation of the reduced form is more stable. For example, the complexation of Fe^{2+} with phenanthroline is more stable than that of Fe^{3+} ($\lg\beta(Fe(phen)_3^{3+})=14.1$, $\lg\beta(Fe(phen)_3^{2+})=21.3$); therefore, electrode potential is increased from 0.77 V to 1.06 V in 1 $mol\cdot L^{-1}$ H_2SO_4.

4. pH Dependence

Hydrogen or hydroxyl ions are involved in many redox half-reactions. The potential of these redox couples can be altered by changing the pH of the solution. If one or both species in a redox couple are weak acids or bases, a change in the pH can directly change the existing forms of the redox couple. Consequently, the electrode potential can also be changed. Consider the following reaction:

$$H_3AsO_4 + 2H^+ + 2I^- \rightleftharpoons HAsO_2 + I_2 + 2H_2O$$
$$\varphi^{\ominus}(H_3AsO_4/HAsO_2) = 0.56\ V,\quad \varphi^{\ominus}(I_2/I^-) = 0.54\ V$$

The standard potentials of the two couples are comparable. The potential of I_2/I^- does not change as pH changes, while the potential of $H_3AsO_4/HAsO_2$ couple is substantially affected by pH. At very acidic pH, the above reaction will occur; however, the reverse reaction will occur as the pH increases.

[Example 6.3] Calculate $\varphi^{\ominus\prime}$ (As(V)/As(Ⅲ)) (25℃) at pH 8 (The effect of ionic strength can be ignored). $pK_{a1} \sim pK_{a3}$ for H_3AsO_4 are 2.1, 6.7, and 11.2; $pK_a(HAsO_2)=9.1$; standard electrode potential for half-reaction $H_3AsO_4 + 2H^+ +$

$2e^- \rightleftharpoons HAsO_2 + 2H_2O$ is $\varphi^\ominus = 0.56$ V.

Answer: The Nernst equation for the half-reaction is

$$\varphi = \varphi^\ominus(H_3AsO_4/HAsO_2) + \frac{0.059}{2}\lg\frac{[H_3AsO_4][H^+]^2}{[HAsO_2]}$$

Substituting $[H_3AsO_4] = c(As(V)) \cdot x(H_3AsO_4)$, $[HAsO_2] = c(As(III)) \cdot x(HAsO_2)$ into the Nernst equation gives

$$\varphi = 0.56 + \frac{0.059}{2}\lg\frac{x(H_3AsO_4)[H^+]^2}{x(HAsO_2)} + \frac{0.059}{2}\lg\frac{c(As(V))}{c(As(III))}$$

Hence
$$\varphi^{\ominus\prime} = 0.56 + \frac{0.059}{2}\lg\frac{x(H_3AsO_4)[H^+]^2}{x(HAsO_2)}$$

The formal potential can be calculated from the pH of the solution. At pH = 8.0, $x(HAsO_2) \approx 1$, and $x(H_3AsO_4) = 10^{-7.2}$. Hence,

$$\varphi^{\ominus\prime} = 0.56 + \frac{0.059}{2}\lg 10^{-7.2-16.0} = -0.12(V)$$

Plot of $\varphi^{\ominus\prime}$ (As(V)/As(III)) versus pH is given in Figure 6.2.

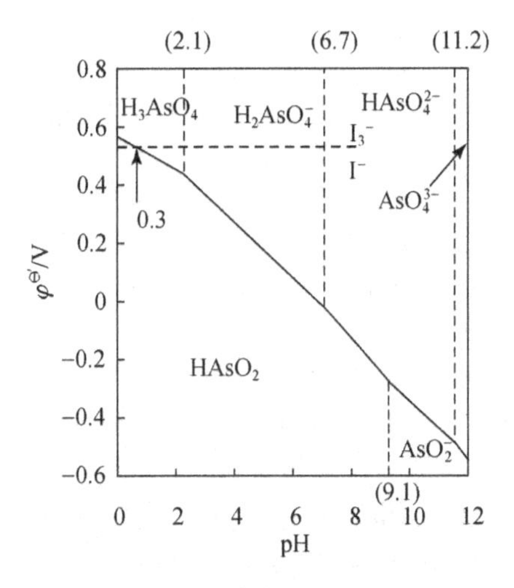

Figure 6.2 pH dependence of $\varphi^{\ominus\prime}$(As(V)/As(III)) and $\varphi^{\ominus\prime}(I_3^-/I^-)$.

At pH ≈ 0.3, $\varphi^{\ominus\prime}$ (As(V)/As(III)) = $\varphi^\ominus(I_3^-/I^-)$; as pH decreases, $\varphi^{\ominus\prime}$(As(V)/As(III)) will increase. As an example, H_3AsO_4 will oxidize I^- to I_2 in 4 mol · L^{-1} HCl; $Na_2S_2O_3$ can be used to titrate I_2 generated to determine indirectly As(V); when the pH increases, $\varphi^{\ominus\prime}$ (As(V)/As(III)) is reduced to values less than $\varphi^\ominus(I_3^-/I^-)$. At pH 8, the difference between the potentials of these two redox couples increases, thus As_2O_3 can be used to standardize an I_2 solution.

From the above example, one can tell that pH significantly affects the electrode potential such that the direction of reaction may even be reversed. The optimization of pH can be used to improve the selectivity of analytical method when a complicated sample is analyzed.

6.1.4 The Equilibrium Constant of Redox Reaction

For a redox reaction

$$p_2\mathrm{Ox}_1 + p_1\mathrm{Red}_2 \rightleftharpoons p_2\mathrm{Red}_1 + p_1\mathrm{Ox}_2 \qquad (6\text{-}12)$$

The half-reactions and their Nernst equations at 25℃ are

$$\mathrm{Ox}_1 + n_1\mathrm{e}^- \rightleftharpoons \mathrm{Red}_1 \quad \varphi_1 = \varphi_1^{\ominus\prime} + \frac{0.059}{n_1}\lg\frac{c(\mathrm{Ox}_1)}{c(\mathrm{Red}_1)}$$

$$\mathrm{Ox}_2 + n_2\mathrm{e}^- \rightleftharpoons \mathrm{Red}_2 \quad \varphi_2 = \varphi_2^{\ominus\prime} + \frac{0.059}{n_2}\lg\frac{c(\mathrm{Ox}_2)}{c(\mathrm{Red}_2)}$$

Since the cell voltage is zero at reaction equilibrium, the difference between the two half-reaction potentials is zero (or the potentials are equal), and the Nernst equations can be equated, i. e. $\varphi_1 = \varphi_2$, then

$$\varphi_1^{\ominus\prime} + \frac{0.059}{n_1}\lg\frac{c(\mathrm{Ox}_1)}{c(\mathrm{Red}_1)} = \varphi_2^{\ominus\prime} + \frac{0.059}{n_2}\lg\frac{c(\mathrm{Ox}_2)}{c(\mathrm{Red}_2)}$$

Rearrangement of the above expression gives the general expression of equilibrium constant of redox reaction (6-12)

$$\lg K' = \lg\left[\left(\frac{c(\mathrm{Red}_1)}{c(\mathrm{Ox}_1)}\right)^{p_2} \cdot \left(\frac{c(\mathrm{Ox}_2)}{c(\mathrm{Red}_2)}\right)^{p_1}\right] = \frac{(\varphi_1^{\ominus\prime} - \varphi_2^{\ominus\prime})p}{0.059} \qquad (6\text{-}13)$$

p is the total number of electrons (mathematically the lease common multiple) gained in the reduction (or lost in the oxidation) represented by redox reaction (6-12). $p = n_2 p_1 = n_1 p_2$; when $n_1 = n_2$, $p_1 = p_2 = 1$.

【Example 6.4】 Calculate the equilibrium constant for the reaction below

$$2\mathrm{Fe}^{3+} + \mathrm{Sn}^{2+} \rightleftharpoons 2\mathrm{Fe}^{2+} + \mathrm{Sn}^{4+}$$

$\varphi^{\ominus\prime}(\mathrm{Fe}^{3+}/\mathrm{Fe}^{2+}) = 0.70$ V and $\varphi^{\ominus\prime}(\mathrm{Sn}^{4+}/\mathrm{Sn}^{2+}) = 0.14$ V in 1 mol · L^{-1} HCl.

Answer:

$$\lg K' = \lg\left[\left(\frac{c(\mathrm{Fe}^{2+})}{c(\mathrm{Fe}^{3+})}\right)^2 \cdot \frac{c(\mathrm{Sn}^{4+})}{c(\mathrm{Sn}^{2+})}\right] = \frac{(0.70 - 0.14)\times 2}{0.059} = 18.98$$

$$K' = 10^{18.98}$$

At stoichiometric point,

$$\frac{c(\mathrm{Fe}^{2+})}{c(\mathrm{Fe}^{3+})} = \frac{c(\mathrm{Sn}^{4+})}{c(\mathrm{Sn}^{2+})}, \quad K' = \left(\frac{c(\mathrm{Fe}^{2+})}{c(\mathrm{Fe}^{3+})}\right)^2 \cdot \frac{c(\mathrm{Sn}^{4+})}{c(\mathrm{Sn}^{2+})} = 10^{18.98}$$

Then the concentration ratio of product to reactant of either redox couple, $c(\mathrm{Fe}^{2+})/$

$c(Fe^{3+}) = c(Sn^{4+})/c(Sn^{2+}) = 10^{6.3}$, indicates completeness of the redox reaction. Percent of Fe^{3+} (Sn^{2+}) is

$$\frac{c(Fe^{3+})}{c(Fe^{3+}) + c(Fe^{2+})} = 10^{-6.3} = 10^{-4.3}\%$$

Volumetric titrimetry requires that completeness should be greater than 99.9%. Substituting the above concentration ratio into Equation (6-13) yields the potential difference required to assure the completeness of a volumetric titration.

For $n_1 = n_2 = 1$, if $c(Red_1)/c(Ox_1) \geqslant 10^3$ and $c(Ox_2)/c(Red_2) \geqslant 10^3$ at stoichiometric point, then

$$K' = \frac{c(Red_1)}{c(Ox_1)} \cdot \frac{c(Ox_2)}{c(Red_2)} \geqslant 10^6$$

Hence $$\varphi_1^{\ominus\prime} - \varphi_2^{\ominus\prime} = \frac{0.059}{p}\lg K' \geqslant 0.059 \times 6 = 0.35(V)$$

For $n_1 = n_2 = 2$, we have

$$K' = \frac{c(Red_1)}{c(Ox_1)} \cdot \frac{c(Ox_2)}{c(Red_2)} \geqslant 10^6$$

Hence $$\varphi_1^{\ominus\prime} - \varphi_2^{\ominus\prime} \geqslant \frac{0.059}{2} \times 6 = 0.18(V)$$

Generally speaking, a redox titration is feasible when the potential difference between the two redox couples is greater than 0.4 V. There is a wide selection of strong oxidizing agents for redox titrations. By altering the formal potential and selection of reaction media, it is easy to assure the completeness of the redox titration. Therefore, titration error is not a major problem and will not be discussed. The kinetics for redox reactions can be more of a problem thus the kinetics will be discussed in the following section.

6.2 FACTORS AFFECTING THE REACTION RATE

Electrode potentials will predict the probability that a given reaction will occur, but this potential indicates nothing about the rate of the reaction. A reaction rate should be fast enough for a titration. In some cases, the reaction rate may be so slow that equilibrium will be reached only after a very long time and thus not favorable for a titration method. The oxidation of arsenic (Ⅲ) with cerium (Ⅳ) in dilute sulfuric acid is one example. The reaction is

$$H_3AsO_3 + 2Ce^{4+} + H_2O \rightleftharpoons H_3AsO_4 + 2Ce^{3+} + 2H^+$$

The electrode reactions and the formal potentials are

$$Ce^{4+} + e^- \rightleftharpoons Ce^{3+} \qquad \varphi^{\ominus\prime} = 1.44\ V$$

$$H_3AsO_4 + 2H^+ + 2e^- \rightleftharpoons H_3AsO_3 + H_2O \qquad \varphi^{\ominus\prime} = 0.56\ V$$

Then $\lg K = \frac{2(\varphi^{\ominus\prime}(Ce^{4+}/Ce^{3+}) - \varphi^{\ominus\prime}(H_3AsO_4/H_3AsO_3))}{0.059} = \frac{2\times(1.44-0.56)}{0.059} = 29.8$

The large equilibrium constant indicates that the reaction is thermodynamically favorable for titration, but several hours are required for the reaction to reach the equilibrium without a catalyst. The rate of a redox reaction can be dictated by the mechanism of reaction. The rate of a redox reaction will be fast, if the reaction involves only electron transfer. Examples are $Ce^{4+} + e^- \rightleftharpoons Ce^{3+}$ and $Fe^{3+} + e^- \rightleftharpoons Fe^{2+}$. The rate of a redox reaction will be slower, if the reaction involves breaking chemical bonds. Examples are $NO_3^- + 2H^+ + 2e^- \rightleftharpoons NO_2^- + 2H_2O$ and $SO_4^{2-} + 2H^+ + 2e^- \rightleftharpoons SO_3^{2-} + 2H_2O$.

For compounds involving the higher valence of an element, the redox reaction rate is generally slow. For example, reducing hypochlorous acid (HClO) is fast in either acidic or basic pH; chloric acid ($HClO_3$) can only be reduced in strong acidic solution; the reduction of perchloric acid ($HClO_4$) is even more difficult, and heating is required in addition to a strong acidic solution.

Most of the redox reactions involve more than one elementary step. In the reactions, one slow elementary step will dictate the overall reaction rate, and thus is called a rate determining step. The rate for redox reactions follows the general rule of kinetics. The effects of reactant concentration, temperature, catalysts, and induced reactions are to be discussed.

6.2.1 Concentrations

The order of a reaction is the sum of the exponents of all the concentrations that appear on the right-hand side of the rate equation. However, a redox reaction may involve more than one elementary steps and the rate determining step dictates the overall reaction rate. Therefore, the overall equation of a reaction does not enable us to predict its order or the equation of its rate law. Generally, the reaction rate can be increased by increasing the concentration of reactants. As an example, the reaction rate of $Cr_2O_7^{2-} + 6I^- + 14H^+ \rightleftharpoons 2Cr^{3+} + 3I_2 + 7H_2O$ can be increased by increasing the concentration of I^- and H^+. The reaction can reach equilibrium in 5 min with $c(H^+) = 0.4\ mol \cdot L^{-1}$ and concentration of KI 5 times greater than the stoichiometric concentration.

6.2.2 Temperature

In most cases, the reaction rate can be improved by elevating the reaction temperature, e. g. , elevating the temperature 10℃ increases the rate 2～3 fold. The oxidation of $C_2O_4^{2-}$ by MnO_4^- is slow at room temperature but much more rapid at elevated temperature; therefore, the titration of $H_2C_2O_4$ by $KMnO_4$ is done at 70～80℃.

6.2.3 Catalysts and Reaction Rate

In redox titrations, catalysts are used to significantly increase the rate of some reactions. The oxidation of arsenic (Ⅲ) with cerium (Ⅳ) in dilute sulfuric acid involves two elementary steps:

$$As(\text{III}) \xrightarrow[\text{Slow}]{Ce^{4+}} As(\text{IV}) \xrightarrow[\text{Fast}]{Ce^{4+}} As(\text{V})$$

Step 1 is the rate determining step. The addition of I^- can catalyze the first elementary step via the following catalytic reactions:

$$Ce^{4+} + I^- \longrightarrow I^0 + Ce^{3+}$$

$$2I^0 \longrightarrow I_2$$

$$I_2 + 2H_2O \rightleftharpoons HOI + H^+ + I^-$$

$$H_3AsO_3 + HOI \longrightarrow H_3AsO_4 + H^+ + I^-$$

Therefore, Ce^{4+} can be standardized by As_2O_3 with I^- as a catalyst.

For the oxidation reaction of $H_2C_2O_4$ by $KMnO_4$ in acidic pH:

$$2MnO_4^- + 5C_2O_4^{2-} + 16H^+ \rightleftharpoons 2Mn^{2+} + 10CO_2 + 8H_2O$$

The initial reaction rate is slow even at elevated temperature (e. g. 80℃) unless Mn^{2+} is present as a catalyst. As more Mn^{2+} is generated, the reaction proceeds more rapidly as a result of autocatalysis. Autocatalysis is a type of catalysis in which the product of a reaction catalyses further reaction, causing the rate of reaction to increase as the reaction proceeds.

6.2.4 Induced Reaction

In some cases, a redox reaction is slow, but other reactions in the same solution can speed up the redox reaction. For example, the oxidation of Cl^- to chlorine by $KMnO_4$ is extremely slow, but the oxidation of Fe^{2+} by $KMnO_4$ can speed up the first reaction. The redox reaction between $KMnO_4$ and Cl^- in this case is induced by the oxidation of Fe^{2+} with $KMnO_4$, thus this is called an induced reaction. Therefore, titration errors occur when permanganate titrations of Fe (Ⅱ) are performed in the presence of chloride ion.

To obtain an accurate titration result, Zimmermann-Reinhardt (Z-R) reagent is usually used to prevent the oxidation of the chloride ion to chlorine when Fe(Ⅱ) is titrated with MnO_4^-. The Mn(Ⅱ) in Z-R reagent with sulfuric acid and phosphoric acid reduces the potential of the MnO_4^-/Mn^{2+} couple sufficiently so that MnO_4^- will not oxidize chloride ion. Phosphoric acid in Z-R reagent is added to complex Fe(Ⅲ) and decrease the potential of the Fe^{3+}/Fe^{2+} couple. When Fe(Ⅲ) is removed from the solution as it is formed, the equilibrium of the titration reaction is shifted to the right to give a sharp endpoint. When the phosphate-Fe(Ⅲ) complex forms, the color of chloro complex in chloride medium fades, thus the endpoint color change is sharper.

6.3 REDOX TITRATIONS

6.3.1 Constructing Redox Titration Curves

In redox titrations, the concentration for oxidizing and reducing forms of both redox couples will change with the addition of titrant as will the electrode potentials. Because most redox indicators respond to changes in electrode potential, the ordinate in oxidation/reduction curves is generally the electrode potential rather than the logarithmic *p*-functions used for precipitation, complexation and neutralization titration curves. However, there is a logarithmic relationship between the electrode potential and the concentration of the analyte or titrant; as a result, the redox titration curves are similar to the *p*-function plot of other types of titrations.

We now proceed with the titration of 0.1000 mol · L^{-1} $FeSO_4$ with 0.1000 mol · L^{-1} $Ce(SO_4)_2$ in 1 mol · L^{-1} H_2SO_4 (at 25℃). The titration reaction is

$$Ce^{4+} + Fe^{2+} \rightleftharpoons Ce^{3+} + Fe^{3+}$$

$$\varphi^{\ominus\prime}(Ce^{4+}/Ce^{3+}) = 1.44\ V, \quad \varphi^{\ominus\prime}(Fe^{3+}/Fe^{2+}) = 0.68\ V$$

It is important to realize that, when the reaction has come to equilibrium, all half-cells in the analyte solution will be at the same potential, thus the potential difference between any of redox couples is zero. The reaction is rapid and reversible so that the system is at equilibrium at all times throughout the titration. Therefore, the system potential at any equilibrium point in the titration can be calculated at either electrode potential in the solution.

$$\varphi = \varphi^{\ominus\prime}(Fe^{3+}/Fe^{2+}) + 0.059\lg\frac{c(Fe^{3+})}{c(Fe^{2+})}$$
$$= \varphi^{\ominus\prime}(Ce^{4+}/Ce^{3+}) + 0.059\lg\frac{c(Ce^{4+})}{c(Ce^{3+})}$$

1. Initial Potential

The solution contains no cerium species before titrant is added. The quantity of Fe^{3+} is also unknown, thus there is not enough information to calculate the initial potential of the system.

2. Pre-stoichiometric Point Potentials

Most of the added Ce^{4+} is reduced to Ce^{3+}, so the amount of Ce^{4+} is vanishingly small; thus calculating the electrode potential of the system using Fe^{3+}/Fe^{2+} redox couple is more convenient.

For example, when 99.9% Fe^{2+} is titrated,

$$c(Fe^{3+})/c(Fe^{2+}) = 999/1 \approx 10^3$$

Hence,
$$\varphi = \varphi^{\ominus\prime}(Fe^{3+}/Fe^{2+}) + 0.059\lg(c(Fe^{3+})/c(Fe^{2+}))$$
$$= 0.68 + 0.059\lg 10^3 = 0.86(\mathrm{V})$$

3. Stoichiometric Point Potential

At the stoichiometric point (SP), a problem is encountered. If the reaction truly goes to completion, the concentration of Ce^{4+} and Fe^{2+} are small, and difficult to calculate; therefore, neither of the couples can be used to calculate the system potential. However, combining the Nernst equations of Ce^{4+}/Ce^{3+} and Fe^{3+}/Fe^{2+} couples, we can calculate the potential of the system at stoichiometric point.

At SP, system potential, φ_{sp}, equals to either of the following potentials.

$$\varphi_{sp} = 0.68 + 0.059\lg(c(Fe^{3+})/c(Fe^{2+}))$$
$$\varphi_{sp} = 1.44 + 0.059\lg(c(Ce^{4+})/c(Ce^{3+}))$$

The sum of the above two equations gives

$$2\varphi_{sp} = 0.68 + 1.44 + 0.059\lg\frac{c(Fe^{3+})\cdot c(Ce^{4+})}{c(Fe^{2+})\cdot c(Ce^{3+})}$$

At the stoichiometric point, $c(Fe^{3+})=c(Ce^{3+})$ and $c(Fe^{2+})=c(Ce^{4+})$, Therefore,

$$\lg\frac{c(Fe^{3+})\cdot c(Ce^{4+})}{c(Fe^{2+})\cdot c(Ce^{3+})} = 0$$

$$\varphi_{sp} = (0.68 + 1.44)/2 = 1.06(\mathrm{V})$$

4. Post-stoichiometric Point Potentials

Almost all Fe^{2+} is oxidized to Fe^{3+}. The amount of Fe^{2+} is small, therefore calculating the electrode potential of the system is more convenient using Ce^{4+}/Ce^{3+} redox couple.

For example, when 0.1% excess of Ce^{4+} is added, $c(Ce^{4+})/c(Ce^{3+})=1/10^3$. Hence,

$$\varphi=\varphi^{\ominus\prime}(Ce^{4+}/Ce^{3+})+0.059\lg(c(Ce^{4+})/c(Ce^{3+}))$$
$$=1.44+0.059\lg10^{-3}=1.26(V)$$

The system potentials at different equilibrium points are given in Table 6.1, and the titration curve is plotted as shown in Figure 6.3. It can be seen that the system potential is independent of of both analyte and titrant concentrations.

Table 6.1 Electrode Potentials in the Titration of 0.1000 mol · L^{-1} $FeSO_4$ with 0.1000 mol · L^{-1} $Ce(SO_4)_2$ in 1 mol · L^{-1} H_2SO_4

Percent Titration/%		φ/V
	$c(Fe^{3+})/c(Fe^{2+})$	
9	10^{-1}	0.68−1×0.059=0.62
50	10^{0}	0.68+0 =0.68
99	10^{2}	0.68+2×0.059=0.80
99.9	10^{3}	0.68+3×0.059=0.86
100		1.06
	$c(Ce^{4+})/c(Ce^{3+})$	
100.1	10^{-3}	1.44−3×0.059=1.26
101	10^{-2}	1.44−2×0.059=1.32
110	10^{-1}	1.44−1×0.059=1.38
200	10^{0}	1.44

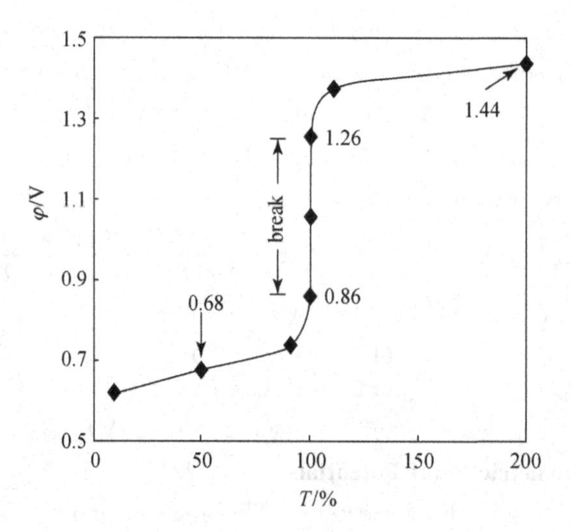

Figure 6.3 The titration curve of 0.1000 mol · L^{-1} $FeSO_4$ with 0.1000 mol · L^{-1} $Ce(SO_4)_2$ in 1 mol · L^{-1} H_2SO_4.

For the Ce^{4+}-Fe^{2+} titration reaction, the titration curve is symmetrical to the stoichiometric point because the number of electrons involved in the half-reaction is 1, the same for both redox couples. In general, redox titration curves are symmetrical to the stoichiometric point when the oxidizing agent and the reducing agent react in a 1 : 1 molar ratio.

For a redox reaction with different number of electrons, n_1 and n_2, involved for both redox couples, respectively:

$$p_2 Ox_1 + p_1 Red_2 \rightleftharpoons p_2 Red_1 + p_1 Ox_2$$

the half-reactions and formal potentials are

$$Ox_1 + n_1 e^- \rightleftharpoons Red_1 \quad \varphi_1^{\ominus'}$$

$$Ox_2 + n_2 e^- \rightleftharpoons Red_2 \quad \varphi_2^{\ominus'}$$

At SP,

$$\varphi_{sp} = \varphi_1^{\ominus'} + \frac{0.059}{n_1}\lg\frac{c(Ox_1)}{c(Red_1)}$$

$$\varphi_{sp} = \varphi_2^{\ominus'} + \frac{0.059}{n_2}\lg\frac{c(Ox_2)}{c(Red_2)}$$

Sum and rearrangement of the above Nernst equations yield

$$(n_1 + n_2)\varphi_{sp} = n_1\varphi_1^{\ominus'} + n_2\varphi_2^{\ominus'} + 0.059\lg\left(\frac{c(Ox_1)}{c(Red_1)} \cdot \frac{c(Ox_2)}{c(Red_2)}\right)$$

At SP,

$$\frac{c(Ox_1)}{c(Red_2)} = \frac{c(Red_1)}{c(Ox_2)} = \frac{p_2}{p_1}$$

Hence

$$\lg\left(\frac{c(Ox_1)}{c(Red_1)} \cdot \frac{c(Ox_2)}{c(Red_2)}\right) = 0$$

Therefore, the general expression for system potential at SP, φ_{sp}, is

$$\varphi_{sp} = \frac{n_1\varphi_1^{\ominus'} + n_2\varphi_2^{\ominus'}}{n_1 + n_2} \qquad (6\text{-}14)$$

The titration break is $\left(\varphi_2^{\ominus'} + \frac{3\times0.059}{n_2},\ \varphi_1^{\ominus'} - \frac{3\times0.059}{n_1}\right)$.

If $n_1 \neq n_2$, the titration curve is asymmetrical to the stoichiometric point. The φ_{sp} is close to the redox couple which has higher number of electrons involved in the half-reaction. For example, in the titration of Sn^{2+} with Fe^{3+} in 1 mol · L^{-1} HCl, the titration reaction is

$$2Fe^{3+} + Sn^{2+} \rightleftharpoons 2Fe^{2+} + Sn^{4+}$$

$$\varphi^{\ominus'}(Fe^{3+}/Fe^{2+}) = 0.70\ V, \quad \varphi^{\ominus'}(Sn^{4+}/Sn^{2+}) = 0.14\ V$$

At SP,

$$\varphi_{sp} = (1\times 0.70 + 2\times 0.14)/(1+2) = 0.33(\mathrm{V})$$

and the titration break is 0.23~0.52 V.

The electrode potential of irreversible redox couple (e. g. MnO_4^-/Mn^{2+} and $Cr_2O_7^{2-}/Cr^{3+}$) does not obey Equation (6-14). Equation (6-14) is also not applied to the electrode reaction dependent on the hydronium ion. The calculated titration curve deviates remarkably from the experimentally determined titration curve. For titrations involving such irreversible redox couples, the titration curve is experimentally measured.

6.3.2 Indicators

Two types of chemical indicators are used to indicate endpoints for redox titrations: redox indicators and non-redox indicators.

1. Redox Indicators

Redox indicators are substances that change color upon being oxidized or reduced. The color change of redox indicators depends only on the change in the system potential that occurs as the titration progresses.

The half-reaction for a redox indicator can be written as

$$\mathrm{In(Ox)} + ne^- \rightleftharpoons \mathrm{In(Red)}$$

The Nernst equation for the above reaction is

$$\varphi = \varphi^{\ominus\prime}(\mathrm{In}) + \frac{0.059}{n}\lg\frac{c(\mathrm{In(Ox)})}{c(\mathrm{In(Red)})}$$

As the system potential changes, the concentration ratio of the oxidized to reduced form of indicator ($c(\mathrm{In(Ox)})/c(\mathrm{In(Red)})$) changes, and the color of the solution changes. At $\varphi=\varphi^{\ominus\prime}(\mathrm{In})$, $c(\mathrm{In(Ox)})=c(\mathrm{In(Red)})$. This is potential theoretically considered as the color transition point for the indicator. With $c(\mathrm{In(Ox)})/c(\mathrm{In(Red)})$ in the range from 1/10 to 10/1, the solution has the color between the oxidized and reduced forms; therefore, the theoretical color transition potential range is $\varphi^{\ominus\prime}(\mathrm{In}) \pm 0.059/n$.

Now, we take tris(1,10-phenanthroline) iron (Ⅱ) sulfate and diphenylaminesulfonic acid as examples. Tris (1,10-phenanthroline) iron (Ⅱ) sulfate ($Fe(Phen)_3^{2+}$), commonly called ferroin, undergoes a reversible electrode reaction as

$$\underset{\text{(pale blue)}}{Fe(Phen)_3^{3+}} + e^- \rightleftharpoons \underset{\text{(red)}}{Fe(Phen)_3^{2+}}$$

The standard potential for the ferriin-ferroin couple is +1.06 V and the color change at the endpoint is very sharp, which makes ferroin an ideal indicator for a titration with titrant of high electrode potential, such as cerium(Ⅳ).

Diphenylamine sulfate is widely used as the redox indicator, especially for titration of Fe^{2+} with dichromate ($K_2Cr_2O_7$). This redox indicator is irreversibly oxidized to an intermediate form which is then oxidized reversibly to a purple color at the endpoint. The purple product can be further oxidized irreversibly to pale or colorless; therefore diphenylamine sulfate can not be used for titration of $K_2Cr_2O_7$ with Fe^{2+}. Since the redox indicator consumes the titrant, a reagent blank correction is necessary to obtain an accurate result.

$2\ {}^{-}O_3S-C_6H_4-NH-C_6H_5$

diphenylamine sulfate (colorless)

irreversible ↓ oxidation

${}^{-}O_3S-C_6H_4-NH-C_6H_4-C_6H_4-NH-C_6H_4-SO_3^{-} + 2H^+ + 2e^-$

diphenylbenzidine (colorless)

reversible ⇅ oxidation

${}^{-}O_3S-C_6H_4-NH^+{=}C_6H_4{=}C_6H_4{=}NH^+-C_6H_4-SO_3^{-} + 2e^-$

diphenylbenzidine violet (purple)

Generally, the transition potential of a redox indicator needs to be in the range of titration break to insure titration error $\leqslant 0.1\%$. For example, in the titration of Fe^{2+} with Ce^{4+} in $1\ mol \cdot L^{-1}\ H_2SO_4$, the titration break is from 0.86 V to 1.26 V. Certainly, ferroin ($Fe(phen)_3^{2+}$) can be used for indicating the endpoint. The titration error will be greater than 0.1%, if diphenylamine sulfate ($\varphi^{\ominus\prime}=0.85$ V) is used for endpoint indication. The error can be reduced within 0.1%, if the titration is carried out in $1\ mol \cdot L^{-1}\ H_2SO_4 + 0.5\ mol \cdot L^{-1}\ H_3PO_4$. In this situation, $\varphi^{\ominus\prime}(Fe^{3+}/Fe^{2+}) = 0.61$ V, and the system potential at 0.1%, the pre-SP is $\varphi = 0.61 + 0.059 \times 3 = 0.79(V)$; thus, diphenylamine sulfate can be used to indicate the endpoint.

For redox titrations, titration error is less important because strong oxidizing agents to ensure thermodynamically the completeness of redox reactions are easy to find; however, the success and feasibility of a redox titration still rely on the rate of the redox reaction.

2. Non-redox Indicators

For reactions involving colored analyte or titrant with colorless reaction product, there is no need for an additional indicator because the color change of the solution will indicate the endpoint. For example, permanganate is dark purple and the Mn^{2+} ion is almost colorless. MnO_4^- is a strong oxidizing agent that can be reduced to Mn^{2+} in acidic solution. In this case, an indicator is not necessary because the purple MnO_4^- can serve as an indicator. After the stoichiometric point, a slight excess of the unreacted MnO_4^- at 2×10^{-6} mol · L^{-1} can produce a detectable pink color.

Non-redox indicators are chemicals that can react with the titrant or the analyte to produce color. Starch is the best-known non-redox indicator which forms a dark blue complex with the triiodide ion at 1×10^{-5} mol · L^{-1}. This complex signals the endpoint in titrations in which iodine is either produced or consumed. Potassium thiocyanate can be used in the titration of Sn^{2+} with Fe^{3+}. The appearance of red color of iron(Ⅲ)/thiocyanate complex indicates the endpoint.

6.3.3 Auxiliary Oxidizing and Reducing Agents

When samples are dissolved, the chemical entity to be analyzed is usually in a mixed oxidation state or in an oxidation state that is not titrable. A pre-oxidizing or pre-reducing step is used to convert the sample into a single titrable oxidative state prior to titration.

The criteria to select the pre-oxidizing or pre-reducing agent are: (1) must react quantitatively and selectively with the analyte of interest; (2) excess pre-oxidizing or pre-reducing agent must be easily removed before the titration because these reagents usually interfere with the titration.

Ammonium persulfate ($(NH_4)_2S_2O_8$) is a very powerful oxidizing agent that can be used in the oxidization of chromium(Ⅲ) to dichromate (Ⅵ), vanadium(Ⅳ) to vanadium (Ⅴ), cerium (Ⅲ) to cerium (Ⅳ), as well as manganese (Ⅱ) to permanganate. The oxidation is carried out in hot acid solution with a small amount of silver (Ⅰ) as a required catalyst. The excess of persulfate is destroyed by boiling. For the determination of Mn^{2+} and Cr^{3+} in a sample, an oxidizing titrant with electrode potentials higher than those of MnO_4^-/Mn^{2+} ($\varphi^\ominus = 1.51$ V) and $Cr_2O_7^{2-}/Cr^{3+}$ ($\varphi^\ominus = 1.33$ V) are difficult to find, the realistic approach is to pre-oxidize Mn^{2+} and Cr^{3+} to MnO_4^- and $Cr_2O_7^{2-}$. Then a redox titration can be done using a Fe^{2+} standard solution.

Because most reductants are susceptible to air oxidation, most redox titrations employ oxidizing agents as titrants. Therefore, reduction of an analyte's oxidative state before titration is usually accomplished with a metal reductor column is used for this purpose. As shown in Figure 6.4, the analyte solution is poured through a glass column filled with granules (or coarse powder) of the metal. An amalgam of metal ensures that the reaction with the H^+ of the acid to form H_2 is very slow, thus the use of zinc as a reducing agent with minimal H_2 production is possible. The perforated plate and glass wool prevent particles of the metal granules from reaching the exit flask containing the analyte. For example, when a mixture of Fe^{2+} and Fe^{3+} is present in a iron ore sample, the zinc in a Jones reductor will reduce all the Fe^{3+} to Fe^{2+} so that only one standard solution, e.g., $K_2Cr_2O_7$ is needed for titration.

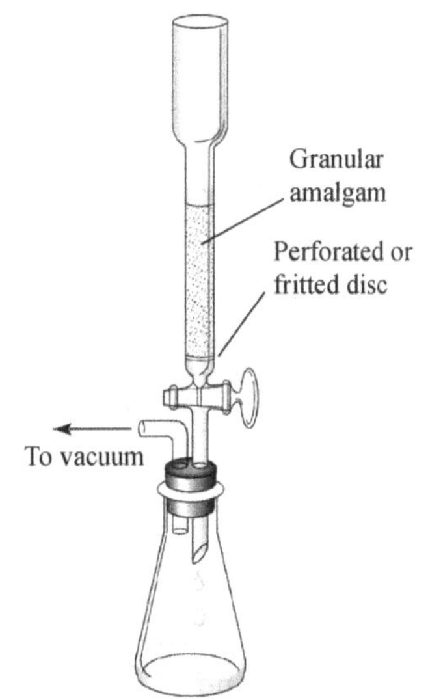

Figure 6.4 A Jones reductor.

$$2Fe^{3+} + Zn \rightleftharpoons 2Fe^{2+} + Zn^{2+}$$

6.4 EXAMPLES OF REDOX TITRATIONS

6.4.1 Potassium Permanganate ($KMnO_4$)

1. Half-reactions of Permanganate

Potassium permanganate is widely used for the determination of inorganic and organic chemicals, because the permanganate ion is a very strong oxidizing agent having a fast reaction rate with most reductants. Permanganate acts as a self-indicator for endpoint detection because the MnO_4^- is purple. Problems with permanganate standard solution include poor stability, complicated reaction mechanisms, side reactions, and poor selectivity; however, such flaws can be overcome with careful solution preparation and storage as well as controlling the titration conditions.

The electrode potential and half-reaction of permanganate varies as the pH changes. In very strong basic solution, MnO_4^- is reduced to MnO_4^{2-} according to the following reaction:

$$MnO_4^- + e^- \rightleftharpoons MnO_4^{2-} \quad \varphi^\ominus = 0.56\ V$$

This reaction is faster in basic solution than in acidic solution when the reaction is used to titrate organic compounds via back titration.

In a very strongly acidic solution, $KMnO_4$ is reduced to Mn^{2+} according to the following reaction:

$$MnO_4^- + 8H^+ + 5e^- \rightleftharpoons Mn^{2+} + 4H_2O \quad \varphi^\ominus = 1.51\ V$$

Thermodynamically, MnO_4^- can not co-exist with Mn^{2+} because the equilibrium constant for the reaction of MnO_4^- and Mn^{2+} ($2MnO_4^- + 3Mn^{2+} + 2H_2O \rightleftharpoons 5MnO_2(s) + 4H^+$) is about 10^{47}. Fortunately, the reaction rate is very slow and the concentration of MnO_4^- near the endpoint is very low thus this reaction will not occur during the titration. Reductant (e. g., Fe^{2+}) cannot be used to titrate MnO_4^-, because MnO_4^- and the generated Mn^{2+} will react to produce MnO_2, reducing the reaction rate with the reductant titrant (Fe^{2+}), thus the endpoint is difficult to detect. For the titration of MnO_4^-, back titration is used by adding an excess of the standard reductant solution (e. g., Fe^{2+}) to reduce MnO_4^- to Mn^{2+}. Then, the MnO_4^- standard solution is used for titration of the excess of the reductant.

2. Preparation of the $KMnO_4$ Standard Solution

The purity of commercially available $KMnO_4$ is 99%~99.5% with MnO_2 as the major contaminant. The trace amount of organic compound in distilled water will react slowly with $KMnO_4$ to produce $MnO(OH)_2$, which acts as a catalyst for the further decomposition of the permanganate. Therefore, $KMnO_4$ standard solution can not be prepared by simply dissolving $KMnO_4$ in water. The moderately stable $KMnO_4$ standard solution can be prepared using the following procedure:

- Weigh a slight excess of the theoretically calculated amount of $KMnO_4$ and dissolve in a volume of distilled water near the final volume desired for the $KMnO_4$ standard solution.
- Boil the solution for about 1 h to reduce all the reductants in solution.
- Remove the $MnO(OH)_2$ by filtering through a sintered-glass filter (filter paper cannot be used because the paper is reductive).
- Store the filtered $KMnO_4$ in a brown bottle kept in the dark to avoid light catalyzed decomposition.

Diluted $KMnO_4$ is less stable, so a diluted solution should be freshly prepared from a more concentrated stock solution and then standardized. After standardization of the $KMnO_4$ standard solution, one should continually look for precipitation. If $MnO(OH)_2$ occurs, the solution should be filtered and restandardized.

$KMnO_4$ is not a primary standard. Any of the following primary reagents can be used to standardize $KMnO_4$: $H_2C_2O_4 \cdot 2H_2O$, $Na_2C_2O_4$, $(NH_4)_2Fe(SO_4)_2 \cdot 6H_2O$, As_2O_3 and Fe. Anhydrous $Na_2C_2O_4$ that is stable and ready for use after drying for 2 h at 105~110℃ is the most commonly used primary standard for $KMnO_4$. In acidic solution, the titration reaction is

$$2MnO_4^- + 5C_2O_4^{2-} + 16H^+ \rightleftharpoons 2Mn^{2+} + 10CO_2 + 8H_2O$$

A few reaction parameters must be controlled to assure that the reaction is carried out quantitatively and stoichiometrically. The solution must be heated to 70~80℃ to obtain a rapid reaction rate. If temperature is greater than 90℃, $H_2C_2O_4$ will decompose and positive error will occur.

$$H_2C_2O_4 \rightleftharpoons CO_2 + CO + H_2O$$

The titration must be carried out in a carefully controlled acidic solution with the initial H_2SO_4 concentration at 1 mol • L^{-1}. $H_2C_2O_4$ will decompose at higher acidity; and MnO_4^- will be reduced to MnO_2 at lower acidity. The reaction in H_2SO_4 instead of HCl avoids the catalytic oxidation of Cl^-.

The titration must start slowly to avoid a negative error, because the reaction rate for MnO_4^- with $C_2O_4^{2-}$ is slow so that $KMnO_4$ may decompose before reacting with the $C_2O_4^{2-}$.

$$2MnO_4^- + 12H^+ \rightleftharpoons 4Mn^{2+} + 5O_2 + 6H_2O$$

A small amount of $MnSO_4$ can be added to the titrand to accelerate the initial rate of titration reaction.

3. Methods and Applications

(1) Direct titration. The following reductants can be directly titrated with $KMnO_4$ standard solution in acidic pH, Fe^{2+}, As(Ⅲ), Sb(Ⅲ), $C_2O_4^{2-}$, H_2O_2, and NO_2^-. For example, the titration reaction for H_2O_2 can be carried out easily at room temperature.

$$2MnO_4^- + 5H_2O_2 + 6H^+ \rightleftharpoons 2Mn^{2+} + 5O_2 + 8H_2O$$

A small amount of Mn^{2+} catalyst may be added to accelerate the reaction at the beginning of titration.

(2) Indirect titration. Metal ions such as Ca^{2+}, Sr^{2+}, Ba^{2+}, and Th^{4+} can be determined indirectly by forming oxalate precipitate (e. g., CaC_2O_4). The oxalate precipitate is dissolved in acid and the solution is titrated with $KMnO_4$, thus the

amount of metal ion can be calculated from the amount of oxalic acid ($H_2C_2O_4$) determined.

It is desirable to keep a straight 1 : 1 stoichiometry between Ca^{2+} and $C_2O_4^{2-}$, as well as to obtain good crystalline CaC_2O_4 for complete removal of the calcium oxalate precipitate. First, excess of $(NH_4)_2C_2O_4$ is added to the acidified analyte solution and then a precipitate is produced by slowly adding diluted ammonia until the solution changes from red to yellow with methyl orange (MO) indicator. Precipitation in neutral or weakly basic solution will result in the formation of $Ca(OH)_2$ or basic calcium oxalate to give a negative error. To optimize removal of foreign ions and obtain large crystalline particles, the CaC_2O_4 precipitate is kept in a hot water bath for 1~2h. The amount of $C_2O_4^{2-}$ adsorbed on the precipitate is then washed away using cold distilled water.

(3) Back titration. Some oxidants, such as MnO_2, PbO_2, $Cr_2O_7^{2-}$, and MnO_4^- can be back titrated using $KMnO_4$. Using the determination of MnO_2 in an ore as an example, first a measured slight excess of $Na_2C_2O_4$ is added to reduce the oxidant in the presence of H_2SO_4 at an elevated temperature. The remaining $Na_2C_2O_4$ is then titrated by $KMnO_4$ standard solution. The reaction of MnO_2 with $C_2O_4^{2-}$ in acidic solution is

$$MnO_2 + C_2O_4^{2-} + 4H^+ \rightleftharpoons Mn^{2+} + 2CO_2 + 2H_2O$$

6.4.2 Potassium Dichromate ($K_2Cr_2O_7$)

1. Dichromate Half-reaction

In acidic solution, the half-reaction for $Cr_2O_7^{2-}$ is

$$Cr_2O_7^{2-} + 14H^+ + 6e^- \rightleftharpoons 2Cr^{3+} + 7H_2O \quad \varphi^\ominus = 1.33\ V$$

There are a few advantages of $Cr_2O_7^{2-}$ over $KMnO_4$. $K_2Cr_2O_7$ is available at very high purity (99.99%) and is stable in solution; thus, the potassium dichromate can be used as a primary standard. The standard solution can be prepared directly from $K_2Cr_2O_7$ after drying for 2 h at 150~180℃. $Cr_2O_7^{2-}$ can be a more selective oxidizing agent, because it is a slightly weaker oxidizing agent than MnO_4^-. $Cr_2O_7^{2-}$ does not oxidize Cl^- in a solution with a HCl concentration less than 3 $mol \cdot L^{-1}$; thus the titration of Fe^{2+} can be carried out in HCl. Diphenylamine sulfonate is usually used as the indicator for dichromate titration.

2. Applications

(1) Analysis of an iron ore by titration with potassium dichromate.

Dichromate titration is a Chinese National Standard method for determination of total iron in iron ore that uses $SnCl_2$-$HgCl_2$-$K_2Cr_2O_7$ as a reagent. The $SnCl_2$ reduces Fe^{3+} to Fe^{2+}. The $HgCl_2$ is used to oxidize the excess $SnCl_2$; however, $HgCl_2$ is toxic and not environmentally friendly. The $SnCl_2$-$TiCl_3$ method is now adopted as a Chinese National Standard (GB 6730.5—86) for iron ore. For an iron ore analysis, a specimen is dissolved in hot HCl and most of the Fe(Ⅲ) is reduced to Fe(Ⅱ) by the stannous chloride ($SnCl_2$). The remaining Fe(Ⅲ) is then reduced by titanium chloride ($TiCl_3$) with sodium tungstate as the indicator. When the excess of $TiCl_3$ reduces sodium tungstate to tungsten blue, all the Fe^{3+} would have been quantitatively reduced. The excess of tungsten blue can be oxidized with 1~2 drops of $K_2Cr_2O_7$ standard solution. The excess of $TiCl_3$ is all in the oxidized state and will not interfere with titration of Fe(Ⅱ). About 100 mL water is added to the resulting solution to reduce the darkness of the solution. H_2SO_4-H_3PO_4 as well as diphenylamine sulfonate indicator are added and the titration with $K_2Cr_2O_7$ is carried out immediately. At the endpoint, the solution changes color from light green (Cr^{3+}) to purple.

The addition of H_3PO_4 is to reduce the electrode potential of Fe^{3+}/Fe^{2+} so that the transition point of diphenylamine sulfate will be within in the titration break to form the colorless $Fe(HPO_4)_2^-$ and to make the endpoint judgment easier.

(2) Determination of oxidants such as NO_3^- and ClO_3^- using the $Cr_2O_7^{2-}$-Fe^{2+} titration.

It is advantageous to indirectly determine some oxidizing agents using the $Cr_2O_7^{2-}$-Fe^{2+} titration: the reaction is fast; the stoichiometry is specific; there are no side reactions; and color change at the endpoint is sharp. For determination of NO_3^- and ClO_3^-, an excess of Fe^{2+} is added to the analyte solution. After the reaction is complete, $K_2Cr_2O_7$ is used to titrate the remaining Fe^{2+}. Strong reductants such as Ti(Ⅲ) are not stable in the air, so the oxidized form of titanium (Ti(Ⅳ)) is usually reduced after passing through a Jones reductor. In the receiving flask, Ti(Ⅲ), the reduced form of Ti(Ⅳ), is oxidized by Fe^{3+} standard solution.

$$\text{Ti(III)} + Fe^{3+} \rightleftharpoons \text{Ti(IV)} + Fe^{2+}$$

Then the $K_2Cr_2O_7$ titration is carried out to determine Fe^{2+} produced in the above reaction.

6.4.3 Iodine: Iodimetry and Iodometry

Iodine is an oxidizing agent that can be used to titrate fairly strong reducing agents.

On the other hand, the iodide ion is a mild reducing agent and serves as the basis for determining strong oxidizing agents.

1. Iodimetry

Iodine is a moderately strong oxidizing agent and can be a more selective titrant than the strong oxidizing agents. Titrations with I_2 are called iodimetric methods that are usually performed in neutral, mildly alkaline (pH 8) or weakly acidic solutions. The endpoint is the appearance of the blue starch-iodine color.

In strongly alkaline solutions, I_2 will be converted to hypoiodite and iodide:

$$I_2 + 2OH^- \rightleftharpoons IO^- + I^- + H_2O$$

The iodimetric titration that can be carried out in neutral solution may not be feasible in strongly acidic solution because the electrode potential of the reducing agent will be increased when hydronium ions are involved in the redox half-reaction. For example, the electrode potential of As(Ⅴ)/As(Ⅲ) is low enough to reduce I_2 in neutral solution, but the potential will be higher than that of I_2/I^- couple in very strongly acidic solutions thus the following titration is not useful.

$$H_3AsO_3 + I_2 + H_2O \rightleftharpoons H_3AsO_4 + 2I^- + 2H^+$$

Another reason iodimetry can not be carried out in strongly acidic solution is that the I^- produced in the reaction tends to be oxidized by dissolved oxygen in acidic solutions:

$$4I^- + O_2 + 4H^+ \rightleftharpoons 2I_2 + 2H_2O$$

Additionally, starch tends to hydrolyze or decompose in strong acid, making endpoint detection difficult.

The pH for the titration of arsenic (Ⅲ) with I_2 can be maintained at about 8 by adding $NaHCO_3$. The bubbling action of the CO_2 formed also removes the dissolved oxygen and maintains a blanket of CO_2 over the solution to prevent air oxidation of the I^-.

Although high purity I_2 can be obtained by sublimation, iodine solutions are usually standardized with a primary standard reducing agent such as arsenous oxide (As_2O_3). Because arsenous oxide is not soluble in acid, the reagent standard is dissolved in sodium hydroxide and the pH of the solution made neutral after the added reagent dissolves.

Iodine has a low solubility in water but the I_3^- complex is very soluble, thus iodine solutions are prepared by dissolving I_2 in a concentrated solution of potassium iodide:

$$I_2 + I^- \rightleftharpoons I_3^-$$

I_3^- is the actual titrant species used in the titration.

2. Iodometry

Iodide ion is a weak reducing agent and will reduce strong oxidizing agents. Iodide ion not used as a titrant mainly because the speed of the reaction is slow and a convenient visual indicator is not available.

When an excess of iodide is added to a solution of an oxidizing agent, I_2 is produced in an amount equivalent to the oxidizing agent present. This I_2 can be titrated with a reducing agent to get the same result as titrating the oxidizing agent directly. Analysis of an oxidizing agent in this way is called an iodometry method. The titrating agent often used is sodium thiosulfate ($Na_2S_2O_3$). The reaction is $I_2 + 2S_2O_3^{2-} \rightleftharpoons 2I^- + S_4O_6^{2-}$ with 1 mole I_2 reacting with 2 moles of $Na_2S_2O_3$.

The reasons that oxidizing agents are not directly titrated with the thiosulfate include: (1) strong oxidizing agents oxidize thiosulfate to oxidation states higher than that of tetrathionate ($S_4O_6^{2-}$), e.g., SO_4^{2-}, but the reaction is generally not stoichiometric; and (2) several oxidizing agents (e.g., Fe^{3+}) form mixed complexes with thiosulfate. In the reaction of iodide with a strong oxidizing agent an equivalent amount of I_2 is produced, which then reacts stoichiometrically with thiosulfate.

The titration should be performed rapidly to minimize the air oxidation of the iodide.

$S_2O_3^{2-}$ can be decomposed in acidic solution: $S_2O_3^{2-} + 2H^+ \rightleftharpoons H_2SO_3 + S\downarrow$. H_2SO_3 further reacts with I_2: $I_2 + H_2SO_3 + H_2O \rightleftharpoons SO_4^{2-} + 4H^+ + 2I^-$. In this case, the stoichiometric ratio of I_2 to $S_2O_3^{2-}$ is 1 : 1 instead of 1 : 2, thus thorough stirring is necessary to prevent local excess of $S_2O_3^{2-}$. As long as $S_2O_3^{2-}$ is added slowly with sufficient stirring, an accurate result can still be obtained even though the titration is carried out in relatively strong 3~4 mol · L^{-1} acid solution.

Sodium thiosulfate solution is standardized iodometrically against a primary standard grade oxidizing agent such as $K_2Cr_2O_7$, KIO_3, and $KBrO_3$. For standardization with $K_2Cr_2O_7$, the titration is carried out in acidic solution with the reaction $Cr_2O_7^{2-} + 6I^- + 14H^+ \rightleftharpoons 2Cr^{3+} + 3I_2 + 7H_2O$.

The reaction rate between $K_2Cr_2O_7$ and I^- is slow. The rate can be increased by adding an excess of KI and increasing the acidity. Generally, the 5 min reaction can be carried out in 0.4 mol · L^{-1} acid solution in the dark. Before titration with $S_2O_3^{2-}$, the solution is diluted with distilled water to reduce the acidity and lighten the color of generated Cr^{3+} to make endpoint detection easier.

Starch is not added at the beginning of the titration when the iodine concentration is high. Instead, starch is added just before the endpoint when the dilute iodine color of the titration solution becomes pale yellow. The reasons for the later addition of starch in an iodometric titration are: (1) the iodine-starch complex is slowly dissociated, thus a diffuse endpoint may result when most of the iodine is adsorbed on the starch; and (2) to minimize the hydrolysis of starch in the acidic solutions typically used for iodometry.

The reaction of KIO_3 with I^- is fast so the titration can start immediately after the KIO_3 standard solution that contains excess I^- is added to the analyte in acidic pH.

3. Examples of Iodimetry and Iodometry

(1) Iodometric determination of copper

The determination of copper by iodometry involves the following reactions:

$$2Cu^{2+} + 5I^- \rightleftharpoons 2CuI\downarrow + I_3^-$$

$$I_3^- + 2S_2O_3^{2-} \rightleftharpoons 3I^- + S_4O_6^{2-}$$

Copper (Ⅱ) reacts with excess of KI to produce a CuI precipitate and an equivalent amount of I_3^-, which is then titrated with $Na_2S_2O_3$ standard solution. The iodine adsorbed on the surface of the cuprous iodide precipitate will slowly react with the thiosulfate titrant to produce a diffuse endpoint to give a negative error. Potassium thiocyanate (KSCN) is added near the endpoint to displace the adsorbed I_2 on the CuI by forming a layer of copper thiocyanate (CuSCN). One must be aware that potassium thiocyanate should be added near the endpoint, because the thiocyanate is slowly oxidized by iodine to sulfate to give a negative error in the titration.

For copper ore with other metals including iron, arsenic, and antimony, all the metals in the sample will be oxidized in solution to their highest oxidation states. Excess amounts of Fe(Ⅲ), As(Ⅴ), Sb(Ⅴ) and HNO_3 can oxidize I^- to result in positive titration errors. The oxides of nitrogen are removed by adding H_2SO_4 and boiling to fuming solution. The solution is then neutralized with NH_3 and NH_4HF_2 ($NH_4F + HF$) is added to buffer the solution to pH 3～4. As(Ⅴ) and Sb(Ⅴ) will not oxidize I^- at this pH, and the F^- forms complex with Fe(Ⅲ) to reduce the oxidation power of Fe(Ⅲ) so that the I^- will not be oxidized.

The solution must be buffered to about pH 3. If the pH is too high, copper (Ⅱ) will hydrolyze and cupric hydroxide will precipitate. If the pH is too low, air oxidation of iodide occurs, because the oxidation of the iodine ion is catalyzed by copper.

(2) Iodimetric determination of water with Karl Fischer reagent

In 1935, Karl Fischer developed a method for the iodimetric determination of water. In the method, one mole of iodine reacts with one mole of water in the presence of pyridine (C_5H_5N) and methanol (CH_3CH_2OH):

$$C_5H_5N \cdot I_2 + C_5H_5N \cdot SO_2 + C_5H_5N + H_2O \rightleftharpoons 2C_5H_5N \cdot HI + C_5H_5N \cdot SO_3$$

$$C_5H_5N \cdot SO_3 + CH_3OH \rightleftharpoons C_5H_5NHOSO_2OCH_3$$

The overall reaction is the sum of these two reactions. Each mole of water requires 1 mole of iodine, 1 mole of sulfur dioxide, 3 moles of pyridine, and 1 mole of methanol. This combination is called the Karl Fischer reagent. The Karl Fischer reagent has a dark brown color which changes to a yellow color upon reaction with water. The color at the endpoint changes from yellow to dark brown. Karl Fischer reagent is a very active desiccant. The apparatus used to store and dispense the Karl Fischer reagent and the titration flask assembly must be designed to protect the Karl Fischer reagent from moisture in the air.

Karl Fischer method can be used to determine the water content in a sample as well as the water generated or consumed in reactions.

6.4.4 Potassium Bromate ($KBrO_3$)

Potassium dichromate is a very strong oxidizing agent ($\varphi^{\ominus\prime}(BrO_3^-/Br_2)=1.5$ V) that can be easily purified by recrystallization from water and drying at 180℃. The bromate method is primarily applied to titration of organic compounds as well as inorganic reductants such as As(Ⅲ), Sb(Ⅲ), and Sn(Ⅱ).

When $KBrO_3$ standard solution is mixed with excess of potassium bromide (KBr), the following reaction will happen in acidic pH:

$$BrO_3^- + 5Br^- + 6H^+ \rightleftharpoons 3Br_2 + 3H_2O$$

to generate a bromine (Br_2) standard solution($\varphi^{\ominus\prime}(Br_2/Br^-)=1.07$ V). Br_2 is not stable and can not be prepared in a standard solution as a titrant. However, $KBrO_3$-KBr standard solution is stable. It is comparable to a freshly prepared Br_2 when acid is added to $KBrO_3$-KBr standard solution. For determination of organic compounds, bromine can react with phenol and aromatic amines to form brominated substitution products. The bromination of phenol is an example. A measured amount of $KBrO_3$-KBr standard solution is added to acidified phenol solution. The following reaction will occur after the reagents are mixed.

$$C_6H_5OH + 3Br_2 \longrightarrow C_6H_2Br_3OH + 3HBr$$

Bromine can also react with many organic compounds with the addition of bromine to measure the degree of unsaturation. The bromination of ethylene is an example.

$$H_2C{=}CH_2 + Br_2 \longrightarrow H_2CBr{-}CH_2Br$$

Bromine is very volatile. After the reaction is completed, excess KI is usually added to convert Br_2 to I_2. This reaction must be carried out in iodine flask sealed with H_2O or KI solution. A reagent blank determination can be carried out exactly in the same way to estimate the systematic errors from apparatus, reagent and bromine volatization.

6.4.5 Ceric Sulfate ($Ce(SO_4)_2$)

Ceric sulfate ($Ce(SO_4)_2$) is a strong oxidizing agent with an oxidizing strength comparable to that of permanganate ($\varphi^{\ominus\prime}=1.44$ V in 1 mol · L^{-1} H_2SO_4). Ce^{4+} may be used in most titrations in which permanganate is used. Ce^{4+} may be preferred over $KMnO_4$ due to indefinite stability. The Ce^{4+} ion in sulfuric acid does not oxidize the chloride ion so the titration can be carried out in concentrated HCl. Another advantage of Ce^{4+} is that reduction is simple and no intermediate products form during the reduction of Ce^{4+}. Ceric sulfate is more than 25 times the cost of permanganate, thus the ceric sulfate is less widely used as permanganate. The speed of the reaction of Ce^{4+} with some reductants is not favorable for titration; therefore, heating is required for titration of $C_2O_4^{2-}$ and a catalyst is required for titration of As (Ⅲ).

Ce^{4+} standard solutions can be directly prepared from primary standard grade cerium (Ⅳ) ammonium sulfate ($Ce(SO_4)_2 \cdot (NH_4)_2SO_4 \cdot 2H_2O$). More commonly, Ce^{4+} standard solutions are prepared from less expensive reagent-grade cerium (Ⅳ) salt followed by standardization using As_2O_3 or $Na_2C_2O_4$. Ce^{4+} is prone to hydrolysis and must be prepared in acidic pH. Ceric titration with ferroin as an indicator is also carried out in acidic solution.

Chapter 6 Questions and Problems

6.1 Write the half-cell reaction and the Nernst equation for the Hg_2Cl_2/Hg couple.

6.2 Can ferroin be used for indicating the endpoint for the titration of Fe^{2+} with Ce^{4+}?

6.3 When and how should the reagent blank correction be performed in the titration of Fe^{2+} with $K_2Cr_2O_7$ using diphenylamine sulfate to indicate the endpoint?

6.4 How should one prepare the $KMnO_4$ and $Na_2S_2O_3$ standard solutions?

6.5 One of the methods for standardization of $Na_2S_2O_3$ is based on the stoichiometrical oxidization of I^- to I_2 with a primary standard $K_3Fe(CN)_6$ in a strong acidic solution. Calculate $\varphi^{\ominus\prime}(Fe(CN)_6^{3-}/Fe(CN)_6^{4-})$ in 2 mol · L^{-1} HCl.

Constants: $\varphi^{\ominus}(Fe(CN)_6^{3-}/Fe(CN)_6^{4-})=0.36V$; $H_3Fe(CN)_6$ is a strong polyprotic acid; for $H_4Fe(CN)_6$, $K_{a3}=10^{-2.2}$, $K_{a4}=10^{-4.2}$; the effect of ionic strength can be neglected.

6.6 Calculate formal electrode potential for Fe^{3+}/Fe^{2+} at pH 3.0, and the concentration of EDTA in solution not binding with Fe^{3+} is 0.010 mol · L^{-1} (The effect of ionic strength can be neglected).

Constants: $\varphi^{\ominus}(Fe^{3+}/Fe^{2+})=0.77$; $\lg K(Fe^{III}Y)=25.1$, $\lg K(Fe^{II}Y)=14.3$; at pH 3.0, $\alpha_{Y(H)}=10^{10.8}$.

6.7 Equal aliquot of 0.40 mol · L^{-1} Fe^{2+} is mixed with 0.10 mol · L^{-1} Ce^{4+}, and the $c(H_2SO_4)$ in the final solution is 0.5 mol · L^{-1}. Calculate $[Ce^{4+}]$.

$\varphi^{\ominus\prime}(Fe^{3+}/Fe^{2+})=0.68$ V(this value is $\varphi^{\ominus\prime}(Fe^{3+}/Fe^{2+})$ in 1 mol · L^{-1} H_2SO_4 and is used for calculation); $\varphi^{\ominus\prime}(Ce^{4+}/Ce^{3+})=1.45$ V(0.5 mol · L^{-1} H_2SO_4).

6.8 For titration of Sn^{2+} using Fe^{3+} in 1 mol · L^{-1} HCl, calculate system potentials when titration reaches the following points ($T/\%$): 50, 99.9, 100.0, 100.1, 200(%).

$$2Fe^{3+} + Sn^{2+} \rightleftharpoons 2Fe^{2+} + Sn^{4+}$$

In 1 mol · L^{-1} HCl, $\varphi^{\ominus\prime}(Fe^{3+}/Fe^{2+})=0.70$ V, $\varphi^{\ominus\prime}(Sn^{4+}/Sn^{2+})=0.14$ V.

6.9 A measured amount of $KHC_2O_4 \cdot H_2C_2O_4 \cdot 2H_2O$ can be oxidized by $KMnO_4$ of a measured volume and the same amount of $KHC_2O_4 \cdot H_2C_2O_4 \cdot 2H_2O$ can also be neutralized by 0.2000 mol · L^{-1} NaOH with exactly half the volume of $KMnO_4$, calculate the concentration of $KMnO_4$.

6.10 A K^+ sample is quantitatively precipitated as $K_2NaCo(NO_2)_6$. The precipitate then is dissolved by acid and titrated using $KMnO_4$ ($NO_2^- \longrightarrow NO_3^-$, $Co^{3+} \longrightarrow Co^{2+}$). Calculate the molar ratio of K^+ to MnO_4^- in $n(K^+) : n(KMnO_4)$.

6.11 A 0.3216 g Mn ore was weighed and mixed with 0.3685 g analytical grade $Na_2C_2O_4$. After addition of H_2SO_4, the solution was heated to allow the reaction to take place. After the reaction is completed, the remaining $Na_2C_2O_4$ is then titrated using 11.26 mL of 0.02400 mol · L^{-1} $KMnO_4$. Calculate the percentage of Mn as MnO_2.

6.12 A 0.5000 g phenol sample was weighed and dissolved. A 25.00 mL of 0.1000 mol · L^{-1}

$KBrO_3$ (with excess KBr) was added to the solution. After HCl was added, the resulted solution was put aside until the reaction is completed. Then KI was added. A 29.91 mL of 0.1003 mol • L^{-1} $Na_2S_2O_3$ was used to titrate I_2 generated in the reaction. Calculate the percentage of phenol.

6.13 A 0.5000 g KI sample was dissolved in water and oxidized to IO_3^- by Cl_2. The excess Cl_2 was removed by boiling the solution. After an excess KI is added to the solution, the generated I_2 is titrated with 21.30 mL of 0.02082 mol • L^{-1} $Na_2S_2O_3$. Calculate the percentage of KI.

6.14 A 1.234 g sample containing PbO and PbO_2 was weighed and mixed with 20.00 mL of 0.2500 mol • L^{-1} oxalic acid to reduce PbO_2 in the sample into Pb^{2+}. The solution was then neutralized with ammonia to precipitate Pb^{2+} in the form of PbC_2O_4. The precipitate was then filtered and made acidic. A 30.00 mL of 0.0400 mol • L^{-1} $KMnO_4$ was used to titrate the resulted $C_2O_4^{2-}$ from dissolution of precipitation. The filtrate was made acidic and titrated by 10.00 mL of 0.0400 mol • L^{-1} $KMnO_4$. Calculate the percentage of PbO and PbO_2.

6.15 A 1.000 g sample containing sodium iodate ($NaIO_3$) and sodium periodate ($NaIO_4$) was dissolved in a 250 mL volumetric flask. A 50.00 mL of the test solution was pipetted into an Erlenmeyer flask. After the solution was adjusted to a weakly basic pH, excess KI was added to reduce IO_4^- into IO_3^- (IO_3^- does not oxidize I^- at this pH), I_2 produced can be titrated by 10.00 mL of 0.04000 mol • L^{-1} $Na_2S_2O_3$. A 20.00 mL of the test solution ($NaIO_3$ and $NaIO_4$) was adjusted to acidic pH by HCl and excess KI was added. A 30.00 mL of 0.04000 mol • L^{-1} $Na_2S_2O_3$ was used to titrate resulted I_2. Calculate the percentage of $NaIO_3$ and $NaIO_4$.

6.16 The carbon in oxalic acid ($H_2C_2O_4$) is easily oxidized to CO_2 by permanganate (MnO_4^-) in acidic pH. A 21.68 mL of $KMnO_4$ standard solution is required to completely oxidize 25.00 mL oxalic acid to CO_2. The $KMnO_4$ was standardized separately by reaction with Fe^{2+}. A 32.85 mL of 0.1385 mol • L^{-1} Fe^{2+} is required to standardize 25.00 mL of the $KMnO_4$. What is the concentration of the oxalic acid solution?

6.17 Write the equations and conditions for the analysis of each component in the mixture by listing the titrand, the titrant, the indicator and the major experimental conditions as well as the equations for calculation.

(1) $Sn^{2+} + Fe^{2+}$; (2) $Sn^{4+} + Fe^{3+}$; (3) $Cr^{3+} + Fe^{3+}$; (4) $H_2O_2 + Fe^{3+}$; (5) $As_2O_3 + As_2O_5$; (6) $H_2SO_4 + H_2C_2O_4$; (7) $MnSO_4 + MnO_2$.

6.18 How to determine the amount of Ba^{2+} in a solution by iodometry?

CHAPTER 7

PRECIPITATION EQUILIBRIUM, TITRATION, AND GRAVIMETRY

7.1 Precipitation Equilibria and Solubility

7.1.1 Solubility of Precipitates in Pure Water

7.1.2 Ionic Strength and the Solubility of Precipitates

7.1.3 Common Ion and the Solubility of Precipitates

7.1.4 pH and the Solubility of Precipitates

7.1.5 Complexing Agents and the Solubility of Precipitates

7.2 Precipitation Titrations

7.2.1 Titration Curves

7.2.2 Examples of Methods Classified by Endpoint Indication

7.2.3 Preparation of Standard Solutions

7.3 Precipitation Gravimetry

7.3.1 Classification of Gravimetric Methods of Analysis

7.3.2 General Procedure and Requirements for Precipitation

7.3.3 Precipitate Formation

7.3.4 Obtaining High Purity Precipitates

7.3.5 Experimental Considerations

7.3.6 Examples of Organic Precipitating Reagents

7.1 PRECIPITATION EQUILIBRIA AND SOLUBILITY

Precipitation equilibrium is very important for precipitation titration and precipitation gravimetry. Both are required for a complete precipitation of the species to be determined. When dissolved, most sparingly soluble ionic compounds are essentially completely dissociated in a saturated aqueous solution. For a precipitate compound (MA), M stands for a metal ion and A stands for an anion. When excess MA is equilibrated with water, the dissociate equilibrium is described by Equation (7-1).

$$MA(s) \rightleftharpoons M + A \tag{7-1}$$

and the equilibrium constant is given in Equation (7-2).

$$K^{\ominus} = \frac{a(M) \cdot a(A)}{a(MA)_{(s)}} \tag{7-2}$$

where $a(MA)_{(s)}$ represents the activity of MA in the solid which is a separate phase in contact with the saturated solution. Because the concentration of a compound in its solid phase is constant regardless how much excess solid is present, Equation (7-2) can be rewritten as

$$K_{sp}^{\ominus} = K \cdot a(MA)_{(s)} = a(M) \cdot a(A) \tag{7-3}$$

$K_{sp}^{\ominus}$ is the **solubility-product constant**, or the **solubility product**. It is important to understand that this equilibrium is independent of the amount of the solid (MA) as long as the solution is saturated. The solubility product ($K_{sp}^{\ominus}$) changes only as the temperature changes.

7.1.1 Solubility of Precipitates in Pure Water

The **solubility** (S) of a slightly soluble precipitate (MA) is $S = [MA]+[A]=[MA]+[M]$, where [MA] is intrinsic solubility of molecule MA, and is often negligible as compared to [M] and [A] because MA is completely ionized in water. Therefore, solubility of MA is given by

$$S = [M] = [A] = (K_{sp})^{1/2} \tag{7-4}$$

7.1.2 Ionic Strength and the Solubility of Precipitates

In the precipitation process, the buffer and the addition of precipitating reagents will increase the ionic strength of the solution. The molar concentrations of the sparingly soluble salts are of interest to the analytical chemist. Then,

$$K_{sp}^{\ominus} = a(M) \cdot a(A) = \gamma(M)\gamma(A)[M][A]$$

and
$$K_{sp} = [M][A] = \frac{K_{sp}^{\ominus}}{\gamma(M)\gamma(A)} \quad (7\text{-}5)$$

K_{sp} is the **concentration solubility-product constant**, whose value changes with the ionic strength of the solution, because $\gamma(M) \leqslant 1$ and $\gamma(A) \leqslant 1$, $K_{sp} \geqslant K_{sp}^{\ominus}$. Evidently an increase in the concentration of an indifferent electrolyte (one neither reacting chemically with the precipitate nor having a common ion) will increase the solubility product and subsequently the solubility of the precipitate by decreasing the activity coefficients of the component ions. The K_{sp} values of selected sparingly soluble salts are provided in Appendix D8. Other than in pure water, K_{sp} rather than $K_{sp}^{\ominus}$ is used for solubility calculation.

7.1.3 Common Ion and the Solubility of Precipitates

The molar solubility of a precipitate MA is reduced when excess precipitating ions, M or A, are present. The **common ion** is often added to reduce the solubility of the precipitate in gravimetry. For example, in the gravimetric determination of sulfate (SO_4^{2-}), when the amount of added barium ion (Ba^{2+}) is at the stoichiometric ratio to SO_4^{2-}, the solubility $S=(K_{sp})^{1/2}=(6\times10^{-10})^{1/2}=2.4\times10^{-5}$ ($mol \cdot L^{-1}$). Assuming the volume of the final saturated solution is 250 mL, the mass amount of $BaSO_4$ lost from dissolution is $233.4\times(2.4\times10^{-5}\times250/1000)=1.4$ (mg). If excess precipitating reagent (Ba^{2+}) is added to make a final concentration of 0.01 $mol \cdot L^{-1}$, the solubility $S=6\times10^{-10}/0.01=6\times10^{-8}$ ($mol \cdot L^{-1}$), then the mass amount of $BaSO_4$ lost from dissolution is $233.4\times(6\times10^{-8}\times250/1000)=0.004$ (mg). The excess Ba^{2+} insures the completeness of the precipitation reaction and thus the error from loss of precipitate is reduced or eliminated.

7.1.4 pH and the Solubility of Precipitates

The solubility of many precipitates is strongly dependent on the pH of the solution, when the precipitate contains an anion with basic properties or a cation with acidic properties or both. At acidic pH, the anion will be protonized to produce the acid form of the anion; at basic pH, the cation easily forms hydroxo complex with hydroxide ion, thus the solubility of the precipitate will increase in both cases. For example, the dissociation equations related to the dissociation of calcium oxalate (CaC_2O_4) in excess oxalic acid are:

$$CaC_2O_4(s) \rightleftharpoons Ca^{2+} + C_2O_4^{2-}$$
$$\downarrow H$$
$$HC_2O_4^-$$
$$H_2C_2O_4$$

The formation of bioxalate ion and oxalic acid increases the solubility of calcium oxalate.

In practical precipitation reactions, there exists common ion. The anion may be protonized to become the weak acid form, and the cation may form hydroxo complex or react with other complexing agents. All these side reactions as summarized in the following chart may increase the solubility of the precipitate.

$$MA(s) \rightleftharpoons M + A$$
$$M \xrightarrow{OH} MOH \cdots, \quad M \xrightarrow{L} ML \cdots, \quad A \xrightarrow{H} HA \cdots$$

The analytical concentration of M is

$$[M'] = [M] + [ML] + [ML_2] + \cdots + [MOH] + [M(OH)_2] + \cdots$$

The analytical concentration of A is

$$[A'] = [A] + [HA] + [H_2A] + \cdots$$

Substitution of $[M] = [M']/\alpha(M)$ and $[A] = [A']/\alpha(A)$ into Equation (7-5) yields

$$K_{sp} = [M][A] = \frac{[M'][A']}{\alpha(M) \cdot \alpha(A)} = \frac{K'_{sp}}{\alpha(M) \cdot \alpha(A)}$$

i. e.

$$K'_{sp} = [M'][A'] = K_{sp} \cdot \alpha(M) \cdot \alpha(A) \tag{7-6}$$

K'_{sp} is the conditional solubility-product constant. Because $\alpha(M) \geqslant 1$ and $\alpha(A) \geqslant 1$, $K'_{sp} > K_{sp}$. The occurrence of side reactions increases the solubility of the precipitate. K'_{sp} is the product of the analytical concentration of the anion and the cation of the dissociated precipitate compound at the equilibrium between solid and aqueous phases, thus K'_{sp} is a practical measure of the completeness of a precipitation reaction.

【Example 7.1】 Calculate the solubility of CaC_2O_4

(1) in pure water;

(2) in pH 1.0 HCl solution;

(3) in a pH 4.0 oxalate solution with analytical concentration of non-precipitated oxalate of 0.10 mol · L^{-1}.

Answer:

(1) In pure water,

$$S = \sqrt{K^{\ominus}_{sp}} = \sqrt{10^{-8.6}} = 10^{-4.3} (\text{mol} \cdot \text{L}^{-1})$$

(2) At pH 1.0, $[H^+] = 0.1\ mol \cdot L^{-1}$,

$$\alpha(C_2O_4(H)) = 1 + [H^+]\beta_1 + [H^+]^2\beta_2 = 1 + 10^{-1.0+4.0} + 10^{-2.0+5.1} = 10^{3.4}$$

$$K'_{sp}(CaC_2O_4) = K_{sp}(CaC_2O_4) \cdot \alpha(C_2O_4(H)) = 10^{-7.8+3.4} = 10^{-4.4}$$

Thus, $$S = [Ca^{2+}] = [(C_2O_4^{2-})'] = \sqrt{K'_{sp}} = 10^{-2.2}(mol \cdot L^{-1})$$

(3) At pH=4.0,

$$\alpha(C_2O_4(H)) = 1 + [H^+]\beta_1 + [H^+]^2\beta_2 = 1 + 10^{-4.0+4.0} + 10^{-8.0+5.1} = 10^{0.3}$$

$$K'_{sp}(CaC_2O_4) = K_{sp}(CaC_2O_4) \cdot \alpha(C_2O_4(H)) = 10^{-7.8+0.3} = 10^{-7.5}$$

In excess precipitating agent $C_2O_4^{2-}$, $S = [Ca^{2+}]$,

$$[(C_2O_4^{2-})'] = S + 0.1 \approx 0.10(mol \cdot L^{-1})$$

$$S = [Ca^{2+}] = K'_{sp}(CaC_2O_4)/[(C_2O_4^{2-})'] = \frac{10^{-7.5}}{10^{-1.0}} = 10^{-6.5}(mol \cdot L^{-1})$$

If the anion of the precipitate compound has a large protonation constant, the dissociation of precipitate in pure water may affect the pH of the solution when the K_{sp} is not small enough. In this case, the solubility approximately equals the concentration of hydroxide ion in solution.

【Example 7.2】 Calculate solubility of the following precipitates in pure water.

(1) Copper sulfide (CuS);

(2) Manganese sulfide (MnS);

Constants for calculation: $K^{\ominus}_{sp}(CuS) = 10^{-35.2}$; $K^{\ominus}_{sp}(MnS) = 10^{-12.6}$; pK_a values for hydrogen sulfide (H_2S): $pK_{a1} = 7.1$ and $pK_{a2} = 12.9$.

Answer:

(1) The small solubility product (K_{sp}) of CuS means that the solubility is very low. As a result, the amount of hydroxide ion generated from protonation of sulfide ion can be neglected compared to the amount of hydroxide ion from dissociation of water, thus, the pH of the CuS saturated solution can be assumed to be about 7.0. At this pH,

$$\alpha(S(H)) = 1 + 10^{-7.0+12.9} + 10^{-14.0+20.0} = 10^{6.3} \gg 1$$

Thus $[S^{2-}]$ in solution is much less than solubility (S). Therefore,

$$S = [S^{2-}] + [HS^-] + [H_2S] = [S'] = [Cu^{2+}]$$

$$[Cu^{2+}][S'] = K'_{sp}(CuS) = S^2$$

$$S = \sqrt{K'_{sp}(CuS)} = \sqrt{K^{\ominus}_{sp}(CuS) \cdot \alpha(S(H))} = \sqrt{10^{-35.2+6.3}} = 10^{-14.5}(mol \cdot L^{-1})$$

(2) The solubility product of manganese sulfide is relatively large, thus the solubility is large. In this case, the amount of hydroxide ion almost equals to the amount of HS^- generated from dissociation of MnS followed by protonation of

sulfide. The solubility (S) can be calculated from the dissociation reaction of MnS.

$$MnS(s) + H_2O \rightleftharpoons Mn^{2+} + HS^- + OH^-$$
$$S \quad S \quad S$$

The expression of the equilibrium constant for the above reaction is

$$K = [Mn^{2+}][HS^-][OH^-] = \frac{[Mn^{2+}][S^{2-}][H^+][OH^-]}{K_{a2}}$$

$$= \frac{K^{\ominus}_{sp}(MnS) \cdot K_w}{K_{a2}} = \frac{10^{-12.6} \times 10^{-14.0}}{10^{-12.9}} = 10^{-13.7}$$

Thus, $S = \sqrt[3]{10^{-13.7}} = 10^{-4.6}(mol \cdot L^{-1})$

In the solution, $[OH^-] = S = 10^{-4.6}\ mol \cdot L^{-1}$, that is $[H^+] = 10^{-9.4}\ mol \cdot L^{-1}$,

$$\alpha(S(H)) = 1 + 10^{-9.4+12.9} + 10^{-18.8+20.0} = 1 + 10^{3.5} + 10^{1.2}$$

Therefore, $[S^{2-}] : [HS^-] : [H_2S] = 1 : 10^{3.5} : 10^{1.2}$

As can be seen from the above equation, HS^- predominates in the solution. Therefore, it is reasonable to assume $S = [HS^-]$.

7.1.5 Complexing Agents and the Solubility of Precipitates

It is often found that in spite of the common-ion effect, an excess precipitating reagent increases dramatically the solubility of a precipitate or even causes the precipitate to dissolve completely, because the excess reagent forms complexes with the anion or the cation of the precipitate.

【Example 7.3】 Calculate the solubility of lead oxalate (PbC_2O_4) at pH 4.0 when the analytical concentration of oxalate is 0.2 $mol \cdot L^{-1}$ and the analytical concentration of EDTA (Y) not binding lead ion (Pb^{2+}) is 0.01 $mol \cdot L^{-1}$. ($K_{sp}(PbC_2O_4) = 10^{-9.7}$, $\lg K(PbY) = 18.0$)

Answer: The precipitation reaction and the side reactions occurring are summarized in the following chart.

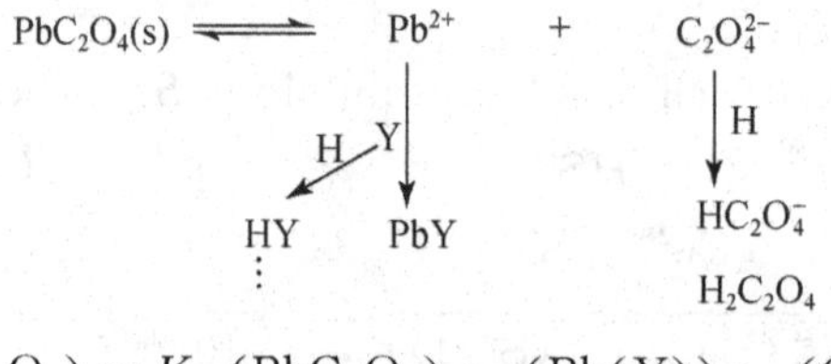

$PbC_2O_4(s) \rightleftharpoons Pb^{2+} + C_2O_4^{2-}$

Pb^{2+} —Y→ PbY; H (from Y) → HY ⋮

$C_2O_4^{2-}$ —H→ $HC_2O_4^-$, $H_2C_2O_4$

$$K'_{sp}(PbC_2O_4) = K_{sp}(PbC_2O_4) \cdot \alpha(Pb(Y)) \cdot \alpha(C_2O_4(H))$$

At pH 4.0, $\alpha(C_2O_4(H)) = 10^{0.3}$ (See Example 7.1) and $\alpha(Y(H)) = 10^{8.6}$. Thus,

$$[Y] = \frac{[Y']}{\alpha(Y(H))} = \frac{10^{-2}}{10^{8.6}} = 10^{-10.6}(mol \cdot L^{-1})$$

$$\alpha(\text{Pb(Y)}) = 1 + [\text{Y}] \cdot K(\text{PbY}) = 1 + 10^{-10.6+18.0} = 10^{7.4}$$

$$K'_{sp}(\text{PbC}_2\text{O}_4) = 10^{-9.7+7.4+0.3} = 10^{-2.0}$$

Therefore, $$S = [\text{Pb}'] = \frac{K'_{sp}}{[(\text{C}_2\text{O}_4^{2-})']} = \frac{10^{-2.0}}{0.2} = 0.05(\text{mol} \cdot \text{L}^{-1})$$

PbC_2O_4 in this situation is not going to be precipitated because the solubility is very high.

Sometimes, the effects of common-ion and complex formation exist at the same time, because the excess precipitating reagent is also a complexing agent for the same cation. In this situation, the solubility is governed by the concentration of precipitating reagent. In precipitation gravimetry, too much excess precipitating reagent may increase the solubility of the analyte, thus reducing the recovery of the analyte and causing an analytical error. For example, in the determination of silver by precipitation of silver ion (Ag^+) with excess potassium chloride (KCl), chloride ion (Cl^-) reacts with Ag^+ to form not only AgCl precipitate but also the $AgCl_2^-$, $AgCl_3^{2-}$, and $AgCl_4^{3-}$ complexes ($\lg\beta_1 \sim \lg\beta_4$ are 2.9, 4.7, 5.0, and 5.9). The equilibrium equations and the constant expressions involved are:

$$\text{AgCl(s)} \rightleftharpoons \text{AgCl(aq)} \rightleftharpoons \text{Ag}^+ + \text{Cl}^- \qquad K_{sp} = [\text{Ag}^+][\text{Cl}^-]$$

$$\text{Ag}^+ + \text{Cl}^- \rightleftharpoons \text{AgCl(aq)} \qquad \beta_1 = [\text{AgCl}_{(aq)}]/[\text{Ag}^+][\text{Cl}^-]$$

$$\text{AgCl(aq)} + \text{Cl}^- \rightleftharpoons \text{AgCl}_2^- \qquad \beta_2 = [\text{AgCl}_2^-]/[\text{AgCl}_{(aq)}][\text{Cl}^-]$$

$$\text{AgCl}_2^- + \text{Cl}^- \rightleftharpoons \text{AgCl}_3^{2-} \qquad \beta_3 = [\text{AgCl}_3^{2-}]/[\text{AgCl}_2^-][\text{Cl}^-]$$

$$\text{AgCl}_3^{2-} + \text{Cl}^- \rightleftharpoons \text{AgCl}_4^{3-} \qquad \beta_4 = [\text{AgCl}_4^{3-}]/[\text{AgCl}_3^{2-}][\text{Cl}^-]$$

$$\begin{aligned} S &= [\text{AgCl}_{(aq)}] + [\text{Ag}^+] + [\text{AgCl}_2^-] + [\text{AgCl}_3^{2-}] + [\text{AgCl}_4^{3-}] \\ &= \beta_1[\text{Ag}^+][\text{Cl}^-] + [\text{Ag}^+] + \beta_2[\text{Ag}^+][\text{Cl}^-]^2 + \beta_3[\text{Ag}^+][\text{Cl}^-]^3 + \beta_4[\text{Ag}^+][\text{Cl}^-]^4 \\ &= [\text{Ag}^+](1 + \beta_1[\text{Cl}^-] + \beta_2[\text{Cl}^-]^2 + \beta_3[\text{Cl}^-]^3 + \beta_4[\text{Cl}^-]^4) \\ &= K_{sp}(1/[\text{Cl}^-] + \beta_1 + \beta_2[\text{Cl}^-] + \beta_3[\text{Cl}^-]^2 + \beta_4[\text{Cl}^-]^3) \end{aligned} \qquad (7\text{-}7)$$

where the $\beta_3[Cl^-]^2$ and $\beta_4[Cl^-]^3$ can be neglected because $AgCl_3^{2-}$ and $AgCl_4^{3-}$ are not predominant forms.

To find the minimum value of solubility, S, the derivative of S was taken with $[Cl^-]$ equal to zero:

$$\frac{dS}{d[\text{Cl}^-]} = K_{sp}\left(-\frac{1}{[\text{Cl}^-]^2} + \beta_2\right) = 0$$

$$[\text{Cl}^-] = 10^{-2.4}(\text{mol} \cdot \text{L}^{-1})$$

Thus, $$S_{min} = 10^{-9.5} \times (1/10^{-2.4} + 10^{2.9} + 10^{4.7-2.4} + \cdots) = 10^{-6.4}(\text{mol} \cdot \text{L}^{-1})$$

The solid curve in Figure 7.1 illustrates the effect of chloride ion concentration on the solubility of silver chloride precipitate with data obtained from Equation (7-7). The crosses in Figure 7.1 represent the experimental data points. At $[Cl^-]$

$< 10^{-2.4}$ mol · L^{-1}, the common-ion effect is dominant; and at $[Cl^-] >$ $10^{-2.4}$ mol · L^{-1}, the complexation effect is dominant. One should be aware that an appropriate excess amount precipitating reagent needs to be added to assure the minimum solubility in precipitation gravimetric analysis.

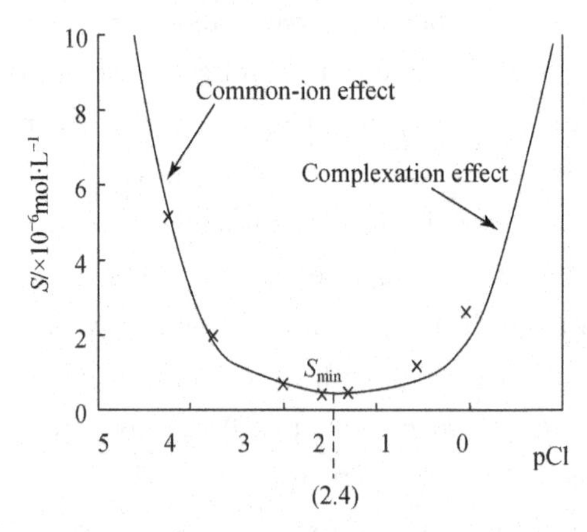

Figure 7.1 The effect of chloride ion concentration on the solubility of AgCl. The solid curve is obtained from Equation (7-7). The crosses are experimental data points.

7.2 PRECIPITATION TITRATIONS

Precipitation titration is a volumetric method based on reactions that produce an ionic product of limited solubility. Although there are many reactions which produce insoluble precipitates, only a limited number of precipitating reagents are successfully used for titration purpose, because (1) the rate for most precipitate formation is slow; (2) there is no specific stoichiometry for most precipitation reactions; (3) the solubility of some precipitates is not small enough to assure the completeness of the titration reaction; and (4) the adsorption of reactant ion on the precipitate causes a titration error. The most commonly used precipitation titrimetry is called **argentometry** which is based on the reaction between silver nitrate ($AgNO_3$) with the halides (Cl^-, Br^- and I^-) and halide-like ions (such as SCN^- and CN^-). In this text, the discussion of precipitation titration is limited to argentometry.

7.2.1 Titration Curves

Consider the titration of 20.00 mL of 0.1000 mol · L^{-1} NaCl with 0.1000 mol · L^{-1} $AgNO_3$. The titration curve can be constructed by plotting pAg against the percent titration ($T/\%$) in a manner similar to that for acid-base titration. To illustrate calculations, the titration is divided into three stages: before the stoichiometric point, at the stoichiometric point, and after the stoichiometric point.

(1) Before the stoichiometric point, the concentration of Ag^+ can be calculated from the concentration of Cl^- and the solubility product of AgCl. The contribution from dissociation of AgCl is negligible. For example, if 19.98 mL of $AgNO_3$ is added to precipitate this NaCl, the $[Cl^-]$ in solution can be calculated as follows:

$$[Cl^-] = 0.1000 \times \frac{0.02}{20.00 + 19.98} = 10^{-4.3}\,(mol \cdot L^{-1})$$

Therefore,

$$[Ag^+] = \frac{K_{sp}}{[Cl^-]} = \frac{3.2 \times 10^{-10}}{5 \times 10^{-5}} = 10^{-5.2}\,(mol \cdot L^{-1})$$

$$pAg = 5.2$$

(2) At stoichiometric point,

$$[Ag^+] = [Cl^-] = \sqrt{K_{sp}} = 10^{-4.7(5)}\,(mol \cdot L^{-1})$$

$$pAg = 4.7(5)$$

(3) After the stoichiometric point, the $[Ag^+]$ can be calculated from the excess $AgNO_3$ added. The contribution from dissociation of AgCl is negligible. For example, when 20.02 mL $AgNO_3$ is added,

$$[Ag^+] = 0.1000 \times \frac{0.02}{20.00 + 20.02} = 10^{-4.3}\,(mol \cdot L^{-1})$$

$$pAg = 4.3$$

Additional data are provided in Table 7.1. The titration curve according to data in Table 7.1 is given in Figure 7.2 with pAg as the y-axis and the percent titration ($T/\%$) as the x-axis.

Table 7.1 pCl and pAg Changes during the Titration of 0.1000 mol · L^{-1} NaCl with $AgNO_3$ at the Same Concentration

Volume of $AgNO_3$/mL (20.00 mL NaCl)	$T/\%$	pCl (0.01 mol · L^{-1})	pAg (0.1 mol · L^{-1})
0.00	0.0	1	
18.00	90	2.3	7.2
19.80	99	3.3	6.2
19.98	99.9	4.3	5.2
20.00	100.0	4.7(5)	4.7(5)
20.02	100.1	5.2	4.3
20.20	101	6.2	3.3
22.00	110	7.2	2.3
40.00	200	8.0	1.5

The profile of the titration curve for this titration is similar to that of acid-base titration. The titration break is roughly symmetrical to the stoichiometric point. The titration break is dependent on the concentration of both the titrant and titrand as well as the solubility of the precipitate. When the concentration of titrant and titrand are increased 10 times, the titration break will be expanded by 2 pAg units with one unit being expanded at each end of the titration break. The solubility of AgI ($K_{sp}=10^{-15.8}$) is much smaller than that of AgCl ($K_{sp}=10^{-9.5}$), and the titration break is expanded by 6.3 pAg (the pK_{sp} difference between AgI and $AgNO_3$) from titration of Cl^- to titration of I^- at the pre-stoichiometric point side of titration. After the stoichiometric point, the curves for the two titrations are the same. The titration curves for 1.000 mol · L^{-1} Cl^-, Br^- and I^- with Ag^+ at the same concentration are illustrated in Figure 7.2.

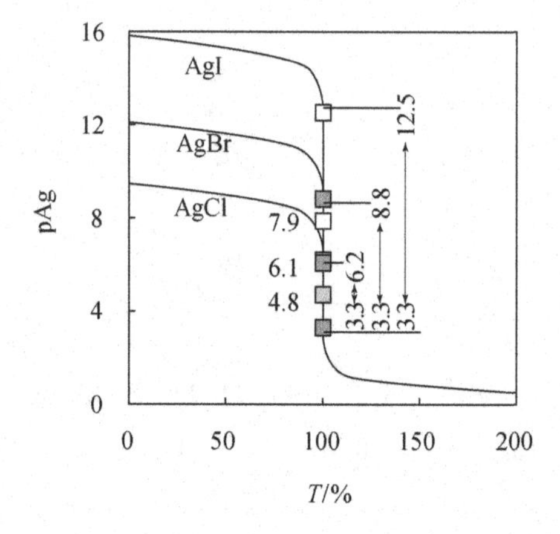

Figure 7.2 Titration curves for 1.000 mol · L^{-1} NaCl, NaBr and NaI using $AgNO_3$ at the same concentration (pK_{sp} for AgCl, AgBr, and AgI are 9.75, 12.31, and 16.08).

7.2.2 Examples of Methods Classified by Endpoint Indication

The visual endpoint detection of precipitation titration can be realized by forming colored precipitate, colored complex and color change resulting from adsorption of fluorescent dyes.

1. The Mohr Method

In **the Mohr method**, the halide ion, such as Cl^-, is titrated using Ag^+ as a titrant. A small amount of K_2CrO_4 is added to the solution to be titrated. After the stoichiometric point, a brick red Ag_2CrO_4 precipitate forms when a small increment of Ag^+ is added, indicating the endpoint. If necessary, an analyte-free reagent blank is titrated to determine the amount of titrant needed for endpoint visualization. The volume for the reagent blank is subsequently subtracted from the experimental endpoint to obtain a true titrant volume for the titrand. Because CrO_4^{2-} is a weak base, the solution usually is maintained in the pH range of 7~10 to ensure the formation of Ag_2CrO_4 precipitate so that a significant error in detecting the endpoint can be avoided. The presence of ammonia may cause a significant error because ammonia forms silver-ammonia complex at the pH for titration. When ammonia is present, the suitable pH for titration is 6.5~7.2 to ensure that no silver-ammonia complex is formed because at pH 6.5~7.2 the ammonium ion is the predominant form. The Mohr method is not suitable for titration of I^- and SCN^-, because the precipitate readily adsorbs the analyte to make the endpoint occur earlier than expected. The Mohr method is prone to interferences from other ions because cations such as Ba^{2+}, Pb^{2+} and Hg^{2+} form precipitates with CrO_4^{2-} and anions such as SO_3^{2-}, PO_4^{3-}, AsO_4^{3-}, S^{2-} and $C_2O_4^{2-}$ form precipitates with the silver ion.

2. The Volhard Method

In **the Volhard method**, Ag^+ is titrated with SCN^- in the presence of Fe^{3+} ($NH_4Fe(SO_4)_2$). At the endpoint of the titration, additional SCN^- forms red colored $Fe(SCN)_2^+$ complex. The titration reaction and the endpoint indicating the completion of the reaction are as follows:

$$Ag^+ + SCN^- \rightleftharpoons AgSCN(s)$$

$$Fe^{3+} + SCN^- \rightleftharpoons Fe(SCN)_2^+$$

The titration must be carried out in a strongly acidic pH.

Back titration is used for determination of halide ions. Usually the analytical sample is prepared in HNO_3. Excess $AgNO_3$ standard solution is then added to the solution, and the NH_4SCN standard solution is used to titrate the excess Ag^+.

In determination of Cl^-, NH_4SCN may slowly convert AgCl to AgSCN precipitate close to the endpoint with the following reaction $AgCl(s) + SCN^- \rightleftharpoons AgSCN(s) + Cl^-$. In this case, more SCN^- is consumed than expected, causing a positive titration error. The following approaches can help to reduce the titration error: (1) Heat the resulted AgCl suspension after the addition of excess $AgNO_3$ to make the AgCl agglomerate thus reducing the adsorption of Ag^+. The precipitate is then filtered and washed using diluted HNO_3. The collected wash solution with the excess Ag^+ is then titrated with NH_4SCN standard solution. (2) After addition of excess $AgNO_3$, 1~2 mL of an organic solvent, such as nitrobenzene or 1,2-dichloroethylene, is added to the suspension to block contact between AgCl and SCN^- so that the AgCl cannot be converted to AgSCN. (3) Increase the concentration of Fe^{3+} to reduce the amount of SCN^- needed for endpoint indication so that the formation of AgSCN is not likely to occur.

As the titration is carried out in HNO_3, PO_4^{3-}, AsO_4^{3-}, and S^{2-} do not form precipitates with silver ion to result in errors. However, attention still needs to be paid when the Volhard method is used: (1) The titration must be carried out in strongly acidic solution, typically greater than 0.3 mol · L^{-1} HNO_3 to avoid the formation of ferric-hydroxo complex. (2) In titration of I^-, excess $AgNO_3$ must be added prior to addition of ammonium ferric sulfate ($NH_4Fe(SO_4)_2$) to avoid oxidation of I^- by Fe^{3+}. (3) Any strong oxidizing agents such as nitrogen oxides, copper ion, and mercury ion must be removed or masked prior to the titration because these species oxidize SCN^-.

3. The Fajans Method

In **the Fajans method**, the adsorption indicator is used for endpoint indication. The adsorption indicator adsorbed on the precipitate is different from the color when the indicator is in solution. For example, in titration of Cl^- with Ag^+, the anionic dye fluorescein (Fl^-) is used as the indicator. Before the endpoint, AgCl precipitate is negatively charged on the surface due to the adsorption of excess Cl^-. Fluorescein is repelled by the precipitate and remains in solution where fluorescein has a greenish yellow color. After the endpoint, the precipitate is positively charged due to the adsorption of excess Ag^+. The anionic fluorescein indicator then adsorbs to the surface of precipitate and its color changes into pink, indicating the endpoint. This process is illustrated as below:

(1) Cl^- in excess

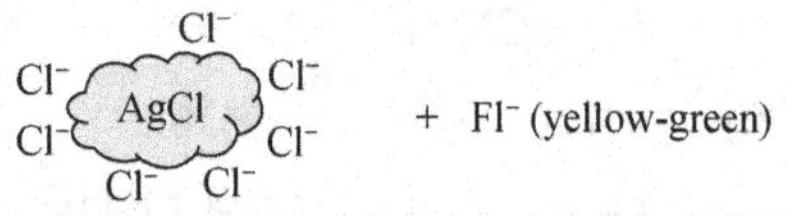

AgCl surrounded by Cl^- ions + Fl^- (yellow-green)

(2) Ag^+ in excess

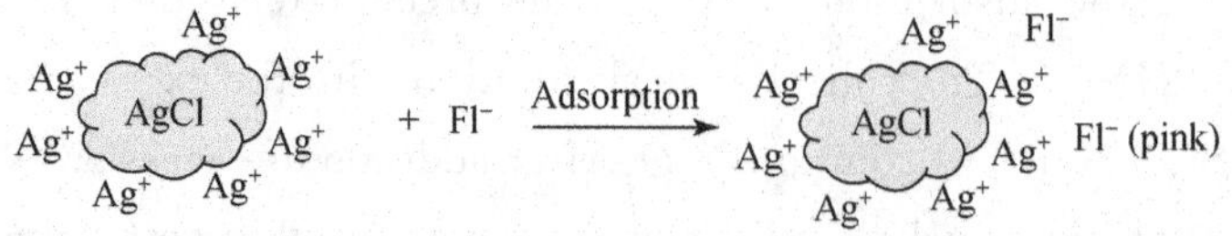

AgCl surrounded by Ag^+ ions + Fl^- —Adsorption→ AgCl surrounded by Ag^+ ions with Fl^- (pink)

Cautions with the use of adsorption indicators are:

- As adsorption occurs at the surface of the precipitate, the larger the surface area of the precipitate the more sensitive is the endpoint. Dextrin is often added to the titration solution to prevent AgCl precipitate to agglomerate so that a large surface area can be maintained to see the endpoint.
- Most of the adsorption indicators are weak organic acids. The pH of titration must be such to keep the indicators negatively charged, that is, pH>pK_a of the indicator.
- The titration is better carried out with mild room lighting because the silver halide precipitate decomposes under irradiation into dark grey colored silver which makes the endpoint more difficult to see.
- The ability of indicator to adsorb to the colloidal surface should be moderately less than the halide ions; otherwise, the endpoint will occur before the stoichiometric point. On the other hand, if the adsorption ability of the indicator toward the precipitate is too small, the endpoint will occur too late, causing a titration error. The adsorption of silver halide colloids toward halide ions and indicators is in the order of $I^- > SCN^- > Br^- >$ eosin $> Cl^- >$ fluorescein. For this reason, fluorescein is selected instead of eosin for the titration of Cl^-. The adsorption indicators for specific applications are given in Table 7.2.

Table 7.2 The Application Examples of Selected Adsorption Indicators

Indicator	Analytes	Titrant	pH for Titration
Fluorescein	Cl^-, Br^-, I^-	$AgNO_3$	7~10
Dichlorofluorescein	Cl^-, Br^-, I^-	$AgNO_3$	4~10
Eosin	SCN^-, Br^-, I^-	$AgNO_3$	2~10
Methyl violet	Ag^+	NaCl	Acidic pH

7.2.3 Preparation of Standard Solutions

$AgNO_3$ standard solution can be prepared from analytical grade $AgNO_3$. $AgNO_3$ standard solution is often prepared by standardization with primary standard grade NaCl with the same experimental conditions as the determination of Cl^-, because some determinate errors can be eliminated by carrying the determination and standardization in the same manner and experimental condition. NaCl primary standard has to be heated in a porcelain crucible with stirring at 500 ~ 600℃ to remove any residual water, because NaCl readily absorbs water from ambient environment. $AgNO_3$ is easily decomposed by normal room lighting and thus should be stored in a brown bottle or in the dark.

7.3 PRECIPITATION GRAVIMETRY

Gravimetric methods of analysis are based on measurement of mass of the reaction product which is stoichiometrically related to the analyte. In the gravimetry, the product chemically related to the analyte is removed from the sample matrix before the mass of the reaction product is measured. The content of analyte in the test sample is calculated based on the mass obtained. An analytical balance with the necessary sensitivity is used for the weighing steps in gravimetric analysis to ensure accurate and precise data.

7.3.1 Classification of Gravimetric Methods of Analysis

Gravimetry methods can be classified into the following categories:

(1) **Volatilization gravimetry**. In the volatilization gravimetry, the analyte is converted to a gas and thus separated from the sample matrix. The mass of the separated gas can be calculated from the difference of the mass of the sample before and after the volatilization. For example, the adsorbed water and the water of crystallization of a sample can be determined in this way. The volatized gas can also be absorbed with a known mass of a drying agent. The mass gained by the drying agent is the mass of the gas.

(2) **Electrogravimetry**. In the electrogravimetry, the analyte is separated by electrolysis and deposited on an electrode. The mass the electrode gained is the mass of the analyte.

(3) **Precipitation gravimetry**. In the precipitation gravimetry, the analyte is separated from a solution of the sample to be analyzed by precipitation. The precipitate is converted to a compound of known composition whose mass can be measured. Precipitation gravimetry will be discussed in this text.

7.3.2 General Procedure and Requirements for Precipitation

In the precipitation gravimetry, after a solution of the sample is prepared, a roughly measured amount of precipitating reagents is added to react with the analyte to form a sparingly soluble precipitate. The precipitate is then filtered, washed to remove impurities, and converted to a product of known composition by drying or ignition and then weighed. The concentration of the analyte can be calculated from the mass of the precipitate. The weighed form and the initial precipitate may or may not have the same composition.

For example, the precipitation gravimetric procedures for the determination of SO_4^{2-} and Mg^{2+} are illustrated as below:

$$SO_4^{2-} + BaCl_2 \longrightarrow \boxed{BaSO_4} \xrightarrow[\text{Wash}]{\text{Filter}} \xrightarrow[\text{Ignition}]{800^\circ C} \boxed{BaSO_4}$$

$$Mg^{2+} + (NH_4)_2HPO_4 \longrightarrow \boxed{MgNH_4PO_4 \cdot 6H_2O} \xrightarrow[\text{Wash}]{\text{Filter}} \xrightarrow[\text{Ignition}]{800^\circ C} \boxed{Mg_2P_2O_7}$$

Analytes | Precipitating reagents | Precipitation forms | Weighing forms

Please note that in the determination of sulfate, the **precipitation form** and the **weighing form** are the same, but in the determination of magnesium ion, the precipitation form and the weighing form are different.

There are requirements for the precipitation forms and the weighing forms in order to assure accuracy of the method as well as easiness of manipulation.

The requirements for a precipitation form: (1) the precipitate should have small solubility to minimize loss of precipitate during precipitation and washing; (2) the precipitate should be easily filtered and washed; (3) high purity of the precipitate is required so that the determinate error from reaction impurities can be avoided.

The requirements for a weighing form: (1) the weighing form must have a known chemical composition in order to give a quantitative measurement of the analyte; (2) the weighing form should be stable, that is, free of influence from H_2O, CO_2 and O_2 in the air; (3) it is desirable that the weighing form has a large molecular weight so that the weighing error can be reduced to assure a high accuracy of the result. This can be illustrated in Example 7.4.

The results of a gravimetric analysis are generally calculated from the mass of the sample (m_s) and the weighing form (m). The mass of the analyte is often needed to be converted from the mass of the weighing form according to the formula mass and stoichiometric ratios of the analytical species to the weighing form. The converting factor, called **gravimetric factor (*F*)**, is given in Equation (7-8) and examples are given in Table 7.3.

$$F = \frac{M(\text{analyte})}{r \cdot M(\text{weighing form})} \tag{7-8}$$

where r is the stoichiometric ratio of the analyte to the weighing form.

Table 7.3 Examples of Gravimetric Factor Calculation

Analytes	Weighing Forms	Gravimetric Factor (F) Calculations	F Values
S	$BaSO_4$	$M(S)/M(BaSO_4)$	0.1374
MgO	$Mg_2P_2O_7$	$2M(MgO)/M(Mg_2P_2O_7)$	0.3622

The percentage of analyte in the analytical sample is calculated by Equation (7-9)

$$w = \frac{mF}{m_s} \times 100\% \tag{7-9}$$

【Example 7.4】 A 0.5000 g sample containing Al is weighed and dissolved. 8-hydroxylquinoline (C_9H_7NO) is added and 0.3280 g $Al(C_9H_6NO)_3$ weighing form is obtained. Calculate Al content in percentage. If the weighing form is Al_2O_3, how many grams of weighing form can be obtained?

Answer: If the weighing form is $Al(C_9H_6NO)_3$,

$$w(Al) = \frac{m(Al(C_9H_6NO)_3) \cdot \dfrac{M(Al)}{M(Al(C_9H_6NO)_3)}}{m_s} \times 100\%$$

$$= \frac{0.3280 \times 0.05873}{0.5000} \times 100\% = 3.853\%$$

If the same amount of Al is converted to weighing form Al_2O_3,

$$w(Al) = \frac{m(Al_2O_3) \cdot \dfrac{2M(Al)}{M(Al_2O_3)}}{m_s} \times 100\%$$

$$= \frac{m(Al_2O_3) \times 0.05293}{0.5000} \times 100\% = 3.853\%$$

Then $$m(Al_2O_3) = \frac{3.853 \times 0.5000}{0.5293 \times 100} = 0.0364(g)$$

In this case when Al_2O_3 is taken as the weighing form, the weighing error will be seriously greater than expected, thus the accuracy of the method will be substantially compromised.

7.3.3 Precipitate Formation

It is desirable to obtain precipitates with large particle size so that the precipitate is easy to filter and wash to remove impurities. Other than this, precipitates with large particle size usually have higher purity than those with small particle size, because the latter has a larger surface area so that the amount of adsorbed impurities will be greater.

The particle size of precipitates varies enormously, and generally speaking, depends on the specific compound involved and the process of nucleation, crystal growth, and aging. Colloidal particles, such as $Fe_2O_3 \cdot xH_2O$, are usually less than 1 μm. The diameters of the crystalline particles, such as $BaSO_4$, are 1 μm or larger.

The process of precipitate formation starts with the **nucleation.** Nucleation is a process that a minimum number of atoms, ions, or molecules join together to form the smallest particles which are then capable of growth to large particles. For example, a $BaSO_4$ nucleus is composed of 8 ions (or 4 ion pairs). Nucleation can happen spontaneously from the solute in a supersaturated solution which contains more of the dissolved salt than occurs at the equilibrium. Sometimes, nuclei form on the surface of suspended solid contaminants, such as dust particles, or the 5~10 nm glass nuclei on the wall of the container. Further precipitation involves a competition between additional nucleation and particle growth. A precipitate with a large number of small particles results if nucleation takes the lead; large particles are produced if particle growth predominates.

The particle size during precipitation is affected by **relative supersaturation** with the expression given as $(Q-S)/S$, where Q is the instantaneous concentration of the solute (precipitate product) and S is the equilibrium solubility of the precipitate. When $(Q-S)/S$ is large, nucleation is the major precipitation mechanism, and the precipitate tends to be colloidal; when $(Q-S)/S$ is small, the rate of particle growth predominates, a crystalline precipitate is more likely to be produced.

Variables that minimize supersaturation to produce larger particle sizes precipitates include: (1) elevated temperature to increase the solubility of the precipitate; (2) dilute solution to minimize concentration of the solute (Q) at any instant; and (3) slow addition of the precipitating reagent with good stirring to minimize the relative supersaturation. Generally, a precipitate with lower solubility forms as a colloid. For example, many sulfides and hydrous oxides of most heavy metal ions only form colloids, because their solubility is very low. In this situation, S is negligible compared with Q so that nucleation is a dominant process in precipitation.

7.3.4 Obtaining High Purity Precipitates

1. Coprecipitation

It is important to obtain a precipitate with acceptable purity in precipitation gravimetry. With normal experimental conditions, some soluble compounds cannot precipitate by themselves, but can **coprecipitate** with the precipitation of the analyte.

For example, $BaCl_2$ or Na_2SO_4 often coprecipitate with $BaSO_4$. The **coprecipitation** includes **surface adsorption**, **occlusion**, and **mixed-crystal formation**. One needs to understand the mechanism of coprecipitation to minimize impurities or reduce the amount of impurities occurring during precipitation.

(1) Surface adsorption

In a crystalline lattice, the ions are arranged in a fixed pattern. Taking AgCl as an example (Figure 7.3), every Ag^+ is surrounded by 6 negatively charged Cl^-. Vice versa for Cl^- in the crystalline lattice. The charge inside the crystal is balanced. There are alternating Ag^+ and Cl^- ions on the surface. The surface tends to adsorb the ion of the precipitate particle that is in excess in the solution. For example, if $AgNO_3$ is used to precipitate Cl^-, the surface of the crystal will first adsorb Ag^+ from the ions in solution. The adsorption creates a primary adsorption layer that is strongly adsorbed and makes the crystal surface positively charged. Because colloids have larger surface area than crystalline particles, the degree of the surface adsorption is generally greater for colloids. The charged particle will adsorb a layer of ions, called the **counter-ion layer**, with opposite charge (primarily NO_3^-) to balance the charge on the surface of the AgCl precipitate particle. The primary adsorption layer and the counter-ion layer constitute an **electric double layer** that exerts an electrostatic repulsive force to prevent particles from adhering when particles approach one another. When the concentration of electrolytes including the precipitating reagent is diluted, the distance between colloidal particles can become small enough not to coagulate. Heating can reduce adsorption of ions on the surface of the precipitate particles, thus hastening coagulation. Coagulation can also be promoted by adding an electrolyte to the medium. In the preparation of a colloidal precipitate with reasonable particle size and filterability, the precipitation is generally carried out in a hot, concentrated electrolyte medium.

The surface of a precipitate selectively adsorbs ions from the solution. If the concentrations of counter ions are comparable, the precipitate surface first adsorbs ions which can form low solubility compounds with the ions of the crystal. High valence and high concentration ions are more likely to be adsorbed. In the example of precipitating Cl^- with excess $AgNO_3$, in the presence of K^+, Na^+, Ac^- in addition to the precipitating reagents, the primary adsorption layer is Ag^+ rather than Na^+ or K^+. The counter layer is Ac^- and not NO_3^-, because AgAc is less soluble than $AgNO_3$, which results in adsorption coprecipitation of AgAc with AgCl.

Surface adsorption is the major source of contamination for colloidal precipitates. As the surface adsorption occurs on the surface of the precipitate, the most effective way to reduce the surface adsorption error is to wash the precipitate after filtering.

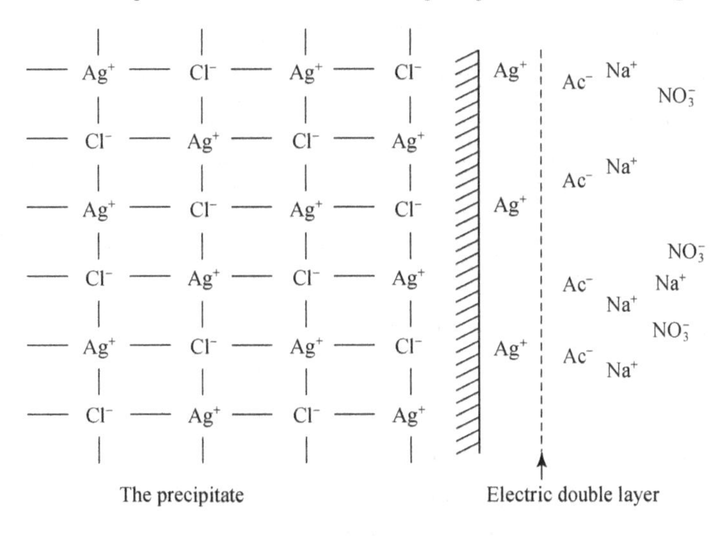

Figure 7.3 Illustration of surface adsorption on AgCl colloids.

(2) Occlusion

If a crystal grows rapidly during precipitation, foreign ions in the counter-ion layer may be trapped or occluded within the growth of the crystal. This type of coprecipitation is called **occlusion**, which also results from surface adsorption and usually occurs with crystalline precipitates. The amount of occlusion is associated with the general mechanism of surface adsorption as discussed above. Therefore, reducing the relative supersaturation is one way to reduce the amount of occlusion. Aging or re-precipitation at elevated temperature can effectively open up the pockets and allow the impurities to be released from the crystals.

(3) Mixed-crystal formation

Mixed-crystal formation is a type of coprecipitation in which a foreign ion replaces an ion in the lattice of a crystal. The requirements for the formation of mixed-crystals are: the two ions have the same charge; their sizes differ by no more than 5%; and the two salts must belong to the same crystal class. Examples are $BaSO_4$-$PbSO_4$ and AgCl-AgBr.

The formation of mixed-crystal is a chemical equilibrium. The extent of mixed-crystal contamination is governed by the equilibrium constant and the law of mass action.

Changing precipitation experimental conditions, washing, aging and reprecipitation are not effective in reducing the extent of mixed-crystal formation. The most effective way to avoid mixed-crystal formation is to separate interfering ions before the precipitation step. Alternatively, masking or using a different precipitating reagent that does not give mixed-crystals with the analyte may be employed.

2. Postprecipitation

According to the IUPAC definition, **postprecipitation** is the subsequent precipitation of a chemically different species upon the surface of an initial precipitate usually, but not necessarily, including a common ion (IUPAC Compendium of Chemical Terminology, O. B. 86, 2nd Edition, 1997). The postprecipitation does not occur by itself but occurs only with the precipitation of species of interest. The longer the time the postprecipitating compound is with the desired precipitate, the more postprecipitation is obtained on the desired precipitate. For example, it is not possible to prepare MgC_2O_4 precipitate because the solubility ($pK_{sp}=3.3$, $I=0.1$) is not small enough to allow MgC_2O_4 to precipitate. In the preparation of CaC_2O_4 precipitate in the presence of Mg^{2+}, MgC_2O_4 does not precipitate initially even if the solution reaches supersaturation. Adsorbed Mg^{2+} can be easily washed away. If CaC_2O_4 is not filtered after the precipitation and the CaC_2O_4 precipitate stays with the mother liquor of Mg^{2+}, subsequent MgC_2O_4 precipitation can be observed. The amount of subsequent MgC_2O_4 precipitation increases with time. The effective approach to avoid postprecipitation is to minimize the time the precipitate is in mother liquor.

3. Analytical Errors from Coprecipitation and Postprecipitation

Coprecipitation and postprecipitation can bring impurities to the precipitate, thus may cause either positive or negative errors in an analysis. If the contaminant is not a compound of the ion being determined, a positive error will always occur. In the determination of chloride ion by AgCl precipitation gravimetry, the adsorption of silver nitrate on AgCl colloidal particles will always cause a positive error. If the contaminant contains the ion being determined, either positive or negative errors may be observed. For example, when $BaCl_2$ is coprecipitated in the preparation of $BaSO_4$ precipitate, it may cause a positive error for determination of SO_4^{2-} because coprecipitation increased the mass of the precipitate. It may cause a negative error for determination of Ba^{2+} because the molar mass of $BaCl_2$ is less than that of $BaSO_4$. If H_2SO_4 is occluded by surface adsorption in this process, whether an error is brought by coprecipitation depends on the temperature for ignition of the precipitate and the analyte of the interest. If ignition is carried out at high temperature, a negative error will occur for determination of sulfur because H_2SO_4 will be decomposed into volatile SO_2, and there will be no error for determination of barium. However, if drying and ignition are carried out in microwave, H_2SO_4 may not be completely removed and a positive error occurs for determination of barium, and a negative error for determination of sulfur.

7.3.5 Experimental Considerations

Experimental conditions must be selected based on the type of precipitate to be prepared to obtain a precipitate with high purity.

1. Precipitation

(1) Crystalline precipitates

Generally, **crystalline precipitates** are easy to filter and purify. The particle size and thus the filterability can be controlled to a degree by minimizing the relative supersaturation. The general experimental guidelines for preparing crystalline precipitates are given below using the preparation of $BaSO_4$ precipitate as an example.

- The precipitate has to be prepared in a diluted and hot solution. The diluted precipitating reagents are added drop wise with stirring to minimize the solute concentration and the relative supersaturation. Elevated temperatures can increase the solubility and the diffusion of ions thus decreasing the relative supersaturation to help with crystal growth.
- Adding HCl solution can also increase the solubility of $BaSO_4$, because SO_4^{2-} can be protonated and thus solubility of $BaSO_4$ is increased which is favorable for preparing precipitates with larger particle size. After the majority of the analyte is precipitated, excess precipitating reagents are added to complete the precipitation and to avoid errors from solubility loss.
- After precipitation, the precipitate is aged by heating for 1 h or more with the mother liquor to allow releasing of the impurity from occlusion. The small particles dissolve and finally grow into larger particles.
- Filtering removes most of the supernatant solution; however, residual traces of the supernatant must be removed to avoid errors from contamination. Rinsing the precipitate to remove the residual impurities must be done carefully to avoid significant loss due to the solubility of the precipitate. For example, in the determination of

barium by forming $BaSO_4$ precipitate, diluted sulfuric acid can be used to wash the precipitate initially because low solubility can still be assured by common ion effect and sulfuric acid can then be removed in the ignition step. For the determination of sulfur, water is the only choice to rinse the precipitate.

(2) Colloidal precipitates

It is hard to control the relative supersaturation of a colloidal precipitate because the solubility S is small so that S is always negligible compared to Q, the instantaneous concentration of the precipitate. Therefore, the particles of colloidal precipitates are so small that the precipitates are difficult to be retained by filters. Colloidal particles can be coagulated to become a filterable, amorphous mass. The precipitation of colloids can be prepared by heating, stirring and adding electrolytes to enhance the coagulation. Heating and stirring decrease the number of adsorbed ions. Addition of electrolytes to the precipitation medium can hasten the coagulation. After the coagulation, the colloidal precipitates need to be filtered while the suspension is hot. Coprecipitated impurities, especially those on the surface, can be removed by washing the precipitate after filtering. Colloidal coagulates cannot be washed with pure water because **peptization** occurs. Peptization is a reverse process of coagulation, and the coagulated colloids will return to the dispersed state because of peptization. The problem can be solved by washing the precipitate with a solution containing an electrolyte, e. g. NH_4NO_3 or NH_4Cl, that easily volatiles when the precipitate is dried or ignited.

(3) Precipitation from homogeneous solution

Homogeneous precipitation is to prepare the precipitate by using a slow chemical reaction to generate precipitating reagents. In this way, local, regional excess of reagent do not occur because the precipitating reagents are generated gradually and homogeneously throughout the solution and react instantly with the analyte. Therefore, the relative supersaturation can be kept low during the entire precipitation, and larger particle precipitates are obtained. For example, in the precipitation of CaC_2O_4, excess $H_2C_2O_4$ is added to a Ca^{2+} sample solution at an acidic pH. The pH of the solution is raised by using urea to homogeneously generate hydroxide ion according to the following equation

$$CO(NH_2)_2 + H_2O \xrightarrow{\triangle} 2NH_4^+ + 2OH^- + CO_2$$

The rate of hydroxide ion generation is dependent of temperature, thus rate of precipitation can be controlled by controlling rate of temperature increase. Urea is particularly valuable for the precipitation of hydrous oxides and basic salts, e. g. hydrous oxides of iron(Ⅲ) and aluminum.

2. Filtration, Washing, Drying or Ignition

After precipitation and aging, the precipitate is then separated from the mother liquor by filtration. The sintered glass crucible or ashless filter paper can be used for filtration. The sintered glass crucible contains a sintered glass bottom with fine (F), medium (M) or coarse (C) porosity. Ashless filter paper is generally used when the precipitate is ignited at high temperature. The filter paper will burn away, leaving the precipitate in a suitable form for weighing. Filter paper with fine porosity is used for filtering fine crystalline precipitates such as $BaSO_4$ and CaC_2O_4 so that very small particles of precipitate can be held by the filter paper. Gelatinous or large crystalline precipitates, such as $Fe_2O_3 \cdot xH_2O$ and $Al_2O_3 \cdot xH_2O$, need to be filtered with filter paper of coarse porosity so that filtration can be finished in a reasonably short period of time. The small or medium particle sized crystalline precipitates such as $MgNH_4PO_4 \cdot 6H_2O$ are generally filtered with medium-porosity filter paper.

After transferred to the filter, the precipitate is washed with several small portions of washing liquid to ensure complete removal of the impurity from the mother liquor. The selection of washing liquid has been discussed previously. It is important that the washing step removes all impurities while minimizing the solubility loss of the precipitate.

After the filtering and washing steps, a weighing form is obtained by heating the precipitate to a constant mass. Heating removes the solvent and the volatile species in the precipitate. The temperature required to produce a suitable weighing form varies with the precipitate to be analyzed.

Precipitates with known chemical composition only require drying to remove the adsorbed solvent to produce weighing form. Precipitates such as AgCl, nickel dimethylglyoxime, and potassium tetraphenylborate, can be dried at 105～120℃ to remove adsorbed water to obtain a constant mass. Precipitates with stable crystal water, such as $CaC_2O_4 \cdot H_2O$ and $Mg(C_9H_6ON_2)_2 \cdot 2H_2O$ (magnesium 8-hydroxyquinolate) can be dried at 105～110℃ to obtain a weighing form with the same chemical formula of the precipitate. Some precipitates such as $BaSO_4$ have a known chemical composition but still require a high ignition temperature to remove water of adsorption and occlusion.

Precipitates with unknown chemical composition must be converted into a suitable weighing form via ignition. Hydrated metal oxides, such as iron hydroxide and aluminum hydroxide have various numbers of crystalline water molecules. The crystalline water of $MgNH_4PO_4 \cdot 6H_2O$ is apt to be lost during drying at ~100°C; igniting the precipitate at 1100℃ produces a weighing form of $Mg_2P_2O_7$.

During ignition, the chemical composition of the precipitate may change as temperature changes. The suitable temperature for a desirable weighing form can be obtained by monitoring the weight as the function of temperature with a thermobalance. The thermal profiles for calcium oxalate and barium sulfate are shown in Figure 7.4. With temperature below 135℃, the adsorbed water and the coprecipitate $(NH_4)_2C_2O_4$ cannot be removed although calcium oxalate is in monohydrate form, $CaC_2O_4 \cdot H_2O$. The monohydrate salt can then be converted to anhydrous oxalate CaC_2O_4 at 225℃, but CaC_2O_4 is apt to absorb water from the ambient environment thus is not a suitable weighing form. A weighing form of $CaCO_3$ can be obtained at 500±25℃. Further increasing in the temperature to 850℃ can produce a weighing form CaO. The weighing forms produced at high temperature are more apt to absorb water and CO_2 from ambient environment.

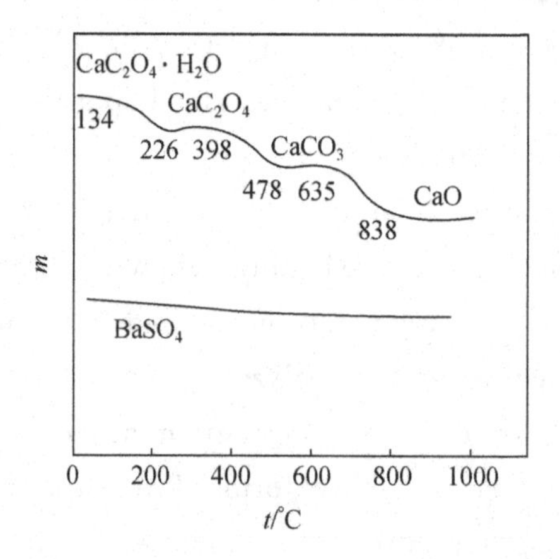

Figure 7.4 The mass of precipitates as a function of temperature.

7.3.6 Examples of Organic Precipitating Reagents

1. The Advantages of Organic Precipitating Reagents

Organic precipitating reagents have wide applications because: (1) organic precipitating reagents are more selective in reactions to form precipitates compared with inorganic species; (2) the solubility of precipitates involving organic precipitating reagents is small; (3) the adsorption of inorganic impurities is small; (4) the precipitating forms usually have known chemical composition and require only drying to produce the weighing forms, which simplifies the experimental procedure; (5) weighing forms with large molar mass are easy to obtain, thus favoring highly accurate analytical results.

2. The Classifications of Organic Precipitating Reagents and Application Examples

There are two types of organic precipitating reagents.

(1) Reagents forming ionic associate precipitates

Organic precipitating reagents having functional groups, such as —COOH, $—SO_3H$, —OH, can form sparingly soluble salts with metal ions via ionic interaction. Sodium tetraphenylborate, $(C_6H_5)_4B^-Na^+$, can form sparingly soluble ionic associate precipitates with K^+, NH_4^+, Tl^+, and Ag^+. Its reaction with K^+ is

$$(C_6H_5)_4B^- + K^+ \longrightarrow K[(C_6H_5)_4B]\downarrow$$

Sodium tetraphenylborate is water soluble and can be used as a near-specific precipitating reagent for K^+ and NH_4^+ to give a stoichiometric composition. The precipitate is easy to be filtered and dried at 105～120℃ to a constant mass.

(2) Reagents forming coordination compound precipitates

Some organic precipitating reagents have coordinating groups such as $—NH_2$, R_2CO, R_2CS, etc., and can form sparingly soluble coordination compounds with metal ions. Generally, this type of coordination compounds have stoichiometric compositions and are used to selectively precipitate metal ions. For example, dimethylglyoxime can specifically precipitate nickel (Ⅱ) from the ammonia solution. The produced precipitate is easy to obtain a constant mass upon drying. This reaction (as given below) is often used for gravimetric analysis of nickel.

$$2\ \begin{array}{c} H_3C-C-C-CH_3 \\ \quad\ \ \| \quad \| \\ \quad\ \ N \quad N \\ HO \qquad\quad OH \end{array} + Ni^{2+} \longrightarrow \begin{array}{ccccc} & H_3C & & CH_3 & \\ & C & & C & \\ & \| & & \| & \\ & O-N & & N-O & \\ H & & Ni & & H \\ & O-N & & N-O & \\ & \| & & \| & \\ & C & & C & \\ & H_3C & & CH_3 & \end{array} + 2H^+$$

8-hydroxyquinoline (oxine) can be used to selectively and quantitatively precipitate Al^{3+}, Zn^{2+} and Mg^{2+} by controlling pH. This reaction with Mg^{2+} is

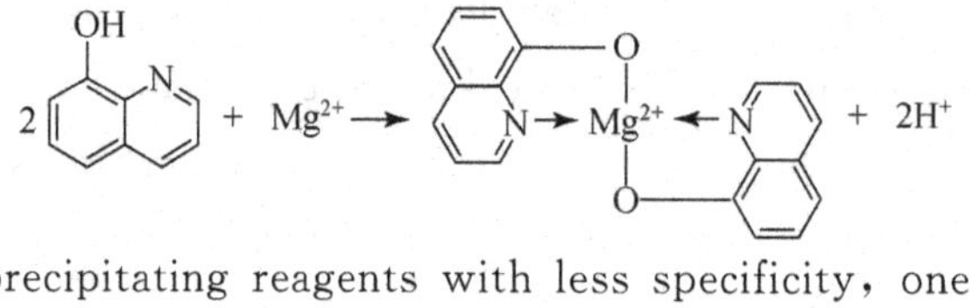

$$2\,C_9H_7NO + Mg^{2+} \rightarrow Mg(C_9H_6NO)_2 + 2H^+$$

When using precipitating reagents with less specificity, one needs to combine pH adjustment or masking approaches to selectively precipitate the metal ion of interest. Sometimes, a derivative compound can be used if the mother compound has low specificity toward metal ions.

The application examples of organic precipitating reagents are given in Table 7.4.

Table 7.4 Selected Gravimetric Methods Based on Precipitation with Organic Precipitating Reagents

Analyte	Precipitating Reagents	Structure	Precipitation Form	Weighing Form
Ni^{2+}	Dimethylglyoxime	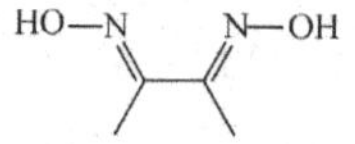 HO—N N—OH	$Ni(C_4H_7O_2N_2)_2$	$Ni(C_4H_7O_2N_2)_2$
Mg^{2+}	8-hydroxyquinoline	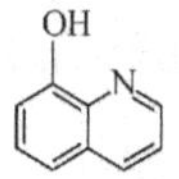 OH N	$Mg(C_9H_6ON_2)_2 \cdot 2H_2O$	$Mg(C_9H_6ON_2)_2 \cdot 2H_2O$
K^+	Sodium tetraphenylborate	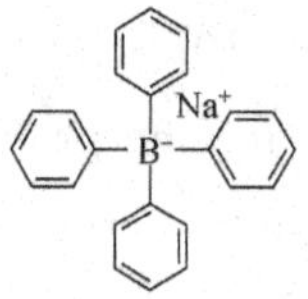 Na^+ B^-	$K[(C_6H_5)_4B]$	$K[(C_6H_5)_4B]$
Co^{2+}	1-nitroso-2-naphenol	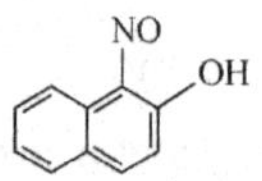 NO OH	$Co(C_{10}H_6O_2N)_3$	Co or $CoSO_4$
Cu^{2+}	Cupron	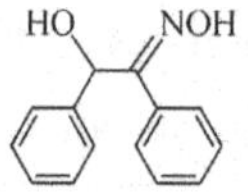 HO NOH	$Cu(C_{14}H_{11}O_2N)$	$Cu(C_{14}H_{11}O_2N)$

Chapter 7 Questions and Problems

7.1 Calculate solubility of the following slightly soluble compounds in a specific media (K_{sp} at $I=0.1$ are used for questions (2)~(5)).

(1) ZnS (α type) in water;

(2) CaF_2 in 0.01 $mol \cdot L^{-1}$ HCl (the amount of HCl for dissolving CaF_2 can be neglected);

(3) AgBr in 0.01 $mol \cdot L^{-1}$ NH_3;

(4) $BaSO_4$ in 0.01 $mol \cdot L^{-1}$ EDTA at pH 7.0;

(5) AgCl in 0.1 $mol \cdot L^{-1}$ HCl.

7.2 In a $MgNH_4PO_4$ saturated solution, $[H^+] = 2.0 \times 10^{-10}$ $mol \cdot L^{-1}$, and $[Mg^{2+}] = 5.6 \times 10^{-4}$ $mol \cdot L^{-1}$. Calculate K_{sp}. ($pK_{a1} \sim pK_{a3}$ for H_3PO_4 are 2.16, 7.21, 12.32; $pK_a(NH_4^+) = 9.25$)

7.3 NaI crystals are slowly added to a solution that is 0.100 $mol \cdot L^{-1}$ $Pb(NO_3)_2$ and 0.100 $mol \cdot L^{-1}$ $AgNO_3$. What will precipitate first, AgI or PbI_2? Will any PbI_2(s) have precipitated when 99.90% of the Ag^+ is precipitated?

7.4 A mixture is composed of KCl and KBr. A 0.3028 g of the mixture is dissolved in water and pH is adjusted. The dissolved solution is then titrated with 30.20 mL of 0.1014 $mol \cdot L^{-1}$ $AgNO_3$ standard solution using the Mohr method. Calculate percentage of KCl and KBr in this mixture.

7.5 A 0.2266 g of chloride salt was mixed with 30.00 mL of 0.1121 $mol \cdot L^{-1}$ $AgNO_3$. Excess $AgNO_3$ is then titrated with 6.5 mL of 0.1158 $mol \cdot L^{-1}$ NH_4SCN standard solution. Calculate percentage of chloride in this salt.

7.6 A pesticides contains arsenic. A 0.2000 g of such pesticide was weighed and oxidized in HNO_3 to be arsenic acid (H_3AsO_4). The resulted solution was adjusted to a neutral pH and the arsenic was precipitated as Ag_3AsO_4 by $AgNO_3$. The precipitate was filtered, washed, and dissolved in diluted HNO_3. Titration was carried out by the Volhard method and 33.85 mL of 0.1180 $mol \cdot L^{-1}$ NH_4SCN was used. Calculate percentage of arsenic as As_2O_3 in the pesticide formulation.

7.7 A sample contains potassium bromate ($KBrO_3$), potassium bromide (KBr) and non-reactive reagents. A 1.000 g sample was dissolved and diluted to 100.0 mL. A 25.00 mL of the solution was pipetted and BrO_3^- was reduced to Br^- by Na_2SO_3 in H_2SO_4 media. After excess Na_2SO_3 was separated, the solution was adjusted to neutral pH and 10.51 mL of 0.1010 $mol \cdot L^{-1}$ $AgNO_3$ was used to titrate total Br^- by the Mohr method. Another aliquot of 25.00 mL sample solution was mixed with H_2SO_4 and Br_2 produced was expelled by heating. The resulted solution was adjusted to neutral pH and 3.25 mL of above $AgNO_3$ standard solution was used to titrate excess Br^- by the Mohr method. Calculate the percentage of $KBrO_3$ and KBr.

7.8 Write the experimental design for analysis of each component in the mixture of HCl+HAc.

7.9 What will the result be if the Mohr method is used to indicate the endpoint in the following titrations with $AgNO_3$ standard solution? State the reasons.

(1) Determine the chloride in a solution at pH 4 or 11;

(2) Determine the chloride in the NH_4Cl solution;

(3) Determine the content of NaCl in a mixture of NaCl and Na_2SO_4.

7.10 What will the result be, if the Volhard method is used to determine Cl^- without the presence of nitrobenzene and if the endpoint is indicated by the Fajans method using eosin?

7.11 Write the experimental design for the analysis of each component in the mixture by listing the titrand, the titrant, the indicator and the major experimental conditions.

(1) NH_4Cl; (2) $BaCl_2$; (3) $FeCl_3$; (4) $CaCl_2$; (5) NaCl + Na_3AsO_4; (6) NaCl+Na_2SO_3.

7.12 List the factors affecting the solubility of a precipitate. How to avoid loss due to solubility in precipitation gravimetric analysis? The K_{sp} values for AgCl and $BaSO_4$ are comparable. Why does $BaSO_4$ crystalline precipitate in a crystalline form, but AgCl precipitate only in s colloidal form?

7.13 Why is dilute HNO_3 solution used instead of H_2O or NH_4NO_3 to wash AgCl precipitate?

7.14 What is the purpose of aging in preparing a crystalline precipitate? Is aging necessary for preparing a colloidal precipitate? State the reason.

7.15 Describe the procedure by which a constant mass is obtained for a weighing form.

7.16 What will the result be in gravimetric analysis of the following analytes, if the following species coprecipitate with $BaSO_4$? (high, low, or no errors)

Analyte \ Coprecipitate	$BaCl_2$	Na_2SO_4	H_2SO_4
S			
Ba^{2+}			

7.17 Calculate the gravimetric factors in the following gravimetric analysis.

Analytes	Weighing Forms
FeO	Fe_2O_3
KCl($\rightarrow K_2PtCl_6 \rightarrow$Pt)	Pt
Al_2O_3	$Al(C_9H_6ON)_3$
P_2O_5	$(NH_4)_3PO_4 \cdot 12MoO_3$

7.18 The purity of the Mohr's salt, $(NH_4)_2SO_4 \cdot FeSO_4 \cdot 6H_2O$, is determined by gravimetry. To make sure that the weighing error for the weighing form, Fe_2O_3, is less than 0.1%, how many grams of the Mohr's salt need to be used if the variability for weighing on an analytical balance is 0.2 mg?

7.19 A 1.2030 g air-dried gypsum sample is dried to obtain 0.0208 g absorbed water and then ignited to obtain 0.2424 g water of crystallization. Calculate the percentage of $CaSO_4 \cdot 2H_2O$ in a dry state before absorbing water.

7.20 Ba^{2+} in a sample is analyzed by $BaSO_4$ precipitation gravimetry. Because $BaSO_4$ precipitate is partially reduced to BaS during ignition, the experimental result for Ba^{2+} is 99% of the truc value. Calculate percentage of BaS in the $BaSO_4$ weighing form.

CHAPTER 8

SPECTROPHOTOMETRY

8.1 Principle of Spectrochemical Analysis

8.1.1 Properties of Electromagnetic Radiation

8.1.2 Interaction of Electromagnetic Radiation with Matter

8.1.3 Beer's Law, the Quantitative Principle of Light Absorption

8.1.4 Limitations to Beer's Law

8.2 Principles of Instrumentation

8.2.1 Instrumentation

8.2.2 Instrumental Errors in Absorption Measurement

8.3 Applications of Spectrophotometry

8.3.1 Single Component Analyses

8.3.2 Multicomponent Analyses

8.3.3 Spectrophotometric Titrations

8.3.4 Studies of Complex Formation in Solutions

8.3.5 Measurements of Dissociation Constants of Organic Acids/Bases

In the former chapters, we have discussed analytical methods based on measurement of mass and volume. In this chapter, we introduce methods based on the measurement of **electromagnetic radiation** produced or absorbed by molecular species of interest, called **spectroscopy** or **spectrochemical methods.** Spectroscopic methods can be classified according to the region of the electromagnetic spectrum involved in the measurements, including γ-ray, X-ray, ultraviolet (UV), visible, infrared (IR), microwave, and radio-frequency techniques. This chapter deals with the basic principles that are necessary to understand measurements made with electromagnetic radiation, particularly with the absorption of UV and visible radiation, the instrumentation for UV and visible spectrum measurement, and also the applications of UV and visible spectrochemical method.

8.1 PRINCIPLE OF SPECTROCHEMICAL ANALYSIS

8.1.1 Properties of Electromagnetic Radiation

Electromagnetic radiation in the UV/visible and sometimes IR region is often called light. Electromagnetic radiation can be described in terms of its **wavelike properties** as well as its **particlelike properties.** The wavelike properties include wavelength, frequency, velocity and amplitude. Light does not require supporting medium for its transmission. With particlelike properties, electromagnetic radiation can be treated as discrete packets of energy or particles called **photons** or **quanta.** The wavelike and particlelike perspectives about the electromagnetic radiation are complementary to each other.

1. Wavelike Properties

The wavelike nature of electromagnetic radiation can be described as a model consisting of perpendicularly oscillating electric and oscillating magnetic fields as shown in Figure 8.1A. The electric field for a single-frequency wave oscillates sinusoidally in space and time as shown in Figure 8.1B. The distance between the adjacent peaks (or troughs) of the wave shown in Figure 8.1B is called the wavelength, λ, that can be given in meters (m), centimeters (cm), micrometers (μm) or nanometers (nm). The frequency of electromagnetic radiation, ν in hertz (Hz), is the rate of number of wave peaks passed by one given point in the space and is independent of the medium traversed.

Wavelength and frequency are related in Equation (8-1).

$$\text{wavelength}(\lambda, \text{cm/peak}) \times \text{frequency}(\nu, \text{peaks/s}) = \text{velocity}(v, \text{cm/s}) \quad (8\text{-}1)$$

where v is the velocity that light travels, depending on the medium traversed. $v=c/n$, where n is the refractive index of the medium and c is the velocity in a vacuum where light travels at its maximum velocity which is 2.998×10^8 m · s^{-1}. The wavenumber is another way to describe electromagnetic radiation. Wavenumber is defined as the number of waves per centimeter (cm^{-1}) and equals $1/\lambda$. It is proportional to the frequency and is given the symbol $\tilde{\nu}$.

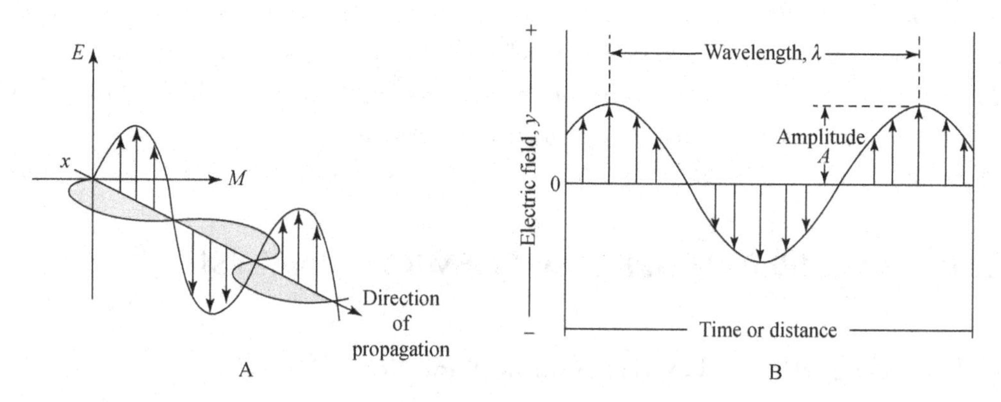

Figure 8.1 Schematic illustration of an electromagnetic wave. A, An oscillating electric field (E axis) and an oscillating magnetic field (M axis) are moving in the x direction with velocity (c). B, the electric field oscillations.

2. Particlelike Properties

Considering the particle nature of the electromagnetic radiation, the energy of a photon associated with an electromagnetic wave is described in Equation (8-2).

$$E = h\nu = h\frac{c}{\lambda} = hc\tilde{\nu} \qquad (8\text{-}2)$$

where h is the Plank's constant (6.626×10^{-34} J · s).

3. Radiant Power and Intensity

The **radiant power** of a beam of electromagnetic radiation is the energy that reaches a given area per unit time. **Intensity** is defined as radiant power from a point source per unit solid angle. Both radiant power and intensity are related to the square of the amplitude of a wave such as that shown in Figure 8.1.

8.1.2 Interaction of Electromagnetic Radiation with Matter

The interactions of electromagnetic radiation with matter include absorption and emission processes, and other types of interactions such as reflection, refraction, elastic scattering, interference, and diffraction. The discussion in this chapter is limited to the absorption processes and quantitation of the analyte based on light absorption.

1. The Electromagnetic Spectrum

The electromagnetic spectrum covers many orders of magnitude in frequency or wavelength and can be used for spectroscopic analysis as shown in Figure 8.2. Spectrochemical methods that use ultraviolet, visible, and infrared radiation are often called optical methods because these methods share many common features of instrumentation and also the similarities in the way we view the interactions of the three types of radiation with matter. The region for ultraviolet, visible, and infrared spectra is given in Table 8.1.

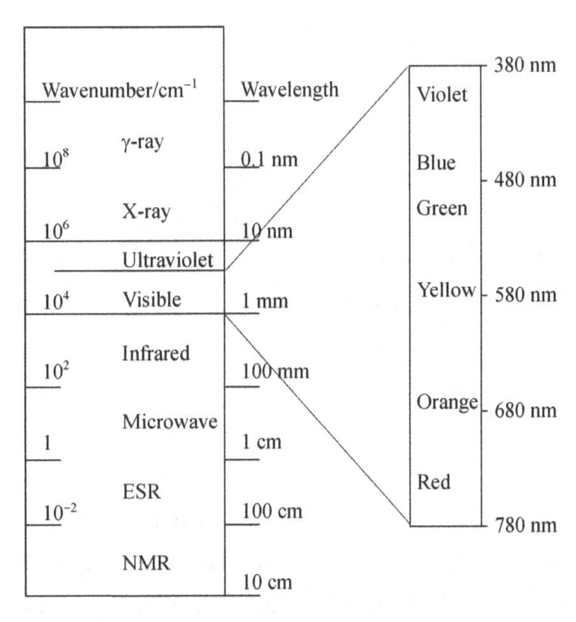

Figure 8.2 The electromagnetic spectrum.

Table 8.1 The Region for Ultraviolet, Visible, and Infrared Spectra

Region	Wavelength Range	Wavenumber Range/cm^{-1}
Ultraviolet (UV)	180～380 nm	$(5.6\sim2.6)\times10^4$
Visible	380～780 nm	$(2.6\sim1.3)\times10^4$
Near-infrared (NIR)	0.78～2.5 μm	$1.3\times10^4\sim4000$
Mid-infrared (IR)	2.5～50 μm	4000～200

2. Light Absorption and Emission

Any signal that can be related to the analyte may form the basis for an analytical method. The **absorption** and **emission** of electromagnetic radiation can be used to establish analytical methods. In light absorption process, the electromagnetic radiation from a source is absorbed by the analyte and results in a decrease of radiant power which reaches a detector. In the light emission process, the electromagnetic radiation emits from the analyte, resulting in an increase in the radiant power that reaches a detector.

The partial energy-level diagram of a polyatomic molecule and the association with the light absorption in infrared, visible and ultraviolet region can be schematically illustrated in Figure 8. 3. Before irradiated by the light, the analyte is predominantly in the lowest energy level state called **ground state** (E_0). The analyte can be promoted to one of the higher energy level states called **excited state** (E_1 or E_2). If the energy of electromagnetic radiation impinging on the molecule equals to the difference in energy between the excited state and the ground state, some of this energy may be absorbed and causes the molecule to reach this excited state. The quantitative and structural information about the analyte can be acquired by measuring the electromagnetic radiation emitted as the molecule returns to the ground state or by measuring the amount of electromagnetic radiation absorbed as a result of excitation.

In addition to the electronic energy (denoted as E_e), a molecule exhibits two other types of radiation-induced transitions: **vibrational transitions** associated with the vibration of chemical bonds and **rotational transitions** associated with the rotational motion of a molecule around its center of gravity and the energy associated with these two types of transitions are denoted as E_v and E_r. Thus the molecule will exhibit various vibrational and rotational energy states as shown in Figure 8. 3. Absorption of a photon can lead to a change in one of these states. The total energy E associated with a molecule is then given by $E = E_e + E_v + E_r$, with the energy-level differences in the order of $\Delta E_e > \Delta E_v > \Delta E_r$. The infrared radiation induces transitions in the vibrational and rotational states associated with the ground electronic state of the molecule irradiated. The ultraviolet and visible radiations can induce transitions in electronic states.

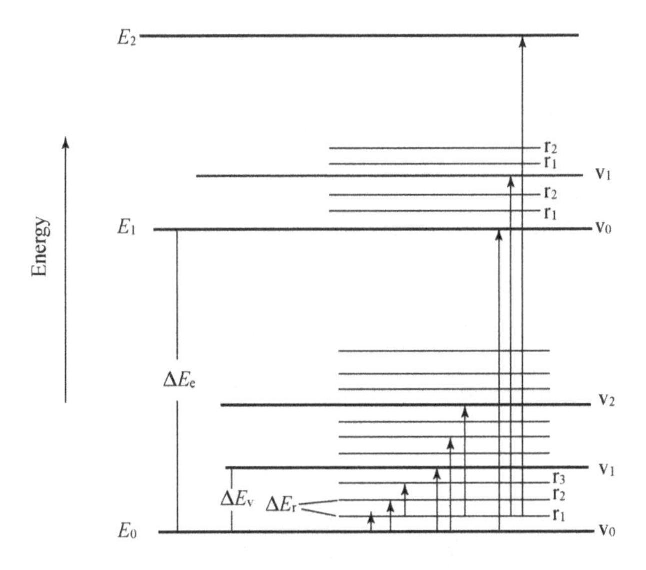

Figure 8. 3 Energy level diagram illustrating energy changes associated with absorption of electromagnetic radiation.

As can be seen from Figure 8. 3, all the three type of transitions are quantized; that is, the energy levels are discretely distributed. Therefore, the molecular absorption spectra in the ultraviolet and visible regions consists of absorption bands composed of closely spaced lines. A real molecule has many more energy levels than are shown in Figure 8. 3, thus the absorption bands consists of a large number of lines. In the solution, the light absorbing species are solvated and this further blur the energy of the quantum states, resulting in seemingly continuous absorption peaks. The spectra of 1, 2, 4, 5-tetrazine in the gas phase, hexane solution and aqueous solution are shown in Figure 8. 4. In the gas phase, the individual transition molecules are sufficiently separated from each other to vibrate or rotate freely, thus some absorption peaks from transitions among certain vibrational and rotational states appear in the spectrum. In hexane solution, the molecules are not able to rotate freely, thus less fine structure are presented in the spectrum. In aqueous solution, because the frequent collisions and electrostatic interactions between tetrazine and water molecules cause the vibrational levels to be modified energetically in an irregular way, the spectrum appears as a single broad peak.

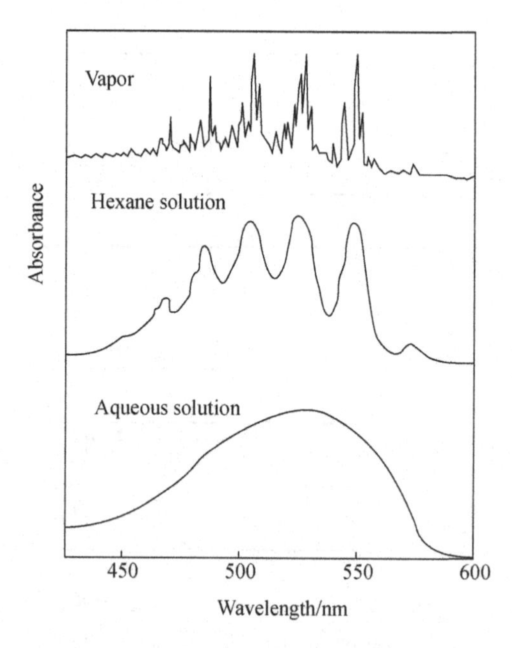

Figure 8.4 The absorption spectra of 1,2,4,5-tetrazine.

White light can be produced by combining all colors of the spectrum at once. It is also possible to make white light by combining only two colors. Any two colors that produce white light, such as blue and yellow, are known as **complementary colors**. When a beam of white light passes through a solution such as $Fe(SCN)^{2+}$, the solution absorbs green from the radiation and exhibits red, its complementary color, which means the red color is transmitted. Therefore, the color of a solution is related to its absorption spectrum. Examples of absorption wavelength and the complementary color are given in Table 8.2.

Table 8.2 Complementary Color in the Visible Spectrum

Wavelength Region Absorbed/nm	Color of Light Absorbed	Complementary Color Transmitted
400～450	Violet	Yellow-green
450～480	Blue	Yellow
480～490	Green-blue	Orange
490～500	Blue-green	Red
500～560	Green	Purple
560～580	Yellow-green	Violet
580～610	Yellow	Blue
610～650	Orange	Green-blue
650～760	Red	Blue-green

8.1.3 Beer's Law, the Quantitative Principle of Light Absorption

1. Measuring the Transmittance and Absorbance

The absorption of electromagnetic radiation can be measured by measuring the radiant power using a detector with or without the analyte solution in the beam of electromagnetic radiation. When place a cell containing solution of analyte in the light path, reflectance losses on the surface of the container, scatting losses in all directions from the surface of large molecules or particles in the solution, and loss by solvent absorption often occur. To compensate for the effects mentioned above, the power of the beam transmitted through a cell containing everything but the analyte is measured and is given the symbol of P_0. Therefore, the experimental **transmittance** T is calculated as

$$T = P/P_0 \qquad (8\text{-}3)$$

where P_0 and P represents the power of the beam that transmitted through the solutions of the blank and the analyte, respectively. The **absorbance** A is defined as

$$A = -\lg T = \lg \frac{P_0}{P} \qquad (8\text{-}4)$$

2. Beer's Law

The Beer's law is that when a parallel beam of monochromatic radiation passes through a solution of analyte, the absorbance is linear to both thickness (**light path**) b of solution and concentration of the analyte c, as expressed by Equation (8-5)

$$A = abc \qquad (8\text{-}5)$$

where a is proportionality constant called the **absorptivity.** Absorbance is a unitless quantity, and if b is in centermeters and c is in moles per liter, the proportionality constant is called the **molar absorptivity** (ε) with the unit of $L \cdot mol^{-1} \cdot cm^{-1}$. Thus,

$$A = \varepsilon bc \qquad (8\text{-}6)$$

When a parallel beam of monochromatic radiation with power P_0 passes through an absorbing solution containing n absorbing particles with a length of b (Figure 8.5), the power of the radiation is decreased to P because absorption occurs. If we divide the absorbing solution into numerous cross-sections with an infinitesimal layer thickness of dx and an area of S, there are dn absorbing particles in this section, thus the light experiences a decrease in power dP (Figure 8.5). If dS is designated to be the total irradiated and absorbing area within the section, the probability of capturing photons within the section can be expressed as the ratio of the capture area to the total area, dS/S. The fraction of power absorbed when radiation pass through the section, $-dP_x/P_x$, equals the probability of capture, where P_x is the power of the beam entering the section and dP_x represents the power reduced within this section. Thus,

$$-\frac{dP_x}{P_x}=\frac{dS}{S} \tag{8-7}$$

dS is the total capture areas for particles within the section, and is proportional to the number of the absorbing particles (dn). Thus,

$$dS=\alpha\, dn \tag{8-8}$$

where α is called the capture cross-section.

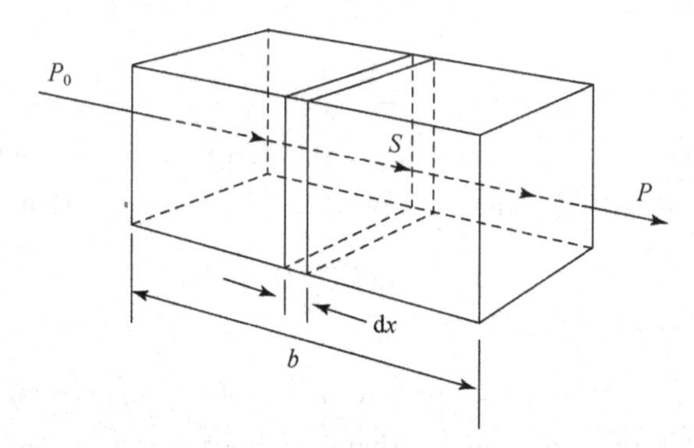

Figure 8. 5 Absorption of radiation with an initial power of P_0 by an analytical solution containing c mol · L^{-1} of absorbing particles and a path length of b cm.

Substituting Equation (8-8) into Equation (8-7) yields

$$-\frac{dP_x}{P_x}=\frac{\alpha\, dn}{S} \tag{8-9}$$

The probability of absorption of an analytical solution with an area S and a thickness of b containing n absorbing particles can be obtained by integration over the interval between 0 and n.

$$-\int_{P_0}^{P}\frac{dP_x}{P_x}=\int_0^n\frac{\alpha dn}{S}$$

that is

$$-\ln\frac{P}{P_0}=\frac{\alpha n}{S} \tag{8-10}$$

or

$$\lg\frac{P_0}{P}=\frac{0.434\alpha n}{S} \tag{8-11}$$

The cross-sectional area S can be expressed in terms of the volume of the solution in cm^3 and its thickness b in cm. Thus

$$S = V/b\ (cm^2) \tag{8-12}$$

The number of absorbing particles n can be expressed in terms of volume of the solution in cm^3 and concentration in $mol \cdot L^{-1}$ as

$$n = 6.022 \times 10^{23} c(V/1000) \tag{8-13}$$

Substituting Equations (8-12) and (8-13) into Equation (8-11) yields

$$\lg \frac{P_0}{P} = \frac{0.434\alpha 6.022 \times 10^{23} c(\cancel{V}/1000)}{\cancel{V}/b}$$

Collecting the constants in the equation into a single term ε gives

$$A = \lg \frac{P_0}{P} = \varepsilon bc \tag{8-14}$$

which is Beer's law. The molar absorptivity ε has the unit of $L \cdot mol^{-1} \cdot cm^{-1}$, and represents the probability that an analyte will absorb a photon of given energy. As a result, both α and ε depend on the wavelength of the electromagnetic radiation.

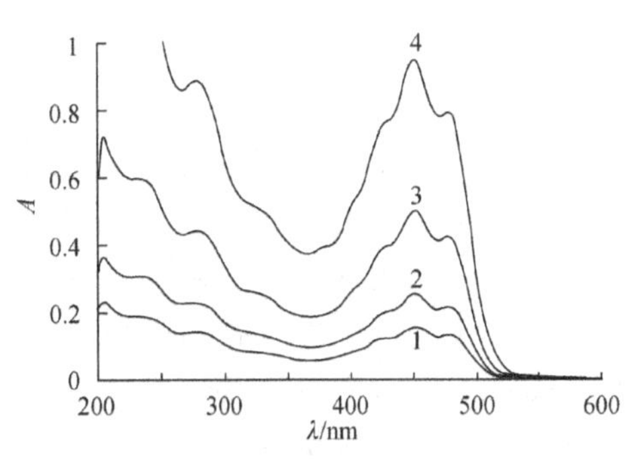

Figure 8.6 The absorption spectra of β-carotene in ethanol solution. The concentrations for spectra 1~4 are 10, 30, 50, and 100 $nmol \cdot L^{-1}$, respectively.

The spectrum of an analyte can be measured by measuring the absorption at every single wave of electromagnetic radiation (single wavelength), then plotting the absorbance against the wavelength as illustrated in Figure 8.6. When the concentration increases, the absorbance at the measured wavelength also increases. A plot of molar absorptivity ε versus wavelength is independent of concentration and is characteristic for a given molecule, which can be sometimes used for identifying or verifying the identity of a species with a particular structural information.

3. Applying Beer's Law to Multicomponent Samples

Beer's law can be extended to solutions containing more than one absorbing entities provided that there is no interaction between the components. The total absorbance of the solution at a given wavelength is the sum of the individual absorbance as expressed in Equation (8-15)

$$A_{total} = A_1 + A_2 + \cdots + A_n = \varepsilon_1 bc_1 + \varepsilon_2 bc_2 + \cdots + \varepsilon_n bc_n \tag{8-15}$$

where the subscripts refer to the absorbing components, 1, 2, $\cdots$, n.

8.1.4 Limitations to Beer's Law

According to Beer's law, A plot of absorbance at a given wavelength versus concentrations of the analyte gives a straight line with an interception of 0 and a slope of εb. However, deviations from linearity are often observed, which are attributed to **fundamental**, **chemical** and **instrumental** sources.

1. Fundamental Limitations

Beer's law is only valid for the dilute solution. At concentrations exceeding about 0.01 $mol \cdot L^{-1}$, the distance between individual absorbing species is reduced, resulting interaction between the absorbing species, thus the ε value is altered.

2. Chemical Deviations

Deviations from Beer's law can also be observed when the absorbing species undergo association, dissociation, reaction with the solvent to give products with absorption spectra different from that of the analyte. In these cases, the concentration of the absorbing species changes and the produced species from the above reaction may not have the same molar absorptivity as the analyte.

For example, for a weak acid (HA) with a total concentration of c, the absorbance at measuring wavelength is $A = \varepsilon_{HA} bc$ if absorbance is measured at $pH \ll pK_a$. However, if measurement is carried out at pH about pK_a, HA undergoes dissociation as

$$HA + H_2O \rightleftharpoons A^- + H_3O^+$$

If both HA and A^- absorb light at measuring wavelength, then

$$A = A_{HA} + A_{A^-} = \varepsilon_{HA} b[HA] + \varepsilon_{A^-} b[A^-]$$

If $\varepsilon_{HA} = \varepsilon_{A^-}$, then $A = \varepsilon_{HA} b([HA] + [A^-]) = \varepsilon_{HA} bc$. Thus, no deviation is observed. If $\varepsilon_{HA} \neq \varepsilon_{A^-}$, then $A \neq \varepsilon_{HA} bc$. Thus, deviation will be observed.

Sometimes when the solution of analyte contains high concentration of other species, particularly the electrolyte, deviation is observed because electrostatic interactions between the analyte and the electrolyte lead to deviation from Beer's law.

3. Instrumental Deviations

There are two principal instrumental limitations to Beer's law. Firstly, Beer's law strictly applies only when the measurements are made with monochromatic radiation. In reality a continuous light source is used in conjunction with a wavelength selector to obtain a narrow band of wavelength to irradiate the analytical solution. However, even the best wavelength selector can only pass radiation with a small, finite bandwidth. Therefore, with polychromatic radiation a negative deviation to Beer's law can be observed as shown in Figure 8.7.

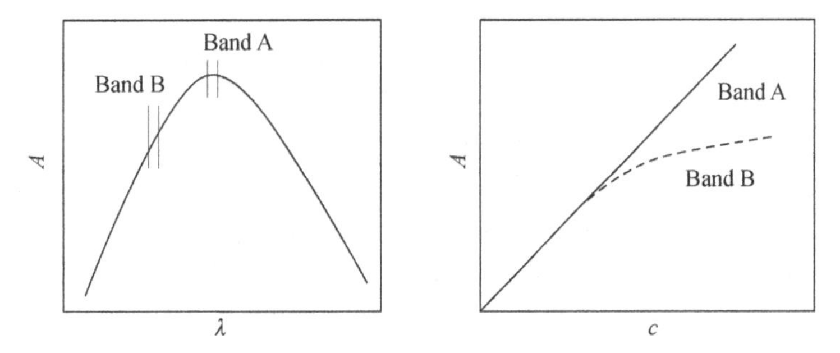

Figure 8.7 The effect of polychromatic radiation on Beer's law.

Stray light is the second source causing deviation to Beer's law. Stray light is the radiation from the instrument that is outside the nominal wavelength band chosen for the measurement. It often results from scattering and reflection off the surface of gratings, lenses or mirrors, filters and windows. With the stray light, the observed absorbance is given by

$$A = \lg \frac{P_0 + P_s}{P + P_s}$$

where P_s is the radiant power of the stray light.

8.2 PRINCIPLES OF INSTRUMENTATION

The basic principles of relating absorption of light to an analyte's concentration are developed. We now need to realize the measurement of light absorption in the laboratory. The section here discusses the instrumentation for measurement of ultraviolet and visible light absorption.

8.2.1 Instrumentation

A spectrophotometer has to have (1) a source providing a stable output of radiant power, (2) a monochromator to allow selection of wavelength, (3) a sample compartment, and (4) a radiant-power detector which converts the optical signal to the electronic signal. A block diagram of components of a spectrophotometer is shown in Figure 8.8.

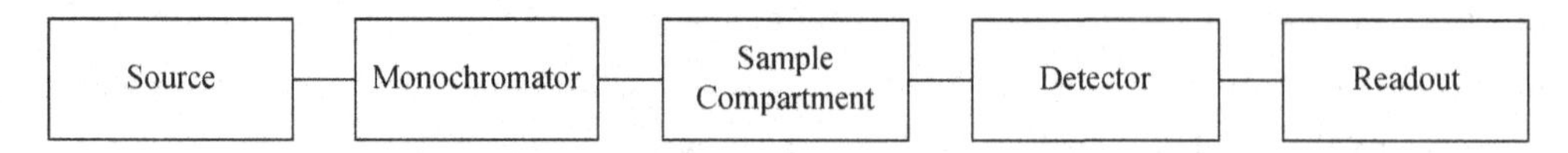

Figure 8.8 The Block diagram of components of a spectrophotometer.

1. Sources

A light source should have high output stability with its radiant-power peak in the wavelength region of interest. The spectra of sources are given in Figure 8.9. As can be seen from Figure 8.9, the xenon arc source has high radiant power over a wide wavelength region although the output drops rapidly at wavelength below 300 nm. The tungsten filament lamp has modest output in the visible region and little in the UV region, but the tungsten lamp is less expensive and can provide a reliable, highly stable source, thus this lamp is widely used for measurements in the visible region. Sublimation of tungsten limited the lifetime of the common tungsten lamp. The lifetime of tungsten/halogen lamp can be more than doubled with iodide in the filament that reacts with sublimed tungsten to yield gaseous tungsten iodide, WI_2 that diffuses back to the hot filament. For measurement in the UV region, electric-discharge sources such as hydrogen and also deuterium lamps are used. The passage of electrons through the gas excites the gaseous molecules. Deuterium lamps have several fold greater output than hydrogen with the same spectral range.

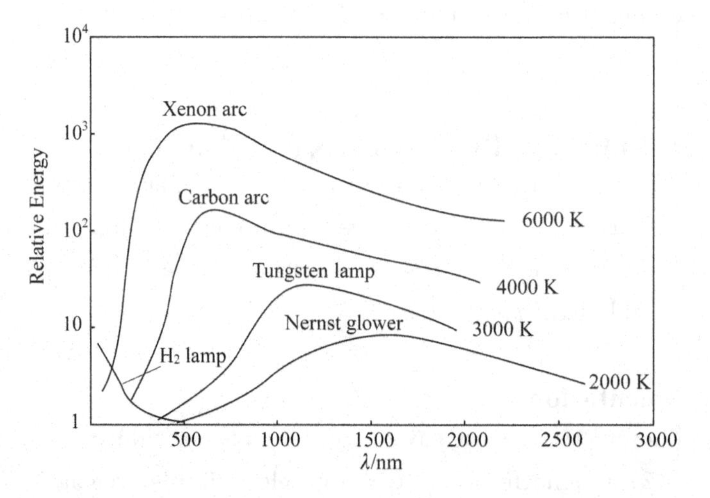

Figure 8.9 Sources of radiant power for spectrophotometry.

2. Monochromators

A monochromator consists of lenses or mirrors to focus the radiation, entrance and exit slits to restrict unwanted radiation and help control the spectral purity of the radiation emitted from the monochromator, and a dispersing element to separate the wavelengths of the polychromatic radiation from the source so that the desired band can be detected and measured. There are two types of monochromators: prism monochromator and grating monochromator. A diffraction grating is generally used to disperse the radiation into separate wavelengths as shown in Figure 8.10. By rotating the grating, different wavelengths of light can be made to pass through an exit slit. Before gratings were available, prisms were used for this purpose with the exit slit being moved along the focal plane (Figure 8.11). The wavelength range passing by a monochromator is called the **spectral band-pass** or effective bandwidth. Many monochromators are equipped with adjustable slits so that bandwidth and the amount of measured radiant power can be controlled by adjusting the slit. A high-quality spectrophotometer will allow bandwidth of less than 1 nm, while less expensive spectrophotometer may have fixed bandwidths of greater than 20 nm.

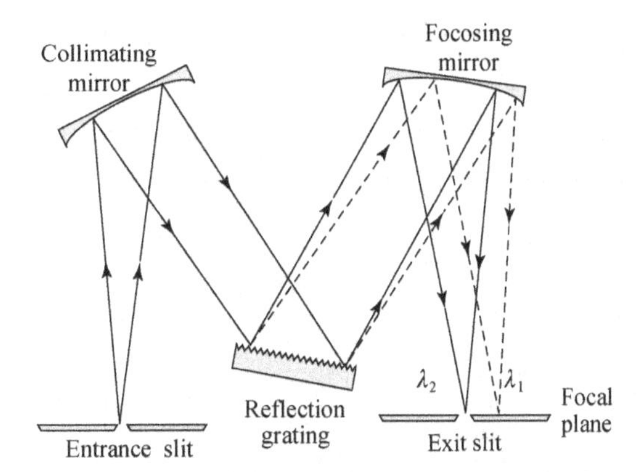

Figure 8.10 Grating type monochromator—a Czerny-Turner design ($\lambda_1 > \lambda_2$).

Gratings are the major means of dispersion in modern spectrophotometry. UV, visible and infrared radiation can be dispersed with either a transmission grating or a reflection grating. The reflection gratings as shown schematically in Figure 8.12 are common in spectrophotometers. The gratings used are often replica gratings consisting of a large number of parallel lines (grooves) produced by depositing a thin film of a metal, e.g. aluminum or by coating a film of an epoxy resin onto a ruled surface. The spacing between grooves d must be comparable to the wavelength of

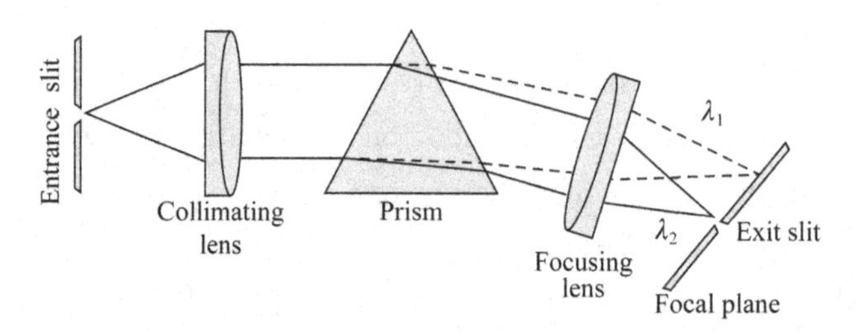

Figure 8.11 Prism type monochromator—a Bunsen design ($\lambda_1 > \lambda_2$).

light. The angle of reflection r is related to the wavelength of the incoming radiation by

$$n\lambda = d(\sin i + \sin r) \tag{8-16}$$

where n is the order of diffraction and is an integer.

According to Equation (8-16), light will be observed at the same angle corresponding to $n=1$ (first order), $n=2$ (second order), and so on. Generally, first-order line is allowed pass through the slit while light from other orders is eliminated by the filters. A grating with 500 ~ 1200 grooves per millimeter is typically used for UV and visible region, and 60~100 grooves per millimeter is used for infrared region.

Holographic gratings are produced by exposing a photo resistant layer on a suitable substrate to the interference pattern produced by two monochromatic laser beams, followed by a photographic development to produce grooves and then a reflective coating process. Holographic gratings have less problems of stray light and "ghosting" (double images) because of smoother lines are produced by laser technology.

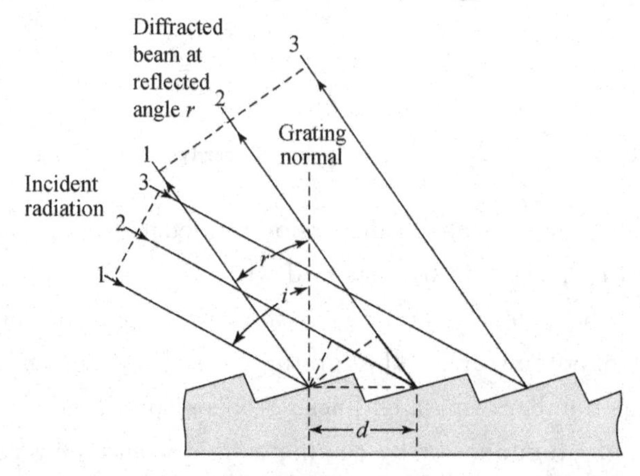

Figure 8.12 Diagram of a reflection grating.

Filters absorb all but a restricted band of radiation from a continuum of light. The bandwidth of wavelength selector is defined as the width of the band of radiation in wavelength units at half-peak height (Figure 8.13). Interference filters and absorption filters used in spectroscopy. The bandwidths of both filters are illustrated in Figure 8.13. Interference filters are used with UV and visible radiation, as well as with wavelengths up to about 14 μm in the infrared region. An interference filter can provide a relatively narrow bandwidth of about 5~20 nm of radiation. Absorption filters are less expensive, but can be used only in visible region. Absorption filters are usually made of colored glass plates which provide an effective bandwidth from 30 to 250 nm.

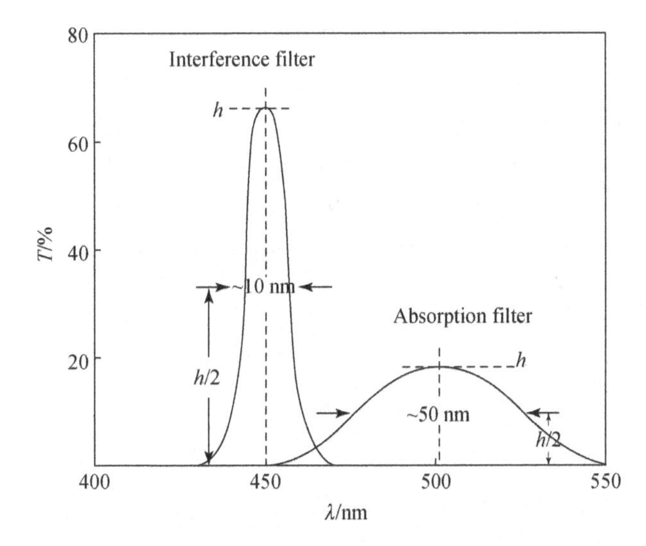

Figure 8.13 The bandwidths of the interference filter with a nominal wavelength of 448 nm and the absorption filter with a nominal wavelength of 500 nm.

3. Cells

Cells for containing the analytical sample must be transparent in the wavelength region measured. For measurements in the UV region, quartz or fused-silica cuvettes are used. For visible region, glass, quartz, or fused-silica can be used. NaCl are frequently used for infrared measurement.

4. Detectors

Radiation detectors generally convert a light signal to an electrical signal by a **transducer.** A major requirement is that the resulting signal S should be directly proportional to the original radiant-power signal, P. This can be expressed as

$$S = kP + d$$

The proportional constant (k) represents the sensitivity of a detector in terms of electrical response per unit of radiant power input. It is desirable that k is high so that low light intensity can be measured. Usually, the detectors exhibit a small constant response called the **dark current** (d) even when no light strikes the detector. Spectrophotometers usually have a means of compensating for dark current so that it can be automatically subtracted to give a response of $S = kP$.

Photon (or photoelectric) detectors such as **phototube** and **photomultiplier tube** (PMT) can be used for converting photon signals in the UV and visible region into electrical signals based on the photoelectric effect. As shown in Figure 8.14, a phototube consists of a photoemissive cathode and an anode. A high voltage is applied between the anode and cathode. When a photo enters the window of the tube and strikes the cathode, an electron is emitted and attracted to the anode, causing photocurrent to flow that can be amplified and measured. The response of photoemissive material (usually an alkali metal or a metal oxide) is wavelength dependent, and different phototubes are available for different region of the spectrum measured.

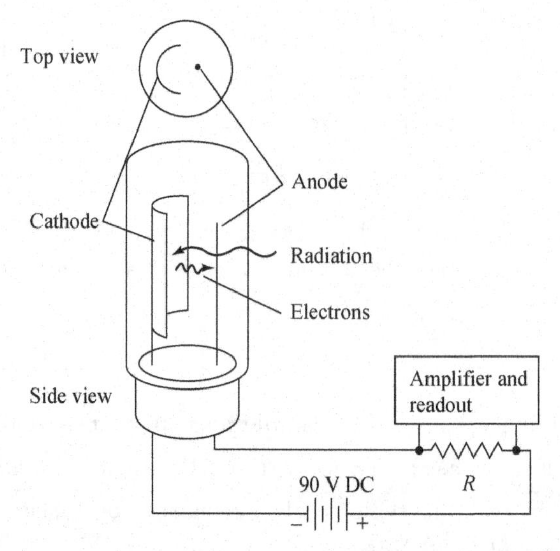

Figure 8.14 Schematic diagram of a phototube with an accompanying circuit. The current due to photons striking the photosensitive cathode is measured by the potential drop in R and then amplified.

Much higher sensitivity can be achieved by using the photomultiplier tubes. In principle, the conversion of light energy to electrical energy is accomplished in the same way as in a phototube. Instead of just collecting the electrons that reach the single anode, PMT has a series of electrodes called **dynodes** as shown in Figure 8.15. The electrons emitted from the cathode are accelerated toward the first dynode, which is about 90V positive respect to the cathode. Each accelerated photoelectron that strikes the dynode surface produces secondary electrons which are then accelerated to dynode 2. Each succeeding dynode must be at a higher voltage (~90 V) than the previous one to accelerate the electrons. One typical PMT utilizes nine of the dynodes before collecting the electrons at the actual anode; therefore, the power supply is typically 900V. If the average number of secondary electrons for each impinging electron is four from each dynode, then the multiplication factor is $4^9 = 2.6 \times 10^5$. Multiplication factors of greater than one million can be achieved by using some PMT.

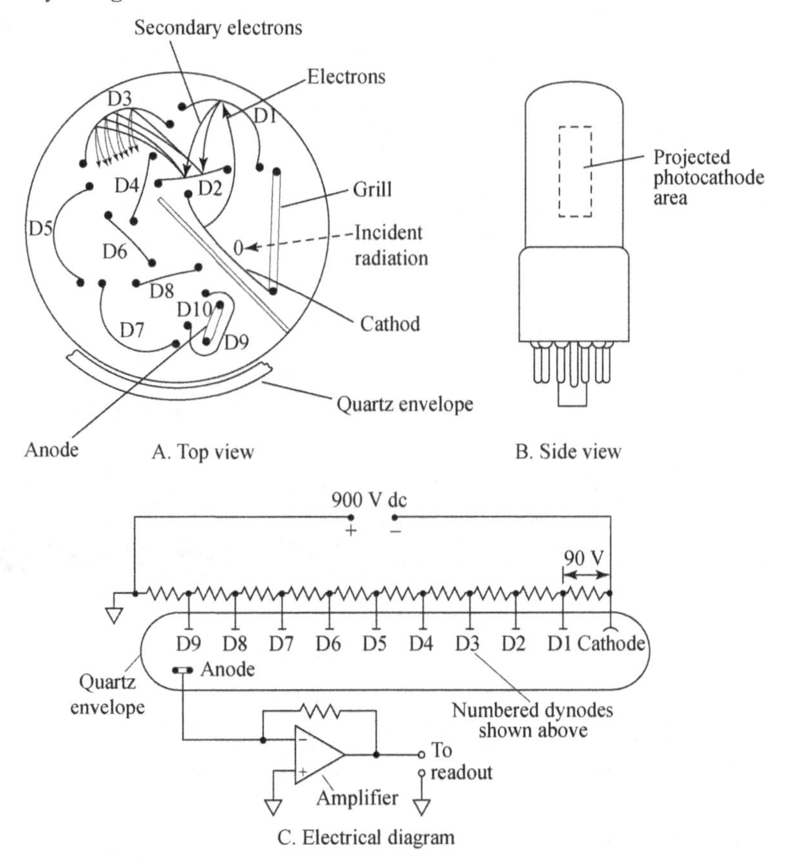

Figure 8.15 Schematic diagram of a photomultiplier tube.

5. Ultraviolet-visible Photometer and Spectrophotometers

We have discussed all the basic components required for spectrochemical measurements. There are different types of instruments to measure the light absorption. An instrument that uses a filter for wavelength selection is a photometer. A **spectrometer** uses a monochromator or a polychromator in conjunction with a detector to convert the radiant power into electrical signals. A spectrophotometer is a spectrometer that allows the measurement of the ratio of the two beams, P_0/P. Both photometers and spectrophotometers can be obtained in single- and double-beam designs.

Figure 8.16 shows the design of a single-beam Model 722 visible spectrophotometer. Before measuring the analytical solution, a 0% T calibration or adjustment has to be made by blocking the detector from receiving radiation. Then a cell containing a reagent blank is placed in the light path, and 100% calibration or adjustment is made. Finally, the analytical solution is placed in the light path to measure the absorbance. The source in this spectrophotometer is a tungsten/halide filament lamp. The spectral range is 330~800 nm with the bandwidth of 6 nm and a wavelength accuracy of ±2.5 nm. The dispersion element in the monochromator is a grating with 1200 grooves per millimeter.

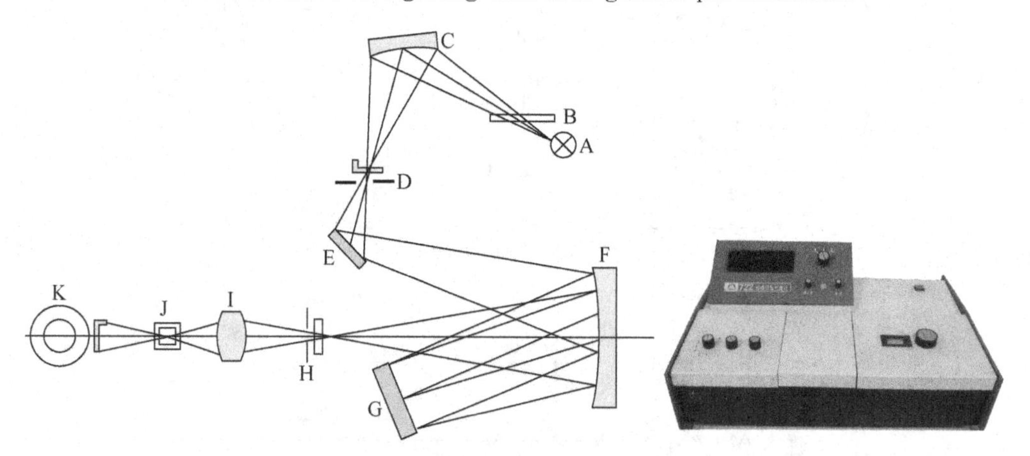

Figure 8.16 The 722 spectrometer.

A—Tungsten lamp; B—Filter; C—Collecting mirror; D—Entrance slit; E—Reflecting mirror; F—Collimating mirror; G—Diffraction grating; H—Exit slit; I—Collecting lens; J—Sample cell; K—Detector.

Many modern spectrophotometer are based on a double-beam design as illustrated in Figure 8.17. The light beam is alternately directed into the reference cell to acquire P_0 and sample cell to acquire P by a rotating sector mirror, and from each to the detector. In such a way, the short-term fluctuations from the source can be compensated. The variation of source power with wavelength can also be compensated.

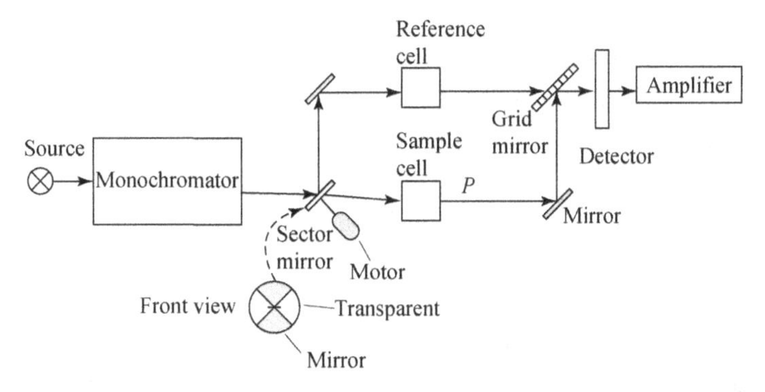

Figure 8.17 A double-beam in-time design for UV/visible spectrophotometer.

There is also another type of instrument which measures the UV/visible absorption of analytical solution with a multichannel detector. This type of instrument adopts a single-beam design with photodiode arrays as the detector. The dispersive system is a grating spectrograph located after the sample cell or the reference cell, and the photodiode array detector is located in the focal plane of the spectrograph (Figure 8.18). The photodiode array consists of more than 1000 diodes, thus the signals in the whole scanning wavelength region can be obtained at the same time usually less than 1 s.

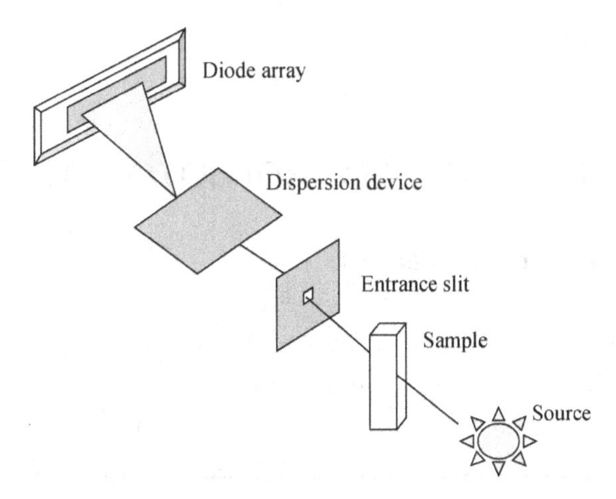

Figure 8.18 Diagram of a multichannel spectrometer with a photodiode array detector.

8.2.2 Instrumental Errors in Absorption Measurement

The accuracy and precision of spectrophotometric results are often limited by the noise associated with the instrument, that is, the random error in reading the radiant power of the light beams, thus the transmittance (T). The concentration random error resulted from the measurement of T can be derived from Beer's law.

Because $A = -\lg T = \varepsilon bc$, we have

$$c = -\frac{0.434}{\varepsilon b}\ln T \tag{8-17}$$

Taking partial derivative of this equation yields,

$$\partial c = -\frac{0.434}{\varepsilon bT}\partial T \tag{8-18}$$

where ∂c can be interpreted as the uncertainty in c resulted from the uncertainty in T. Dividing Equation (8-18) by Equation (8-17) yields

$$\frac{\partial c}{c} = \frac{0.434}{\lg T}\left(\frac{\partial T}{T}\right) \tag{8-19}$$

Replacing ∂c and ∂T with the standard deviation σ_c and σ_T in Equation (8-19) yields

$$\frac{\sigma_c}{c} \approx \frac{0.434}{\lg T}\left(\frac{\sigma_T}{T}\right) \tag{8-20}$$

where σ_c/c and σ_T/T are the relative standard deviations. Equation (8-20) shows that the relative standard deviation of concentration depends on T and reaches a minimum when $T = 0.368$ ($A = 0.434$). Therefore, it is desirable that the final solution of an analyte has a concentration such that the absorbance is about 0.434, Generally, the absorbance of a solution to be measured is better in the range of 0.1~1.

8.3 APPLICATIONS OF SPECTROPHOTOMETRY

8.3.1 Single Component Analyses

Any species which absorbs light in the UV/visible region can be directly determined by spectrophotometry. Organic compounds with functional groups that absorb light in the UV/visible region. Those functional groups are called chromophores. Some examples are given in Table 8.3. Most transition metal ions have absorptions in the visible region but with low absorptivities. However, higher absorptivity can be obtained when the absorption of light results in charge-transfer from the metal ion to a particular ligand (or vice versa), and thus forming the basis for analytical methods.

Table 8. 3 Absorption Data for Some Common Organic Chromophores

Chromophores	Examples	Solvents	λ_{max}/nm	ε_{max}
Ethylene	$CH_2=CH_2$	Vapor	193	10000
Acetylene	$CH\equiv CH$	Vapor	173	6000
Carbonyl	$CH_3C(=O)CH_3$	n-Hexane	186 280	1000 16
	CH_3CHO	n-Hexane	180 293	Large 12
Carboxyl	$CH_3C(=O)OH$	Ethanol	204	41
Amido	$CH_3C(=O)NH_2$	Water	214	60
Azo	$CH_3N=NCH_3$	Ethanol	339	5
Nitroso	C_4H_9NO	Ethyl ether	300 665	100 20
Nitro	CH_3NO_2	Isooctane	280	22
Nitrate	$C_2H_5ONO_2$	Dioxane	270	12
Aromatic	Benzene	n-Hexane	204 256	7900 200

The most commonly used ligands for spectrophotometrical determination of metal ions are organic ligands containing chromophores and complexation functional groups. Examples of commonly used ligands for spectrophotometric determination of metal ions with reasonable sensitivity and selectivity are given in Table 8. 4. The requirements for complexing ligands are:

- The ligand should have a relatively high molar absorptivity.
- The ligand should react selectively with the metal ion of interest.
- The difference between the maximum absorption peaks of the metal-ligand complex and the ligand itself should be greater than 60 nm.
- The complexation reaction should be a stoichiometric reaction and the resulting complex should be stable.

Reaction conditions that may affect the formation of metal-complex, such as temperature, reaction duration of reaction, pH, amount of ligands added, and each potentially interfering entity, must be investigated and optimized so that a high sensitivity and selectivity can be achieved.

Table 8.4 Examples of Spectrophotometric Methods for Metals

Ligands	Analytes	Stoichiometric Ratios	λ_{max}/nm	ε_{max}	Experimental Conditions	References
Chrome Azurol S(CAS)	Al^{3+}	1 : 3	585	5×10^4	pH 5.6	1
CAS+CTMAB[a]	Al^{3+}	Al : CAS : CTMAB =1 : 3 : 2	615	1.3×10^5	pH 5.2~6.0	2
Dimethylglyoxime	Ni^{2+}	1 : 2 or 1 : 4	470	1.3×10^4	pH 11~12, extracted with $CHCl_3$ in the presence of I_2 or H_2O_2	3
Arsenazo Ⅲ	Zr^{3+}	1 : 2	665	1.2×10^5	9 mol · L^{-1}HCl	4,5
Eriochrome Cyanine-R(ECR)	Ga	Ga : ECR=1 : 2	535	5×10^4	pH5.1	6
ECR_CTMAB	Ga	Ga : ECR=1 : 3	588	1.2×10^5	pH5.3	6
PAR + Zeph	Zn	Zn : PAR : Zeph =1 : 2 : 2	505	9.2×10^4	pH 9.7 extracted with $CHCl_3$	7,8
Dithiozone	Pb^{2+}	1 : 2	520	7.0×10^4	pH 8~10 extracted with CCl_4	9,10
Phenanthroline	Fe^{2+}	1 : 3	512	1.1×10^4	pH 2~9	11

CTMAB: Cetyltrimethyl-ammonium bromide; Zeph: Zephiran chloride or tetradecyl dimethyl benzyl ammonium chloride; PAR: 4-(2-pyridylazo) resorcinol. 1. Pakalns P. Spectrophotometric determination of aluminium with chrome azurol S. Analytica Chimica Acta, 1965,32(1):57-&. 2. Marczenko Z, Jarosz M. Formation of ternary complexes of aluminum with some triphenylmethane reagents and cationic surfactants. Analyst, 1982,107(1281):1431-1438. 3. Christopherson H, Sandell EB. The molecular and ionic solubility of nickel dimethylglyoximate. Analytica Chimica Acta, 1954,10(1):1-9. 4. Dupraw W A. Simple spectrophotometric method for determination of zirconium or hafnium in selected molybdenum-base alloys. Talanta, 1972, 19(6): 807-&. 5. Onishi H, Sekine K. Spectrophotometric determination of zirconium, uranium, thorium and rare-earths with arsenazo-iii after extractions with thenoyltrifluoroacetone and tri octylamine. Talanta, 1972, 19(4): 473-&. 6. Marczenko Z, Kalowska H. Sensitive spectrophotometric determination of gallium with eriochrome cyanine-R and cetyltrimethylammonium ions. Mikrochimica Acta, 1979, 2(5-6): 507-514. 7. Ahrland S, Herman R G. Spectrophotometric determination of manganese(Ii) and zinc(Ii) with 4-(2-pyridylazo)resorcinol (Par). Analytical Chemistry, 1975,47(14): 2422-2426. 8. Matsui H. Automated method for determination of trace amounts of metal-ions by Ion-exchange chromatography-determination of zinc(Ii) in waters. Analytica Chimica Acta, 1973,66(1):143-146. 9. Weber O A, Vouk V B. Molar extinction coefficients of dithizone and lead dithizonate in carbon tetrachloride. Analyst, 1960,85(1006):40-45. 10. Jones R A, Szutka A. Determination of microgram quantities of lead by spectrophotometric titration with dithizone. Analytical Chemistry, 1966,38(6):779-&. 11. Harvey A E, Smart J A, Amis E S. Simultaneous spectrophotometric determination of iron(Ii) and total iron with 1,10-phenanthroline. Analytical Chemistry, 1955,27(1): 26-29.

8.3.2 Multicomponent Analyses

Because Beer's law can be applied to a solution with more than one absorbing species, it is possible to determine the concentration of individual species in the solution even if their spectra overlap. For example, if a solution contains x and y absorbing components with the absorption spectra separated from each other, it is easy to obtain calibration curves at their own maximum absorption. Figure 8.19 shows the spectrum of a mixture with absorbing species of x and y as well as the individual spectra of x and y. To analyze x and y component in the sample, the molar absorptivity of each component at λ_1 and λ_2 can be obtained by testing a series of standard solutions of x and y, respectively. Then the absorbance of the mixture is measured at both λ_1 and λ_2. With known molar absorptivities for x and y at both λ_1 and λ_2, the absorbance of the mixture at λ_1 and λ_2 can be expressed in Equations (8-21) and (8-22), respectively, according to the additive property of absorbing components in a mixture as shown in Equation (8-15).

$$A_1 = \varepsilon_{x1} bc_x + \varepsilon_{y1} bc_y \quad (8\text{-}21)$$

$$A_2 = \varepsilon_{x2} bc_x + \varepsilon_{y2} bc_y \quad (8\text{-}22)$$

ε_{x1} and ε_{x2} are the molar absorptivities of component x at λ_1 and λ_2, respectively; ε_{y1} and ε_{y2} are the molar absorptivity of component y at λ_1 and λ_2, respectively. A_1 and A_2 can be measured experimentally. With known ε and b, the concentration of c_x and c_y can be solved by Equations (8-21) and (8-22).

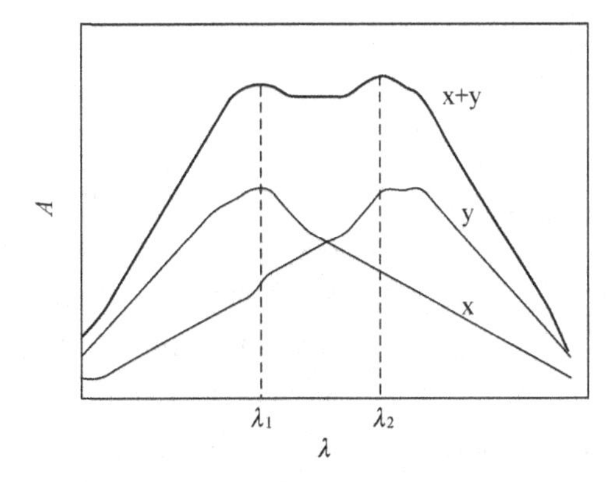

Figure 8.19 Absorption spectrum of a two-component mixture (x+y) and the spectra of the individual components.

8.3.3 Spectrophotometric Titrations

Photometric and spectrophotometric measurements are useful for locating the stoichiometric points of titrations. In the spectrophotometric titrations, the titration curve is plotted with absorbance at a selected wavelength against the volume of titrant added. According to the Beer's law, a linear change in absorbance is observed when the concentration of the absorbing species changes. In a spectrophotometric titration, the concentration of absorbing species changes linearly in the plot of absorbance versus the volume of titrant up to the stoichiometric point. After the stoichiometric point is reached, continued addition of titrant should not affect the concentration of the titration product, thus the slope of absorbance should not change with addition of the titrant. The plot of absorbance versus the volume of titrant beyond the stoichiometric point should also yield a straight line with a different slope. Generally, the titration curve as a plot of absorbance versus the volume of titrant gives two straight lines with different slopes. The endpoint is located at the intersection of extrapolated linear portion of the two lines.

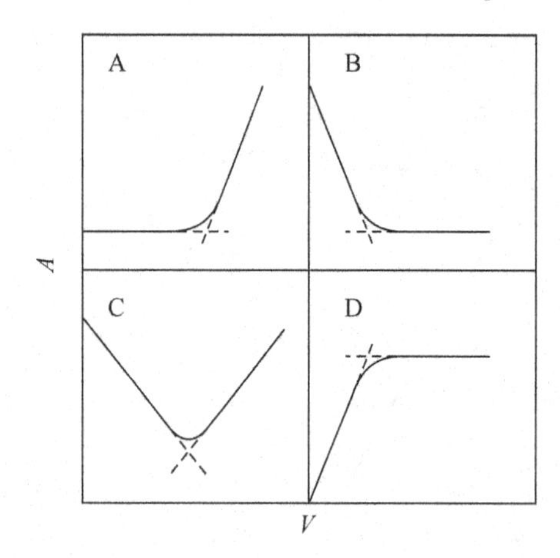

Figure 8.20 Typical spectrophotometric titration curves.

The profile of a titration curve varies with the analyte changes because the absorptivities of the analyte and the titrant are not always the same for all titrations. Figure 8.20A is the curve for the titration of a non-absorbing species with an absorbing titrant that reacts with the analyte to form a non-absorbing product. For example, the titration of Fe(Ⅱ) with $KMnO_4$ follows this profile because the analyte Fe(Ⅱ) and the titration product Mn^{2+} and Fe(Ⅲ) are colorless at the selected wavelength and titrant is the only absorbing species in the entire titration process. Figure 8.20B is the curve for the titration of the absorbing analyte with the non-absorbing titrant to produce a non-absorbing product. The example is the titration of ferrous sulfosalicylic acid complex. The analyte, ferrous sulfosalicylic acid is an absorbing species while the titrant (EDTA) and the products (Fe(Ⅱ)-EDTA) and sulfosalicylic acid, are all colorless. Figure 8.20C is the curve for the titration of an absorbing analyte with an absorbing titrant to yield a non-absorbing product. An example is the titration of Sb^{3+} with $KBrO_3$-KBr standard solution in HCl medium with non-absorbance recorded at 326 nm. Figure 8.20D is the titration of an non-absorbing analyte with a non-absorbing titrant to generate an absorbing product. An example is the titration of *p*-bromophenol with NaOH.

The advantages of spectrophotometric titration over the titrations by visual endpoint observation are:

- The spectrophotometric titration can be applied to titrations that cannot be titrated by visual endpoint observation within an expected titration error. In acid-base titration only the acid with $cK_a \geqslant 10^{-8}$ or the base with $cK_b \geqslant 10^{-8}$ can be titrated with a titration error less than 0.2%. A stepwise titration can only be achieved when $\Delta pK_a \geqslant 5$ for a polyprotic acid with a titration error less than 0.5%. For example, *o*-nitrophenol ($pK_a = 7.15$) and *m*-nitrophenol ($pK_a = 8.39$) cannot be titrated by acid-base titration individually or as a mixture. However, the mixture can be stepwise titrated by spectrophotometry with titration curve constructed based on the absorbance at 545 nm (Figure 8.21).

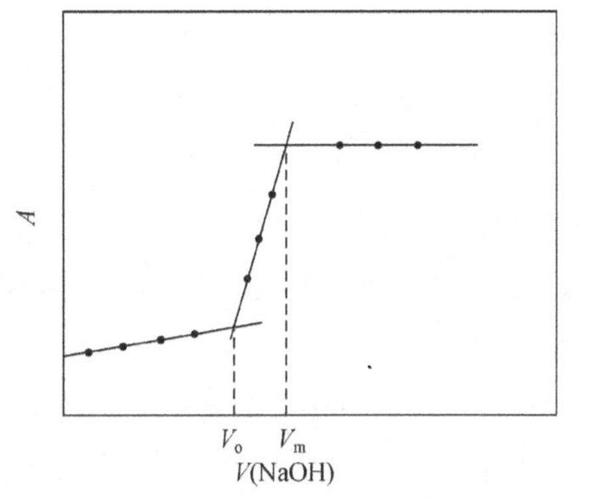

Figure 8.21 The titration curve for *o*-nitrophenol and *m*-nitrophenol mixture by NaOH.

- Because spectrophotometry has very high sensitivity and can measure a very small change in absorbance, spectrophotometric titrations can be applied to some organic acids with poor solubility, but the concentration of these acids in solution may be too small to be titrated with a reasonable titration accuracy based on visual endpoint observation.
- Spectrophotometric titration can be used for titration of colored analytes or colored solutions which are difficult to determine the endpoint by visual observation.

8.3.4 Studies of Complex Formation in Solutions

Spectrophotometry can be used for the determination of the chemical composition and formation constant of a metal ion with a ligand. The two common spectrophotometric methods, the molar ratio method and the method of continuous variation (the Job's method), are introduced here.

1. The Molar Ratio Method

In the molar ratio method, the absorbance of the solutions at a selected wavelength is recorded by keeping the concentration of the metal ion at a constant value while varying the concentration of the ligand in the solution. A plot with absorbance versus c_L/c_M is then constructed. When there is not enough ligand in the solution, the curve increases linearly with increasing concentration of ligand in the solution. When all the metal ions are quantitatively complexed by the ligand, the slope of the curve changes as more ligand is added. If the ligand does not absorb light of the selected wavelength, the slope is zero after metal ions are completely complexed by the ligand. If the ligand absorbs light of the selected wavelength, the slope is greater than zero after metal ions are completely complexed by the ligand. Figure 8. 22 shows the curve of the first case. The composition of the metal complexes can be obtained by the intersection of extrapolated linear portion of the two lines.

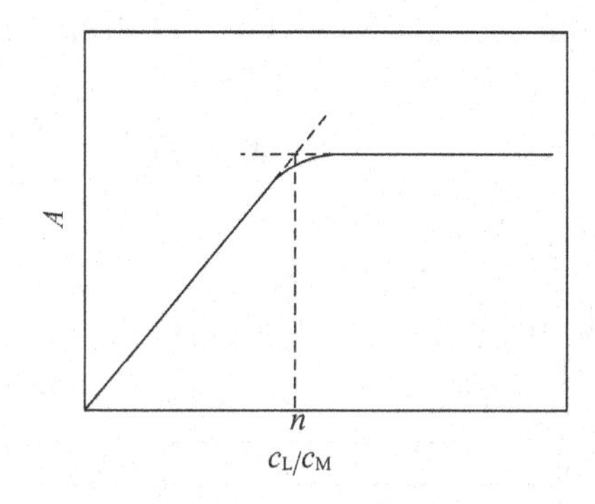

Figure 8. 22 The Molar ratio plot of M complexes with L.

2. The Method of Continuous Variations (The Job's Method)

In the method of continuous variations, metal and ligand solutions with identical concentrations are mixed in a way that the total volume and the total moles of reactants in each mixed solution is constant but the molar ratio of the metal ions to the ligands varies. The absorbance of the mixture at selected wavelength is plotted as a function of the volume fraction of the metal ion or the ligand, that is $V_L/(V_L+V_M)$ or $V_M/(V_L+V_M)$, where V_M is the volume of metal ion and V_L is the volume of the ligand. The typical continuous variations plots are shown in Figure 8. 23. For a 1 : 1 complexes, the maximum absorbance occurs at a $V_M/(V_L+V_M)$ value of 0. 50 for a 1 : 1 metal-ligand complex (Figure 8. 23A) and 0. 33 for a 1 : 2 metal-ligand complex (Figure 8. 23B).

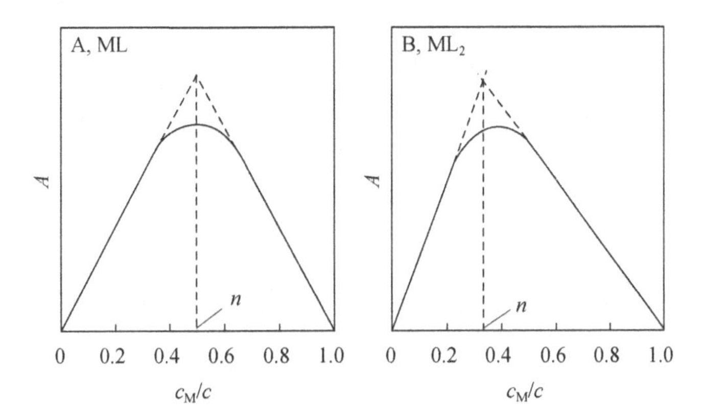

Figure 8.23 Plots for method of continuous variations for ML and ML_2 complexes.

8.3.5 Measurements of Dissociation Constants of Organic Acids/Bases

Many organic ligands and indicators involved in analytical chemistry as well as compounds of pharmaceutical importance are weak acids/bases and also sparingly soluble in water. The dissociation constants of these compounds can be conveniently determined by spectrophotometry.

The monoprotic acid, HL, is taken as an example to derive a mathematical model based on UV/visible absorption data of HL and its conjugate base (L^-). The dissociation reaction and constant expression of HL in aqueous solution are

$$HL \rightleftharpoons H^+ + L^- \quad K_a = [H^+][L^-]/[HL]$$

$$pK_a = pH + \lg \frac{[HL]}{[L^-]} \tag{8-23}$$

With a known ratio of $[HL]$ to $[L^-]$ at a given pH, the pK_a can be calculated according to Equation (8-23). And $[HL]/[L^-]$ can be obtained from their absorbance in solution at a given wavelength where both have absorption and provided that Beer's law is obeyed by the two species.

The absorbance of the HL solution is

$$\begin{aligned} A &= A_{HL} + A_{L^-} = \varepsilon_{HL} \cdot [HL] + \varepsilon_{L^-} \cdot [L^-] \\ &= \varepsilon_{HL} \cdot \frac{[H^+]c}{K_a + [H^+]} + \varepsilon_{L^-} \cdot \frac{K_a c}{K_a + [H^+]} \end{aligned} \tag{8-24}$$

where $c = [HL] + [L^-]$, ε_{HL} at the given wavelength can be determined at a pH where all the HL is in its acid form, HL. At this pH, $A = A_{HL} = \varepsilon_{HL} \cdot [HL] \approx \varepsilon_{HL} \cdot c$. Thus,

$$\varepsilon_{HL} = A_{HL}/c \tag{8-25}$$

In the same way, ε_{L^-} at the given wavelength can be obtained at a pH where all the HL is in its conjugate base form (L^-). At this pH, $A = A_{L^-} = \varepsilon_{L^-} \cdot [L^-] \approx \varepsilon_{L^-} \cdot c$. Thus,

$$\varepsilon_{L^-} = A_{L^-}/c \tag{8-26}$$

Substitution of Equations (8-25) and (8-26) into Equation (8-24) and rearrangement of the obtained equation yield

$$K_a = \frac{[H^+][L^-]}{[HL]} = \frac{A_{HL} - A}{A - A_{L^-}}[H^+] \quad \text{or} \quad pK_a = pH + \lg\frac{A - A_{L^-}}{A_{HL} - A} \tag{8-27}$$

where A_{HL} and A_{L^-} are the absorbance of HL at a given wavelength with the acid completely in its acid and conjugate base form, respectively.

In the experiment, a set of HL solutions of varying pH are prepared with an analytical concentration of c. The absorbance of solutions at the selected wavelength is recorded and the pH is measured with a pH meter. pK_a can be calculated according to Equation (8-27) for each solution. The pK_a can then be obtained by averaging all the pK_a values calculated. Equation (8-27) can be rearranged into

$$\lg\frac{A - A_{L^-}}{A_{HL} - A} = -pH + pK_a \tag{8-28}$$

A plot of $\lg\frac{A - A_{L^-}}{A_{HL} - A}$ versus pH will produce a straight line by linear regression with an intercept of pK_a. The plot for thiazolylazo-5-methoxylphenol is presented in Figure 8. 24 and the pK_a is 7. 87.

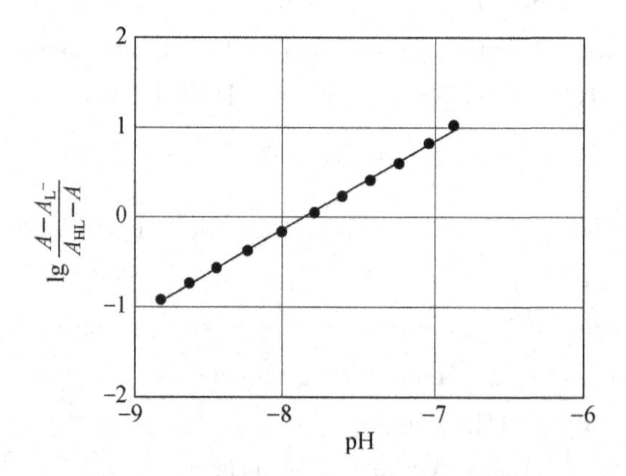

Figure 8. 24 The plot of $\lg\frac{A - A_{L^-}}{A_{HL} - A}$ vs. pH for thiazolylazo-5-methoxylphenol (A_{HL} = 0. 057 and A_{L^-} = 0. 738).

Chapter 8 Questions and Problems

8.1 Why a molecule absorbs light? What is the wavelength region of UV, visible and infrared radiation?

8.2 A solution has color because it absorbs its complementary color composing the white light. If a solution is green, what color this solution absorbs?

8.3 What is the relationship between absorbance and transmittance?

8.4 Why it is better to measure the absorbance of a solution at its maximum absorption wavelength in spectrophotometric analysis?

8.5 State the factors that cause deviation to Beer's law. How to avoid or reduce the influence of these factors?

8.6 State the differences in the design of a single-beam and a dual-beam spectrophotometer.

8.7 If the molar absorptivity of $KMnO_4$ at 345 nm is 2.2×10^3 L·mol^{-1}·cm^{-1}, calculate the transmittance of a 0.002% (m/V) $KMnO_4$ solution placed in a 3.0 cm path cell. If a 1 : 1 dilution is made, calculated the transmittance of the resulting solution.

8.8 Nickel is determined by spectrophotometry using dimethylglyoxime. A solution containing 1.7×10^{-5} mol·L^{-1} nickel-dimethylglyoxime complex has a transmittance of 30.0% in a 2.00 cm cell at 470 nm. Calculate the molar absorptivity of nickel-dimethylglyoxime complex.

8.9 Iron (Ⅱ) is determined by spectrophotometry using phenanthroline. A 0.500 g Fe (Ⅱ) sample is dissolved and phenanthroline is added, the resulted solution is then diluted to 50.0 mL. The solution shows an absorbance of 0.430 in a 1.00 cm cell at 510 nm. If the molar absorptivity (ε) of the Fe (Ⅱ)-phenanthroline complex at 510 nm is 1.1×10^4 L·mol^{-1}·cm^{-1}, calculate the percentage of Fe(Ⅱ). If the final solution is diluted 2 fold, calculate the transmittance of the resulting solution.

8.10 A 1.83×10^{-5} mol·L^{-1} procaine hydrochloride has an absorbance of 0.385 at 288 nm in a 1.00 cm cell. What is the concentration of a solution exhibiting an absorbance of 0.825, and what is the molar absorptivity of procaine hydrochloride?

8.11 Fe(Ⅲ) is determined by spectrophotometry using sulfosalicylic acid. Fe(Ⅲ) standard solution is prepared by dissolving 0.432 g of analytical grade $NH_4Fe(SO_4)_2\cdot12H_2O$ in 500.0 mL water. The calibration curve of Fe(Ⅲ) is constructed by sequential addition of varied volume of Fe(Ⅲ) standard solution and fixed volume of the ligand solution to 50.0-mL volumetric flasks. Then the resulting solutions are diluted to 50.0 mL and the absorbance are measured in a 1.00 cm cell. The volume of Fe(Ⅲ) standard solution added and the corresponding absorbance are given in the table below.

V(Fe(Ⅲ))/mL	1.00	2.00	3.00	4.00	5.00	6.00
A	0.097	0.200	0.304	0.408	0.510	0.618

A 5.00 mL of Fe(Ⅲ) sample solution is pipetted and diluted to 250.0 mL. A 2.00 mL of the prepared solution is withdrawn and an absorbance of 0.450 is obtained by going through the same procedure as that for the calibration curve. Calculate percentage of Fe(Ⅲ).

8.12 The error for transmittance measurement is $\Delta T = \pm 0.010$. Calculate the relative error for concentration determination based on this accuracy for transmittance measurement when the absorbance of an analyte solution at varied concentrations are 0.010, 0.100. 0.200, 0.434, 0.800, and 1.200.

8.13 The spectrophotometric method is used for the determination of two metal-ligand complexes, x and y, in one solution. The absorbance of pure complexes and the test solution are obtained in a 1.00 cm cell and are provided in the table below. Calculate the concentration of x and y in the test solution.

Solutions	c/mol · L^{-1}	A_1 (285 nm)	A_2 (365 nm)
x	5.0×10^{-4}	0.053	0.430
y	1.0×10^{-3}	0.950	0.050
x+y	To be determined	0.640	0.370

8.14 To determine the pK_a of an acid/base indicator (HIn), 1.00 mmol indicator is dissolved to make a 100 mL solution. Aliquots of 2.50 mL are withdrawn, the pH adjusted and diluted to 25 mL. The absorbance at the adjusted pH values measured in a 1.0 cm cell are given in the table below. Calculate the molar absorptivity of In^- and pK_a of this indicator.

pH	1.00	2.00	7.00	10.00	11.00
A	0.00	0.00	0.588	0.840	0.840

8.15 The reaction of compound B with compound C is either $B + C \longrightarrow BC$ or $B + 2C \longrightarrow BC_2$. Only compound B absorbs at 250 nm with the molar absorptivity of 2.00×10^3 L · mol^{-1} · cm^{-1}. Two solutions are prepared by mixing 50.0 mL of B and 50.0 mL of C, and the absorbance of the equilibrated reaction mixtures are then determined at 250 nm using a 1.00 cm cell and the data are:

Solutions	Analytical Concentration of B/ mol · L^{-1}	Analytical Concentration of C/ mol · L^{-1}	Absorbance
1	2.00×10^{-3}	8.00×10^{-3}	0.800
2	2.00×10^{-3}	1.23×10^{-2}	0.400

Determine whether the reaction product is BC or BC_2 and the equilibrium constant, K, for the reaction.

8.16 The iron content of a sample is determined with a photometric titration. A 0.5993 g of the sample is dissolved in HCl diluted to 100.0 mL. A 10.0mL of this solution is taken, salicylic acid is added to form the complex $FeSal_3^{3-}$, and the resulting solution is diluted to V_s = 100.0 mL. This solution is then titrated with 0.01496 mol · L^{-1} EDTA based on the reaction

$$FeSal_3^{3-} + H_3Y^- + 3H^+ \longrightarrow FeY^- + 3H_2Sal$$

Samples of the titration mixture are withdrawn periodically during the titration to measure the absorbance, and the sample is then returned to the titration mixture before adding additional titrant. The experimental data are corrected by compensating volume changes during titration and are displayed in the table below. Calculate the percentage of Fe in the original sample.

V_T/mL	31.22	31.39	31.57	31.82	32.01	32.17	32.38
A_{corr}	0.600	0.537	0.470	0.414	0.353	0.295	0.249
V_T/mL	32.60	32.81	32.97	33.22	33.58	33.97	36.02
A_{corr}	0.187	0.127	0.0751	0.0482	0.0484	0.0485	0.0493

CHAPTER 9

INTRODUCTION TO ANALYTICAL SEPARATION

9.1 General Considerations of Separation Efficiency

9.2 Separation by Precipitation

9.2.1 Inorganic Precipitants

9.2.2 Organic Precipitants

9.2.3 Coprecipitation of Species in Trace Amounts for Separation

9.2.4 Improving the Selectivity of Precipitation Separation

9.3 Separation by Extraction

9.3.1 Principles for Liquid-liquid Extraction

9.3.2 Percent Extraction

9.3.3 Extraction of Inorganic Species

9.3.4 Other Extraction Methods

9.4 Separation by Ion Exchange

9.4.1 Ion Exchange Resins

9.4.2 Cross-linkage and Exchange Capacity

9.4.3 Ion Exchange Equilibria

9.4.4 Applications of Ion Exchange Separation

9.5 Separation by Chromatography

9.5.1 Classification

9.5.2 Chromatogram

9.5.3 Column Chromatography

9.5.4 Planar Chromatography

In Chapters 3 to 8, we discussed some of the quantitative analytical methods as well as the analytical data evaluation based on statistical analysis. These are very important steps in developing an analytical method. In practice, sampling and preparing samples into a form ready to measure is the necessary step for developing an analytical method. Interfering substances have to be separated from the analytical samples. Enrichment of analyte may be necessary, because the quantitative method may not be sensitive enough to detect the analyte without enrichment. Sampling and sample preparation are integral parts of the analytical process and equally important to the other steps in the methods. In this chapter, basic methods for analytical separations are introduced.

9.1 GENERAL CONSIDERATIONS OF SEPARATION EFFICIENCY

An **interference** (or an **interferent**) is a chemical species that causes a systematic error in an analysis by enhancing or attenuating the analytical signal or the background. Separating the interfering substances from the analytical sample is inevitably necessary in developing an analytical method, because few quantitative analytical methods are specific for a species to be determined. Other than sampling, the errors from sample separation are the major source of errors in the analytical process. Categorization of input errors in liquid chromatography is given in Figure 9.1 as an example (Frank Settle. Handbook of Instrumental Techniques for Analytical Chemistry. Prentice Hall PTR, 1997, p20). Attention must be paid to sampling, sample processing and sample separation in order to obtain accurate analytical results. Often sample

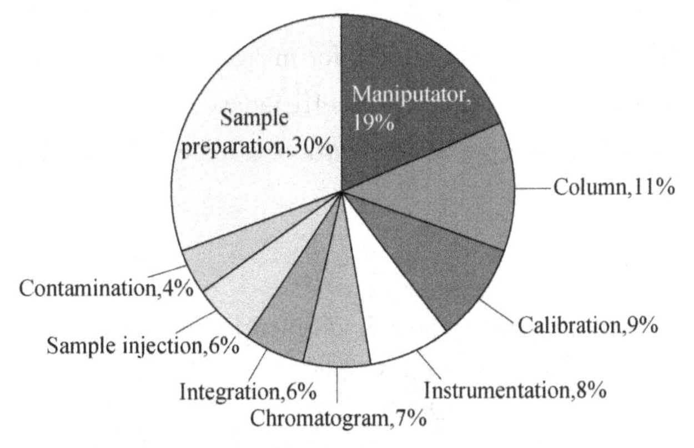

Figure 9.1 Source of errors in a high performance chromatography (HPLC) analysis.

preparation is the most time consuming step to establish methods with high throughput and automation. An effective experimental design for sample preparation can improve the throughput and degree of automation of an analytical method.

It is most desirable that sample preparation meet the following requirements to facilitate an accurate analysis of the analyte:

- The analyte is dissolved in an appropriate solvent for techniques which measure samples in the liquid form.
- Most, if not all, of the interfering substances are removed.
- The concentration of the analyte in the final solution is at a level suitable for quantitation by the selected method.
- The sample preparation procedure is environmentally benign.
- The sample preparation procedure is easy and has the potential to be automated.

The goal of an analytical separation is to remove either the analyte or the interferent from the sample matrix. On one hand, separations that completely remove an interferent may result in the partial loss of the analyte. A separation must assure that the error from loss of the analyte during separation is minimal and can be neglected or the interferences are completely removed. **Recovery** of the analyte (R_A) is often used to evaluate loss of the analyte in a separation procedure. R is defined as

$$R_A = \frac{Q_A(\text{yield})}{Q_A(\text{original})} \times 100\% \qquad (9\text{-}1)$$

where Q_A (yield) is the quantity of the analyte remaining after the separation, and Q_A(original) is the initial quantity of the analyte. A recovery of 100% means that none of the analyte is lost during the separation. In practice, an accurate analytical result requires that $R_A > 99.9\%$ for macroanalysis, $R_A > 99\%$ for analyte with concentration of the analyte greater than 1%, R_A is 95%~105% for microanalysis.

On the other hand, the separation efficiency is affected by the degree of interference removal. The degree of separation can be estimated by the **separation factor**, $S_{I/A}$, defined as

$$S_{I/A} = \frac{R_I}{R_A} \qquad (9\text{-}2)$$

R_I is the recovery of the interferent and is defined in the same manner as R_A,

$$R_I = \frac{Q_I(\text{yield})}{Q_I(\text{original})} \times 100\%$$

For an ideal separation, $R_A = 1$, $R_I = 0$, thus $S_{I,A} = 0$. For microanalysis when the interferent exists in a large amount in the matrix, $S_{I/A}$ is required to be close to 10^{-7}.

For a sample with the analyte and the interference approximately at the same order of magnitude, $S_{I/A}$ is required to be close to 10^{-3}.

Table 9.1 lists the separation methods that are often used in analytical chemistry. In chromatography, quantitation depends upon the efficiency of the separation procedure. In other procedures, the separation step is distinct and independent of the measurement step that follows. In this chapter, separation by precipitation, extraction, ion-exchange and chromatography are introduced.

Table 9.1 Separation Methods

Separation Methods	Basis of Method	Separation Methods	Basis of Method
Size	Filtration Dialysis Size-exclusion chromatography	Change in chemical state	Precipitation Ion-exchange Electrodeposition Volatilization
Mass and density	Centrifugation	Partition between phases	Extraction Chromatography
Change in physical state	Distillation Sublimation Recrystallization		

9.2 SEPARATION BY PRECIPITATION

Separation by precipitation is based on solubility difference between the analyte and potential interferences. Precipitation is often used for quantitative separation of metal ions.

9.2.1 Inorganic Precipitants

Sodium hydroxide ($NaOH$), ammonia (NH_3), and hydrogen sulfide (H_2S) are often used to precipitate metal ions from a solution.

Enormous differences exist in the solubility of the metal hydroxides, thus it is possible that selective precipitation can be achieved by adjusting pH of the solution. Theoretically, the solubility and the pH for the possible formation of metal hydroxide precipitates can be calculated using the solubility products of the metal ions existing in the solution; however, the solubility product may be altered by precipitate changes. For example, the morphology of the crystalline form at initial precipitation and after aging may not be the same, thus solubility product will be different. Other than mononuclear complexes, often polynuclear metal complexes inevitably form during precipitation; therefore, the preferred pH for precipitating a metal ion is often not the same as the value predicted from calculations based on the solubility products.

NaOH is often used for the separation of an amphoteric element (e. g. Zn) from another elements with the latter being precipitated as an oxide and the former staying in the solution as anionic species (e. g. Zn^{2+} forms tetrahydroxozincate ion, $Zn(OH)_4^{2-}$, in strong alkaline solution).

High-valent metal ions, such as Th^{4+} (thallium), Al^{3+} (aluminum), and Fe^{3+} (iron) can be precipitated from a solution with uni- and divalent metal cations by adding ammonia in the presence of ammonium ion (pH 8~9). The coexisting metal ions, such as Ag^{+} (silver), Cu^{2+} (copper), Ni^{2+} (nickel), Zn^{2+} (zinc), and Cd^{2+} (cadmium) form soluble complexes with ammonia.

With the exception of the alkali metals and the alkaline earth metals, most cations form sparingly soluble sulfides whose solubilities differ substantially from one another. By controlling the concentration of S^{2-} (Sulfide ion) via adjusting pH of the saturated H_2S solution, the selective precipitation of heavy metal cations as sulfides can be done. The concentration of S^{2-} can be altered from 0. 1 mol • L^{-1} to 10^{-22} mol • L^{-1} by controlling the pH of a saturated H_2S solution. For example, bubbling H_2S gas continuously through a solution buffered with chloroacetate, Zn^{2+} can be separated from Fe^{2+} (iron), Co^{2+} (cobalt), Ni^{2+} (nickel), and Mn^{2+} (manganese ions) via forming ZnS (zinc sulfide) precipitate. Similarly, Zn^{2+}, Co^{2+}, Ni^{2+} and Fe^{2+} can be separated from Mn^{2+} by forming ZnS, CoS, NiS, and FeS at pH 5~6 buffered with $(CH_2)_6N_4$ (hexamethylenetetramine). Precipitation of divalent metal cations by forming sulfides is often accompanied by coprecipitation; however, the rate of sulfide precipitation is slow, thus the separation of the forming sulfides does not always produce ideal results.

9.2.2 Organic Precipitants

Selected organic reagents for the selective precipitation of inorganic ions were discussed in Chapter 7. There are numerous organic reagents that can be used for selective precipitation of inorganic ions. The precipitates formed are usually large crystals, and the adsorbed water can be easily removed by drying at moderately elevated temperature. Coprecipitation does not occur with the use of organic reagents as the precipitants. These advantages make organic precipitants widely used in precipitation separations.

Dimethylglyoxime can selectively precipitate nickel in ammonia-ammonium buffer with the presence of tartrate.

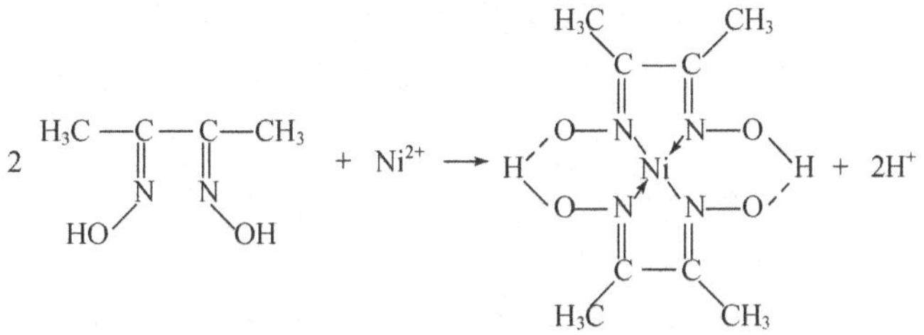

8-hydroxyquinoline can form slightly soluble compounds with many metal cations at different pH. Precipitation separation can be achieved by adjusting pH of the solution together with introducing masking reagents. Selectivity can be improved by introducing more substituent groups into the quinoline structure. For example, both Al^{3+} and Zn^{2+} form precipitates with 8-hydroxyquinoline, while 2-methyl-8-hydroxyquinoline can only form a precipitate with Zn^{2+}, thus effectively separating Zn^{2+} from Al^{3+}.

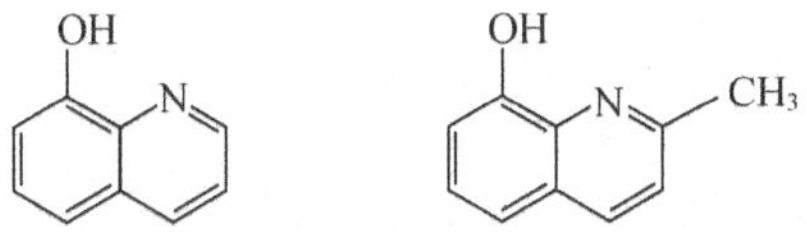

8-hydroxyquinoline 2-methyl-8-hydroxyquinoline

9.2.3 Coprecipitation of Species in Trace Amounts for Separation

For trace amounts of an analyte in a sample, it is nearly impossible to separate the analyte by precipitation, because a very small amount of precipitate is hard to coagulate. In addition, an appreciable fraction of the precipitate may be lost during transfer and filtration. Coprecipitation can be exploited to collect trace amount of species of interest from the sample matrix. In the collection of trace amounts of analyte, a quantity of another ion that also forms a precipitate with the reagent can be added to the solution. The desired minor species is precipitated from the solution by forming a coprecipitate (called a **collector**) with the added ion. For example, the trace amount of lead (Pb) in tap water can be isolated by coprecipitation with $CaCO_3$. Na_2CO_3 is added to the tap water to precipitate Ca^{2+}. The $CaCO_3$ coprecipitates with Pb carries down the trace amount of Pb as $PbCO_3$. Alternatively, $CaCO_3$ is added to the tap water sample and the mixture is shaken vigorously so that $PbCO_3$ can be efficiently collected with $CaCO_3$.

9.2.4 Improving the Selectivity of Precipitation Separation

Selectivity of precipitation separation can be improved by masking via formation of complexes. For example, KCN can be added to complex with Cu^{2+} so that CuS precipitate does not form with the precipitation of Cd^{2+} (cadmium). In a saturated H_2S solution, CdS will form while $Cd(CN)_4^{2-}$ will not form because the latter is less stable.

The selectivity and efficiency of precipitation separation can also be improved by the combination of masking and pH adjustment. The precipitation of CaC_2O_4 (calcium oxalate) in the presence of Pb^{2+} is one illustration. Ca^{2+} cannot be precipitated as CaC_2O_4 while Pb^{2+} stays soluble in the solution, because the solubility of PbC_2O_4 is much less than that of CaC_2O_4. However, adding EDTA (ethylenediaminetetraacetic acid, Y) and adjusting pH to an appropriate range, Ca^{2+} precipitates as CaC_2O_4 leaving Pb^{2+} solution as PbY. The solubility of Ca^{2+} and Pb^{2+} as a function of pH in the presence of EDTA is presented in Figure 9. 2. The solid lines represent the solubility curves without EDTA, and the dashed lines represent the solubility curves in the presence of 10^{-2} mol • L^{-1}EDTA. If at the end of the precipitation, the analytical concentration of $H_2C_2O_4$ is 0. 1 mol • L^{-1} and the total concentration of EDTA not complexed with metal ions is 10^{-2} mol • L^{-1}, then the separation can be accomplished (pCa$\geqslant$5 and pPb$\leqslant$2) by adjusting the pH to a range of 2. 8~4. 9.

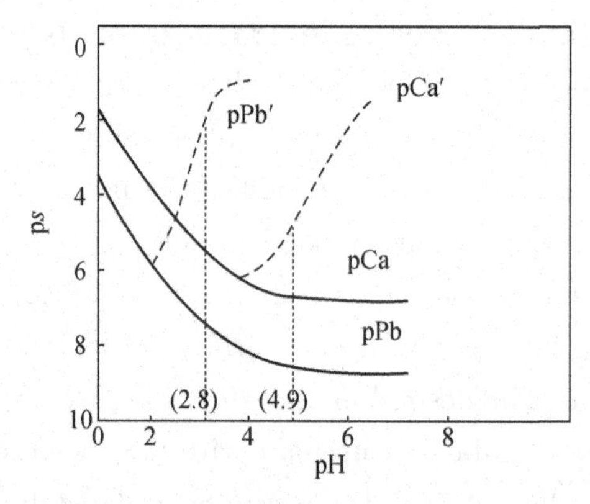

Figure 9. 2 The solubility curves for CaC_2O_4 and PbC_2O_4 as a function of pH with or without EDTA. $c(H_2C_2O_4)$=0. 1 mol • L^{-1}([C_2O_4']=10^{-2} mol • L^{-1}, [Y']= 10^{-2} mol • L^{-1}).

9.3 SEPARATION BY EXTRACTION

An **extraction** is the selective transfer of a compound or compounds from one liquid (usually water) to another immiscible liquid (usually an organic solvent) or from a solid to a liquid. The former process is called liquid-liquid extraction and the latter is called a liquid-solid extraction. More recently, with the development of reactive solid phases, the analyte can be extracted from liquids with a solid. This is called solid-phase extraction. In this section, the distribution law and liquid-liquid extraction are discussed and solid-phase extraction is also introduced.

9.3.1 Principles for Liquid-liquid Extraction

When two immiscible phases are mixed and shaken vigorously to facilitate complete mixing of the two phases, the hydrophobic solutes will stay in the organic phase and the hydrophilic solutes will stay with the aqueous phase, thus the separation by extraction is achieved.

The partition of a solution into two immiscible phases has been used for the extraction of organic compounds. The extraction of inorganic chemicals dates back to the nineteenth century. In 1892, Rothe and Hanroit extracted iron (Ⅲ) in hydrochloric acid with diethyl ether.

The first systematic experimental study of the distribution of a substance between two liquids was made by Berthelot and Jungfleisch in 1872. In 1891, Nernst formulated a **distribution law** for liquid-liquid systems. He stated that the partition coefficient is a constant at a given temperature and pressure for every molecular species present.

If compound A is allowed to distribute between water and an organic phase, the resulting equilibrium may be written as

$$A_{(aq)} \rightleftharpoons A_{(org)}$$

with the letters in parentheses refer to the aqueous (aq) and organic (org) phases, respectively.

Then the ratio of activities for A in the two phases is constant and independent of the total quantity of A at a given temperature and pressure. That is,

$$K_D = \frac{(a_A)_{org}}{(a_A)_{aq}} \tag{9-3}$$

where $(a_A)_{org}$ and $(a_A)_{aq}$ are the activities of A in each of the phases. The equilibrium constant is known as the **distribution constant**, K_D, and also known as the **distribution (or partition) coefficient.** Because only molecules with neutral change can be extracted, and in this case, the activity of the neutral molecules are not greatly affected by the medium, therefore,

$$K_D \approx \frac{[A]_{org}}{[A]_{aq}} \quad (9\text{-}4)$$

$[A]_{org}$ and $[A]_{aq}$ are the molar concentration of A in the organic and aqueous phases, respectively. Remember that distribution law is only applied to the solute of the same form in both phases. In reality, partition between organic and aqueous phases is inevitably accompanied by chemical equilibria such as dissociation, association and complexation. In this case, compound A may be in more than one form in either phase. For analytical purposes, it is more realistic to evaluate the ratio of total concentrations of A in each phase.

$$D = \frac{(c_A)_{org}}{(c_A)_{aq}} = \frac{[A_1]_{org} + [A_2]_{org} + \cdots + [A_n]_{org}}{[A_1]_{aq} + [A_2]_{aq} + \cdots + [A_n]_{aq}} \quad (9\text{-}5)$$

where D is called the **distribution ratio**, and A_1, A_2, $\cdots$, A_n represent all the forms of A in both phases. Because D is not a thermal dynamic constant, it changes as the medium condition changes. D equals to K_D only when A is in the same form in both phases.

The distribution ratio for the partition of iodine (I_2) in the presence of iodide (I^-) between water and carbon tetrachloride (CCl_4) can be calculated as follows

$$D_{I_2} = \frac{(c_{I_2})_{org}}{(c_{I_2})_{aq}} = \frac{[I_2]_{org}}{[I_2]_{aq} + [I_3^-]_{aq}} = \frac{[I_2]_{org}}{[I_2]_{aq}(1 + \beta[I^-])} = \frac{K_{D(I_2)}}{\alpha_{I_2(I^-)}}$$

It can be seen that D is affected by $[I^-]$ in the aqueous phase. When $[I^-]$ is high, $\alpha_{I_2(I^-)} > 1$, then $D_{I_2} < K_D$. When $[I^-]$ is very small, then $\beta[I^-] \ll 1$ and $\alpha_{I_2(I^-)} = 1$, and iodide has the same form (I_2) in both phases; thus $D_{I_2} = K_{D(I_2)}$.

【Example 9. 1】 Seven hundred milligrams of Sb (V) (antimony) were dissolved in 100 mL of HCl, an equal volume of isopropyl ether was added and the system thoroughly shaken. The two phases were separated; and the ether was evaporated. The residue from the ether phase was found to contain 685 mg of Sb (V). Calculate the distribution ratio (D).

Answer:

$$D = \frac{\text{Sb in organic phae(g/mL)}}{\text{Sb in aqueous phase(g/mL)}} = \frac{0.685/100}{(0.700 - 0.685)/100} = 45.7$$

9.3.2 Percent Extraction

For the extraction of compound A from aqueous solution, the extraction efficiency can be evaluated by **percent extraction** (E) as defined

$$E = \frac{\text{Total amount of A in organic phase}}{\text{Total amount of A to be extracted}} \times 100\%$$

That is

$$E = \frac{c_{org} V_{org}}{c_{aq} V_{aq} + c_{org} V_{org}} \times 100\%$$

where c_{org} and c_{aq} are the concentration of compound A in organic and aqueous phase, respectively. V_{org} and V_{aq} are the volume of the organic and the aqueous phases, respectively.

Rearrangement of the above equation by dividing each term with $c_{aq} V_{org}$ yields

$$E = \frac{D}{D + (V_{aq}/V_{org})} \times 100\% \qquad (9\text{-}6)$$

By defining the phase ratio (R) as $R = V_{aq}/V_{org}$, Equation (9-6) can be written as

$$E = \frac{D}{D + R} \times 100\% \qquad (9\text{-}7)$$

Equation (9-7) tells that percent extraction is governed by distribution ratio (D) and phase ratio (R).

Figure 9.3 shows how the percent extraction varies with the change of distribution ratio provided that $R = 1$. Extraction efficiency is high at a larger D. A 99.9% extraction for one extraction requires that $D > 1000$.

Increasing the amount of extraction solvent will improve the percent extraction, which is illustrated in Example 9.2.

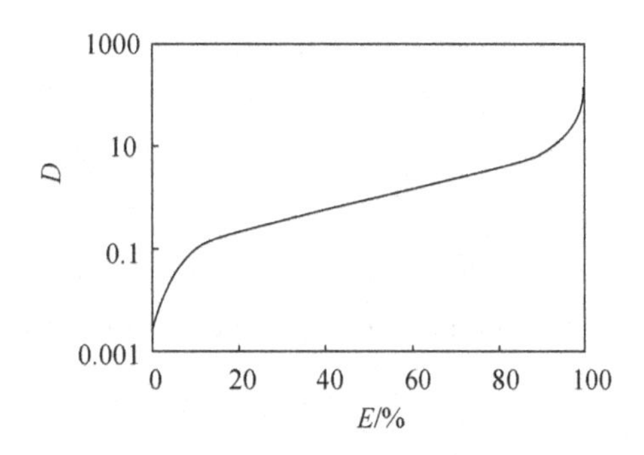

Figure 9.3 A plot of the distribution ratio (D) versus percent extraction (E).

【Example 9.2】 Iron forms a chelate that can be extracted into nitrobenzene with a D of 3. Ten mL of this solution is extracted with 5 mL of nitrobenzene. A second aliquot is extracted with 15 mL of nitrobenzene. Calculate percent extraction for each case.

Answer: For extraction with 5 mL nitrobenzene

$$E = \frac{3}{3 + 10/5} \times 100\% = 60\%$$

And for extraction with 15 mL nitrobenzene

$$E = \frac{3}{3 + 10/15} \times 100\% = 82\%$$

When D is small, percent extraction can be improved by repeating the same extraction several times (called multiple batch extractions). If a V_{aq} (mL) compound A is extracted by a V_{org} (mL) organic solvent, then

$$D = \frac{(c)_{org}}{(c)_{aq}} = \frac{(m_0 - m_1)/V_{org}}{m_1/V_{aq}}$$

where m_0 (g) is the total mass of the compound and m_1 (g) is the mass of compound A remaining in the aqueous solution after one extraction.

Rearrangement of the above equation yields

$$m_1 = m_0[V_{aq}/(DV_{org} + V_{aq})] \tag{9-8}$$

Then percent extraction is

$$E = \frac{m_0 - m_1}{m_0} \times 100\% = \{1 - [V_{aq}/(DV_{org} + V_{aq})]\} \times 100\% \tag{9-9}$$

After n times of extraction, each with a volume of V_{org}, the mass of compound A remaining in the aqueous solution, m_n (g), is

$$m_n = m_0[V_{aq}/(DV_{org} + V_{aq})]^n \tag{9-10}$$

Then percent extraction after n times of extraction is

$$E = \{1 - [V_{aq}/(DV_{org} + V_{aq})]^n\} \times 100\% \tag{9-11}$$

【Example 9.3】 La^{3+} is extracted into chloroform ($CHCl_3$) by forming chelate with 8-hydroxyquinoline at pH 7.0. The distribution ratio is found to be 43. Calculate the percent extraction when 20 mL of 1 mg · L^{-1} La^{3+} is extracted by 10 mL $CHCl_3$

(1) for one time;

(2) for two times, each with 5 mL.

Answer: For one time extraction with 10 mL $CHCl_3$

$$m_1 = 20 \times \left(\frac{20}{43 \times 10 + 20}\right) = 0.89(\text{mg})$$

$$E = \frac{20 - 0.89}{20} \times 100\% = 95.6\%$$

For two time extraction, each time with 5 mL $CHCl_3$

$$m_2 = 20 \times \left(\frac{20}{43 \times 5 + 20}\right)^2 = 0.145(\text{mg})$$

$$E = \frac{20 - 0.145}{20} \times 100\% = 99.3\%$$

It is clear from this example that quantitative extraction is best carried out by multiple batch extractions with a small volume of solvent. This is also the reason that analytical glassware should be washed with multiple portions of solvent or reagent. After the extraction steps, the individual solvent portions are usually combined for completion of the analysis.

9.3.3 Extraction of Inorganic Species

Inorganic species are often extracted by forming uncharged complexes with organic chelating agents.

1. Partition of the Chelating Agent between the Aqueous and Immiscible Organic Phases

Many organic chelating agents are weak acids and their uncharged form is soluble in organic solvent but of limited solubility in water. The dominant forms in the aqueous phase are charged as the result of dissociation or association reaction.

Partition equilibrium for a weak acid (HL) chelating agent between the aqueous and the immiscible organic phases is

$$HL_{(org)} \rightleftharpoons HL_{(aq)}$$

And the distribution ratio is calculated as

$$D = \frac{[HL]_{org}}{[HL]_{aq} + [L]_{aq}} = \frac{[HL]_{org}}{[HL]_{aq}(1 + K_a/[H^+]_{aq})}$$

$$= \frac{K_D}{1 + K_a/[H^+]_{aq}} = K_D \cdot x(HL) \quad (9\text{-}12)$$

It can be seen from Equation (9-12) that $D = \frac{1}{2}K_D$ when $pH = pK_a$; $D \approx K_D$ when $pH < pK_a - 1$ because almost all the HL in aqueous phase is in HL form the same as that in the organic phase; when $pH > pK_a$, D decreases because $x(HL)$ decreases. The partition of acetylacetone ($CH_3COCH_2COCH_3$, $pK_a = 8.9$) between benzene and water ($K_D = 5.9$) as a function of pH is given in Figure 9.4. At $pH \ll pK_a$, acetylacetone is in its acid form, thus $D \approx K_D = 5.9$; at pH 8.9, $x(HL) =$

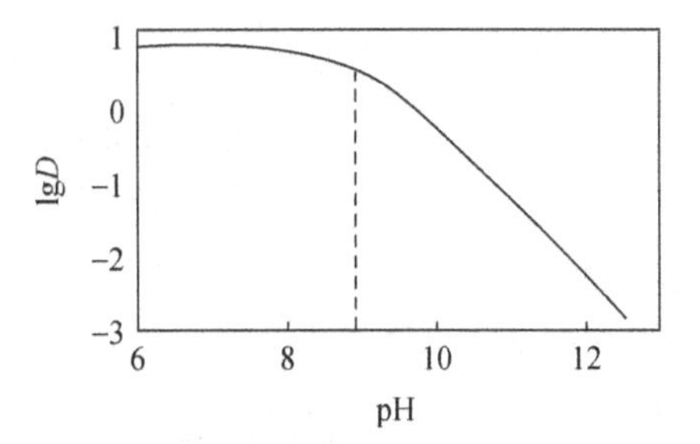

Figure 9.4 Plot of D of acetylacetone versus pH in aqueous phase.

0.5, thus $D=5.9/2\approx 3.0$.

The fractions of all the forms of a chelating agent in both aqueous and organic phases can be calculated. For a monoprotic acid (or base), there are three forms in the two phase system, $HL_{(org)}$, $HL_{(aq)}$ and $L^-_{(aq)}$. The total amount of the chelating agent, n_T, is

$$n_T = [HL]_{org}V_{org} + [HL]_{aq}V_{aq} + [L^-]_{aq}V_{aq}$$
$$= \left(\frac{1}{R}K_D K^H(HL)\cdot[H^+]_{aq} + K^H(HL)\cdot[H^+]_{aq} + 1\right)[L^-]_{aq}V_{aq} \quad (9\text{-}13)$$

Therefore, the molar fraction of each form in the system can be calculated with known K_D, K_a, R, and pH.

$$x(L^-)_{aq} = \frac{[L^-]_{aq}V_{aq}}{n_T} = \frac{1}{\frac{1}{R}K_D K^H(HL)\cdot[H^+]_{aq} + K^H(HL)\cdot[H^+]_{aq} + 1} \quad (9\text{-}14a)$$

$$x(HL)_{aq} = \frac{K^H(HL)\cdot[H^+]_{aq}}{\frac{1}{R}K_D K^H(HL)\cdot[H^+]_{aq} + K^H(HL)\cdot[H^+]_{aq} + 1} \quad (9\text{-}14b)$$

$$x(HL)_{org} = \frac{\frac{1}{R}K_D K^H(HL)\cdot[H^+]_{aq}}{\frac{1}{R}K_D K^H(HL)\cdot[H^+]_{aq} + K^H(HL)\cdot[H^+]_{aq} + 1} \quad (9\text{-}14c)$$

The composition of HL in a water-benzene two phase system as a function of pH is given in Figure 9.5.

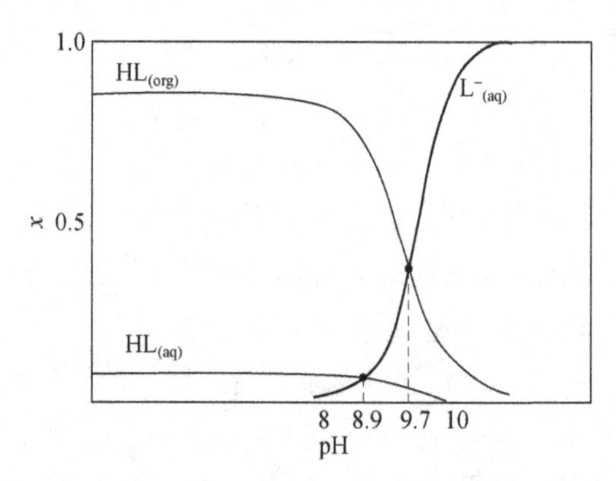

Figure 9.5 Distribution of acetylacetone in water-benzene two-phase system as a function of pH.

2. Extraction of Metal Ions

Because a substance must be neutral to be extracted into an organic phase, metal ions are extracted by forming neutral species by chelate formation or ion-association.

(1) Extraction by forming chelates

The equilibria involved in the solvent extraction of metal chelate are illustrated in Figure 9.6.

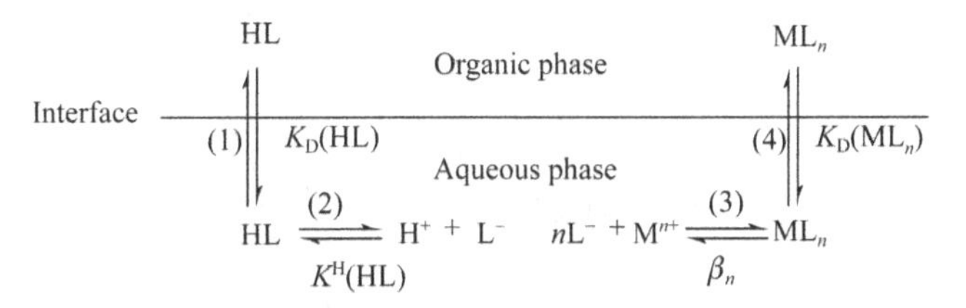

Figure 9.6 Equilibria involved in the extraction of an aqueous metal cation M^{n+} into an immiscible organic solvent containing chelating agent, HL.

The first equilibrium involves distribution of chelating agent (HL) between the aqueous and the organic phases. The second equilibrium is the dissociation of HL to give H^+ and L^- in the aqueous phase. The third equilibrium is the formation of complexes between M^{n+} and L^- in the aqueous phase. The forth equilibrium is the distribution of the formed chelate between the two phases. The overall equilibrium is the sum of these four reactions as given in Equation (9-15)

$$M^{n+}_{(aq)} + nHL_{(org)} \xrightleftharpoons{K_{ex}} ML_{n(org)} + nH^+_{(aq)} \tag{9-15}$$

The equilibrium constant for this reaction, also called extraction constant (K_{ex}) is

$$K_{ex} = \frac{[ML_n]_{org}[H^+]^n_{aq}}{[M^{n+}]_{aq}[HL]^n_{org}} = \frac{K_D(ML_n) \cdot \beta_n}{(K_D(HL) \cdot K^H(HL))^n} \tag{9-16}$$

Assuming that the chelate formed distributes largely into the organic phase and is essentially not dissociated in the organic solvent, the distribution ratio is given by

$$D = \frac{[ML_n]_{org}}{[M^{n+}]_{aq}} = K_{ex}\frac{[HL]^n_{org}}{[H^+]^n_{aq}} \tag{9-17}$$

Generally, the amount of chelating agent in the organic phase is in large excess with respect to M^{n+} in the aqueous phase so that the amount of chelating agent reacted with the metal ion and that in aqueous phases can be neglected, thus $[HL]_{org} \approx c(HL)_{org}$. Equation (9-17) can be simplified as

$$D = K_{ex}\frac{(c(HL)_{org})^n}{[H^+]^n_{aq}} \tag{9-18}$$

Taking logarithmic of Equation (9-18) gives

$$\lg D = \lg K_{ex} + n\lg c(HL)_{org} + n\text{pH}_{aq} \tag{9-19}$$

In addition to the dissociation of HL in aqueous phase and the distribution of chelating agent (HL) between the aqueous and organic phases, there are side reactions associated with the metal ions in aqueous solution as well as the side reaction of HL in the organic phase. Considering the side reactions, the conditional extraction constant (K_{ex}) will be

$$K'_{ex} = \frac{K_{ex}}{\alpha_M \cdot \alpha_{HL}^n} = \frac{[ML_n]_{org}[H^+]_{aq}^n}{[M']_{aq}(c(HL)_{org})^n} \tag{9-20}$$

and distribution ratio is

$$D = \frac{[ML_n]_{org}}{[M']_{aq}} = \frac{K_{ex} \cdot (c(HL)_{org})^n}{\alpha_M \cdot \alpha_{HL}^n \cdot [H^+]_{aq}^n} \tag{9-21}$$

where α_M is calculated in the same way as introduced in complexometric titration, α_{HL} is the total amount of chelating agent in both phases divided by that in organic phase, i.e.,

$$\alpha_{HL} = \frac{[HL]_{org}V_{org} + [HL]_{aq}V_{aq} + [L^-]_{aq}V_{aq}}{[HL]_{org}V_{org}} = \frac{[HL]_{org}/R + [HL]_{aq} + [L^-]_{aq}}{[HL]_{org}/R}$$

$$= 1 + \frac{1}{K_D/R} + \frac{1}{(K_D/R) \cdot K^H(HL) \cdot [H^+]_w} \tag{9-22}$$

When $pH \leqslant \lg K^H(HL)$,

$$\alpha_{HL} \approx 1 + \frac{1}{K_D/R} \tag{9-23}$$

α_{HL} is primarily governed by the distribution constant and the phase ratio, and is close to a constant with $pH \leqslant \lg K^H(HL)$. When $pH > \lg K^H(HL)$, α_{HL} increases drastically as pH increases. The logarithmic value of Equation (9-23) is

$$\lg D = \lg K_{ex} - \lg \alpha_M - n\lg \alpha_{HL} + n\lg c(HL)_{org} + npH_{aq} \tag{9-24}$$

On examining the terms in Equation (9-24), we see that the distribution ratio is associated with the change of concentration of the chelating agent or pH, that is the extraction efficiency can be affected only by changing the concentration of the chelating agent or by changing the pH. Increasing chelating agent concentration or pH can increase the distribution ratio, thus the extraction coefficient. Plots of percent extraction of selected metal ions by 8-hydroxyquinoline as a function of pH are presented in Figure 9.7. Define $pH_{1/2}$ as the pH when $E = 50\%$ with $V_o = V_{aq}$, then the difference of $pH_{1/2}$ between two metal ions needs to be about 3 pH units to achieve a 99.9% separation efficiency. From Figure 9.7, $pH_{1/2}$ for Cu^{2+}, Zn^{2+} and Pb^{2+} are 1.4, 3.3, and 5.1, thus we can conclude that the separations of Cu^{2+}-Zn^{2+} and Zn^{2+}-Pb^{2+} are incomplete and that of Cu^{2+}-Pb^{2+} is satisfactory.

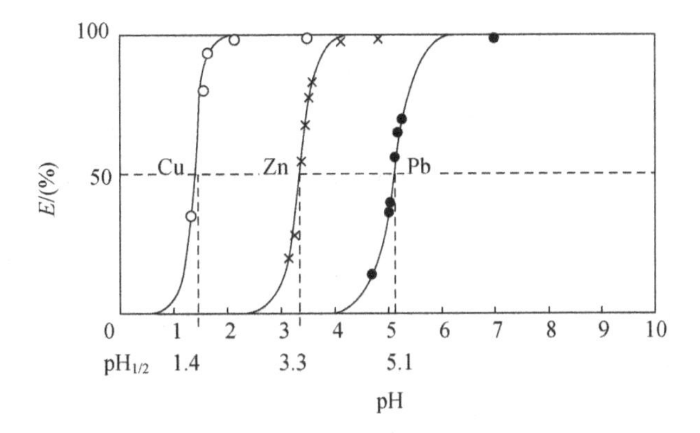

Figure 9.7 Extraction curves for metal-8-hydroxyquinoline complexes in chloroform ($CHCl_3$).

According to Equation (9-7), $E=D/(D+R)$, $D=1$ with $E=50\%$ and $V_o=V_{aq}$. Equation (9-24) can be rearranged into

$$pH_{1/2}=\frac{1}{n}\lg\alpha_M-\frac{1}{n}\lg K_{ex}+\lg\alpha_{HL}-\lg c(HL)_{org} \tag{9-25}$$

In one extraction system, α_{HL} and $c(HL)$ are the same for all metal ions, $pH_{1/2}$ is dictated by the side reaction of the metal ions and the extraction constant (K_{ex}) which is associated with the formation constant of the metal ion with the chelating reagents. To achieve a desirable extraction efficiency which means the desirable $pH_{1/2}$ difference, we can introduce a masking agent to mask the interfering metal ions (increase the $pH_{1/2}$), or select a chelating agent which has much larger formation constant with the metal ions to be separated.

(2) Extraction by ion-association

Ion-association is a chemical reaction whereby ions of opposite electrical charge come together in solution to form an uncharged entity which is insoluble in aqueous solution and can be extracted into an organic phase. For example, diethyl ether can be used to extract $FeCl_4^-$ from a strong acid solution. Oxygen containing solvents with a strong coordination ability, such as diethyl ether, isopropyl ether and methyl isobutyl ketone, form oxonium cations with protons in strong acidic solutions. Metals forming anionic complexes in strong acids can be extracted into this oxygen containing organic solvent as ion pairs with oxonium ions. Fe(Ⅲ) is extracted from 6 mol · L^{-1} hydrochloric acid into diethyl ether as an ion pair $[(C_2H_5)_2O]_3H^+(H_2O)_n$-$FeCl_4^-$ with a distribution coefficient of 100 (Mizuike A. Enrichment Techniques for Inorganic Trace Analysis. Berlin: Springer-Verlag, 1983). Because high extraction efficiency and low selectivity, extractions by forming ion pairs with oxonium ions are often used to remove large amounts of interfering elements from the matrix. In 6~7 mol · L^{-1} HCl, rare earths, U (uranium), Th (thallium), Ba (barium), and Zr (zirconium) are not extractable into diethyl ether, thus the quantitative separation of Fe(Ⅲ) with no loss of the desirable elements is possible (Marin Ayranov, Joaquin Cobos, Karin Popaand Vincenzo V. Rondinella Determination of REE, U, Th, Ba, and Zr in simulated hydrogeological leachates by ICP-AES after matrix solvent extraction. Journal of Rare Earths, 2009, 27(1):123-127).

Dyes containing nitrogen atoms in the molecular structure can be protonized to form large cations. Inorganic anions can be extracted into an organic phase by incorporating with large cations via ion-association. For example, microamounts of boron can be converted into BF_4^- in aqueous solution containing HF and then extracted into benzene or toluene by ion-association with methylene blue. The ion-associate form of methylene blue-boric tetrafluoride (BF_4^-) is

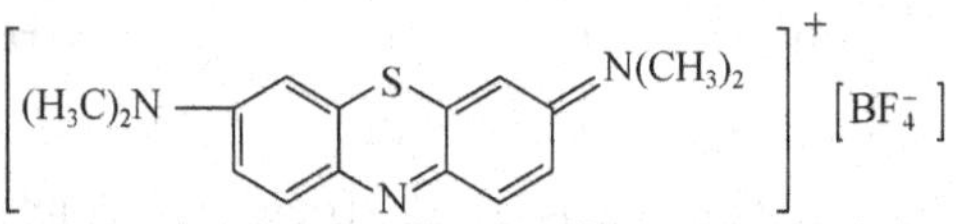

$$\left[(H_3C)_2N\text{—(phenothiazine: S, N)}=N(CH_3)_2\right]^+ [BF_4^-]$$

9.3.4 Other Extraction Methods

1. Continuous Extractions

Many materials of analytical interest have very low distribution ratios that even multiple batch extractions are not practical to use. A continuous extraction with an apparatus setup that can be left unattended for long periods of time can be used to achieve the goals. A continuous extraction setup allows the reuse of small amount of extracting solvent, the large surface area of contact between the two phases, and long contact time so that a greater extraction efficiency is obtained.

As an example, a typical laboratory-scale extractor for use with solvents heavier than water is shown in Figure 9. 8. The solvent is placed in the flask (A) and heated. The vapor rises to B then C, where it is condensed. There is a seal at B so that the liquid cannot drop back in the flask but runs into the extractor (D). The liquid drops through the aqueous layer where the analytical sample is dissolved because the solvent is heavier than water. The solvent containing the extract collects at F and goes back into the flask. With this continuous extraction process, the extract can be quantitatively collected in the flask. Caffeine can be extracted from the cola beverages in one hour by using chloroform ($CHCl_3$) as the solvent.

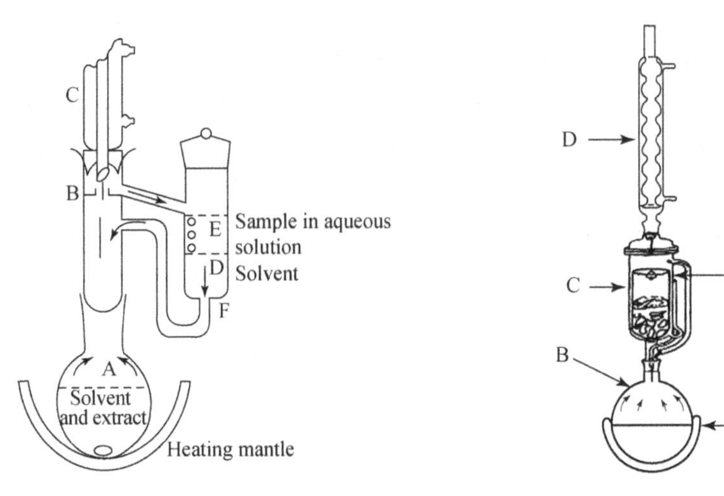

Figure 9.8 A continuous solvent-heavier-than-water extractor.

Figure 9.9 A Soxhlet extractor.

The continuous extractor can also be used for liquid-solid extraction in the cases when the desired component is extracted directly from a solid sample. The problem encountered with extraction from solid sample is that the distribution ratio of desired compound is low, the process is slow, and the large amount of solvent is required, which then has to be removed after extraction. The most convenient laboratory-scale extractor for liquid-solid extraction is the one developed by Franz Ritter von Soxhlet, a German agricultural chemist.

The apparatus (Figure 9.9) consists of a heating device (A) and a solvent reservoir and extract flask (B). An extraction chamber (C) is placed to the top of B. When the solvent is heated, it will go up in the side arm of the extraction chamber, be condensed and drop onto the sample. When the solvent level reaches E, the solvent and the extract will siphon back into A. This process continuously repeats with the fresh solvent dripping onto the sample. After each cycle, the extract becomes a little more concentrated in the flask. The compound of interest can then be quantitatively extracted when long enough extracting period of time is allowed.

2. Solid Phase Extraction (SPE)

Solid phase extraction (SPE) is liquid-solid separation, and the compound of analytical interest is extracted from the liquid phase to the solid phase which can be membranes, small discs, and often small columns or cartridges. A solid phase extraction cartridge is shown in Figure 9.10. A hydrophobic organic compound is coated or chemically bonded to silica particles to form the solid extracting phase. When the sample solution is poured through the cartridge, usually the desired compounds are retained and the impurities pass through. The desired compounds now highly concentrated are eluted with a few milliliters of solvent. The general steps (as shown in Figure 9.11) for a solid phase extraction include: (1) select the proper size SPE tube or disk; (2) condition the tube; (3) load the sample; (4) wash the disk; and (5) elute the compounds of interest. In some solid phase extractions, the impurities are extracted into the solid phase, while compounds of interest pass through unretained.

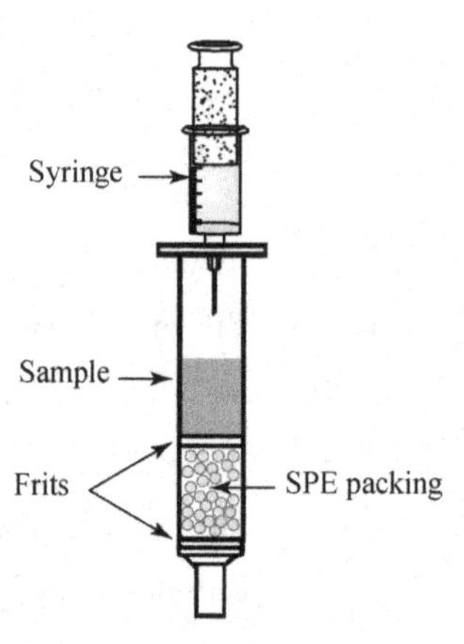

Figure 9.10 A solid phase extraction syringe. The syringe on the top of the cartridge is to apply the pressure to facilitate flow through the sample and the eluent.

Solid phase extraction is advantageous over conventional liquid-liquid extraction when trace components are of interest: The extraction is faster, requires less solvent, reduces the need for time consuming concentration steps, and is easily automatable. The extract can be eluted with a few milliliters of solvent, so the solvent removal for further concentration is nearly or completely eliminated.

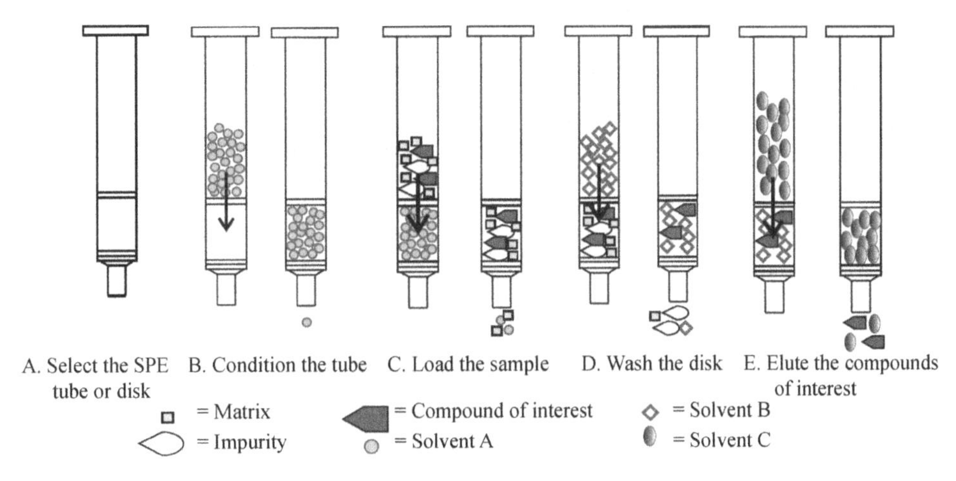

Figure 9.11 A summary of the solid phase extraction process.

9.4 SEPARATION BY ION EXCHANGE

Ion exchange is a process by which ions held on a solid supporting matrix called **resin** are exchanged for ions in a solution brought into contact with the resin. The ion-exchanging capability of different ions in the solution varies with selected resin so that separation can be conducted by sequentially washing the exchanged ions on the resin.

9.4.1 Ion Exchange Resins

Synthetic ion exchange resins are porous structured, cross-linked polymers that contain large numbers of an ionic functional group in the molecules. The resins are usually the divinylbenzene cross-linked polystyrene with covalently attached ionic functional groups, for example the sulfonic group (Figure 9.12). Polystyrene ion exchange resins are stable to about 150℃ and are resistant to strong acids and bases but not to strong oxidants or hydrogen peroxides.

There are basically four types of ion exchange resins used in analytical chemistry. These resins are summarized in Table 9.2. Cation exchange resins contain acidic groups, whereas anion exchange resins contain basic groups. The strong-acid cation exchangers have sulfonic acid groups, $—SO_3H$; and the weak cation exchangers have carboxylic acid groups, $—CO_2H$, sometimes phenolic hydroxy groups, which are partially dissociated. The protons of these groups can exchange with other cations as shown in the reactions below where "res" and "aq" in

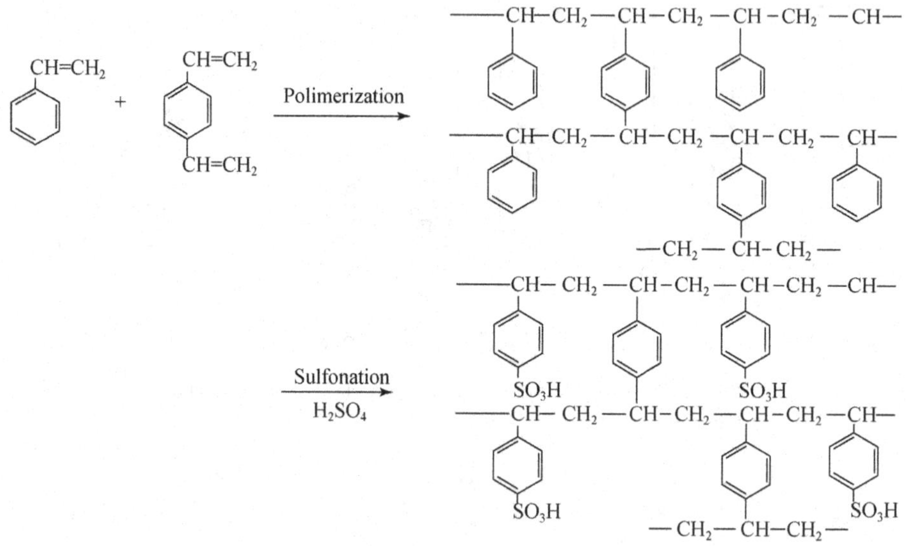

CH=CH$_2$ + CH=CH$_2$ / CH=CH$_2$ —Polimerization→ —CH—CH$_2$—CH—CH$_2$—CH—CH$_2$—CH—

—CH$_2$—CH—CH$_2$—

Sulfonation / H_2SO_4 → —CH—CH$_2$—CH—CH$_2$—CH—CH$_2$—CH— ; SO_3H ; SO_3H ; SO_3H ; SO_3H

—CH$_2$—CH—CH$_2$—

Figure 9. 12 Preparation reaction and structure of a cross-linked polystyrene ion exchange resin containing strong acid cation exchangers.

the parenthesis represent the resin and aqueous phases.

$$nR{-}SO_3H_{(res)} + M^{n+}_{(aq)} \rightleftharpoons (R{-}SO_3)_nM_{(res)} + nH^+_{(aq)}$$

$$nR{-}COOH_{(res)} + M^{n+}_{(aq)} \rightleftharpoons (R{-}COO)_nM_{(res)} + nH^+_{(aq)}$$

Weak acid cation exchange resins are more restricted in pH$>$5 and generally used to separate strongly basic or multifunctional ionic species such as proteins or peptides that are often firmly retained on strong acid exchangers. Strong acid cation exchange resins can be used with pH 1 ~ 14 and are more generally preferred especially for complex mixtures. Cation exchange resins are usually supplied in the hydronium ion form, but they are easily converted to the sodium forms when treated with sodium salts. The sodium salts then exchanges with other cations.

Anion exchange resins contain hydroxyl (—OH) groups and can be exchanged with other anions. The strong-base anion exchangers have quaternary amine groups, $-RN(R)_3^+$; and the weak anion exchangers have amine groups, $-RNH(R)_2$, $-RNH_2R$, or $-RNH_3$. The exchange reactions can be represented by

$$nR{-}NR_3^+OH^-_{(res)} + A^{n-}_{(aq)} \rightleftharpoons (R{-}NR_3)_nA_{(res)} + nOH^-_{(aq)}$$

$$nR{-}NHR_2^+OH^-_{(res)} + A^{n-}_{(aq)} \rightleftharpoons (R{-}NHR_2)_nA_{(res)} + nOH^-_{(aq)}$$

The strong basic exchangers can be used over the pH range of 0 to 12, but the weak base exchangers can only be used at pH of 0 to 9 and preferably for separation of strong acids which can also be done by strong base exchangers.

Table 9.2 Types of Ion Exchange Resins

Types of Exchanger	Functional Groups	Examples
Cation		
Strong acid	Sulfuric acid	$-SO_3^-$, $-CH_2CH_2SO_3^-$
Weak acid	Carboxylic acid	$-COO^-$, $-CH_2COO^-$
Anion		
Strong base	Quaternary amine	$-CH_2N(CH_3)_3^+$, $-CH_2CH_2N(CH_2CH_3)_3^+$
Weak base	Amine	NH_3^+, $-CH_2CH_2NH(CH_2CH_3)_2^+$

Chelating resin is usually perceived as a subgroup of ion exchange resins that is a cross-linked polystyrene containing the active group, $-CH_2N(CH_2COOH)_2$. In chelating resins, counter ions are bound to resin by coordinate covalent bond or by its combination with electrostatic interactions. The selective interaction of the functional group of chelating resin and the metal associated with the nature of the metal allows the selective removal of target metals via suitable functional groups of chelating resins.

9.4.2 Cross-linkage and Exchange Capacity

The degree of **cross-linkage** of the divinylbenzene cross-linked resin depends on the proportions of different monomers used in the polymerization step and is expressed as mass percent of divinylbenzene in the reactants in Equation (9-26).

$$\text{Cross-linkage} = \frac{\text{mass of divinylbenzene}}{\text{total mass of divinylbenzene and polystyrene}} \tag{9-26}$$

Generally, cross-linkage of 8% to 10% is used. The greater the cross-linkage of the resin, the greater the difference in selectivity. Generally, increasing cross-linkage also increases the rigidity of the resin, reduces swelling and porosity, as well as the solubility of the resin and the exchange rate. In practice, the selection of cross-linkage is dependent on the analytes to be separated. To separate amino acids, the resin with 8% cross-linkage is appropriate. For separation of peptides or proteins, the resin with 2%~4% cross-linkage is used. It is always appropriate to use a resin with greater cross-linkage if the exchange rate is reasonable.

The **ion-exchange capacity** is defined as the amount of ions that can be exchanged with 1 gram of dry resins and is dependent on the number of acidic (in the hydronium ion form) or basic (in the chloride form) groups in the porous structured resins. The ion-exchange capacity of the polystyrene ion-exchange resins is typically of 3~6 mmol/g.

9.4.3 Ion Exchange Equilibria

Like any other equilibria, the ion-exchange equilibrium can also be treated by the law of mass action. For example, the equilibrium for a cation exchange resin can be expressed as

$$nR^- A^+_{(res)} + B^{n+}_{(aq)} \rightleftharpoons R_n B^{n+}_{(res)} + nA^+_{(aq)}$$

The equilibrium constant K of the above exchanging equilibrium is

$$K = \frac{[B^{n+}]_{res}[A^+]^n_{aq}}{[A^+]^n_{res}[B^{n+}]_{aq}} \tag{9-27}$$

The K value indicates the ability of the resin to absorb the cation B^{n+}. If K for an ion is large, there is a strong tendency for the resin to retain this ion. With a small K value, the opposite is true.

The distribution coefficients of A^+ and B^{n+} between the resin and aqueous phases, similar to the distribution coefficient in extraction equilibrium, are defined as

$$K_D(A) = \frac{[A^+]_{res}}{[A^+]_{aq}} \quad \text{and} \quad K_D(B) = \frac{[B^{n+}]_{res}}{[B^{n+}]_{aq}}$$

Then Equation (9-27) can be rewritten as

$$K = \frac{K_D(B)}{K_D^n(A)} \tag{9-28}$$

If $n=1$, Equation (9-28) becomes

$$K = \frac{K_D(B)}{K_D(A)} \tag{9-29}$$

K can also be defines as selective coefficient or separation factor of the resin for B respect to A. K is associated with the effective diameters of the hydrated ions as well as the number of charges of the ions. Thus for a sulfonated cation-exchange resin, K values for univalent cations increase in the order $Li^+ < H^+ < Na^+ < NH_4^+ < K^+ < Rb^+ < Cs^+ < Ag^+ < Tl^+$. For divalent cations, the order is $Mg^{2+} < Ca^{2+} < Sr^{2+} < Ba^{2+} < Fe^{2+} < Co^{2+} < Ni^{2+} < Cu^{2+} < Zn^{2+}$. K is bigger for cations with higher charge number, for example, $Na^+ < Ca^{2+} < Fe^{3+} < Th^{4+}$. For a resin containing strong base exchangers, the order for anions is $F^- < OH^- < Ac^- < HCOO^- < H_2PO_4^- < Cl^- < NO_3^- < HSO_4^- < CrO_4^{2-} < SO_4^{2-}$. In exchange process, the ions with bigger K value will be exchanged onto the resin than those with smaller K values. In the elution, the ion with smaller K value is eluted first rather than those with bigger K values.

9.4.4 Applications of Ion Exchange Separation

1. Purification of Water

One of the most important applications of ion-exchange is the deionization of water. Deionized water prepared by ion-exchange is widely used for industrial and research purposes. The water is passed through a mixed-bed ion-exchange resin that contains both a strong acid cation and a strong base anion exchange resins. When water containing a salt such as $CaCl_2$ is passed through the column, the Ca^{2+} ion is exchanged for two H^+ ions and the two Cl^- ions are exchanged for two OH^- ions. The net result is that the salt is exchanged for H_2O, and the water is deionized to provide water with the conductivity less than 0.3 $\mu S/cm$.

2. Concentration of Dilute Electrolytes

The ion-exchange method is effective for collecting ionic substances from large volumes of dilute solutions. Elution with a small volume of eluent can achieve considerable concentration so that the analyte concentration, e.g., copper from milk, is suitable for the analysis.

3. Separation of Interfering Metal Ions

Ion-exchange resins are used to eliminate ions that would interfere with an analysis. In the precipitation gravimetric determination of SO_4^{2-} by formating $BaSO_4$, Fe^{3+} if coexisting with SO_4^{2-} often coprecipitates with $BaSO_4$, thus causes analytical errors. Fe^{3+} can be separated by cation ion-exchange resin before precipitation. Passing the sample through the cation ion-exchange resin, Fe^{3+} is retained and SO_4^{2-} is not. Then effluent containing SO_4^{2-} can be directly precipitated by Ba^{2+}.

4. Separation of Metal Ions

Chromium exists in the environment as the trivalent, Cr(Ⅲ), or hexavalent, Cr(Ⅵ) form. Cr (Ⅲ) is considered to be essential to mammals for the maintenance of glucose, lipid, and protein metabolism, but Cr(Ⅵ) is known to have an adverse affect on the lung, liver, and kidney body organs. The determination of Cr(Ⅲ) and Cr(Ⅵ) in their existing forms as well as total concentration of chromium in drinking water, groundwater and industrial wastewater are very important.

The analytical procedure involves the use of ion-exchange columns. Cation-exchange columns are used for the collection of chromium(Ⅲ), and anion exchange columns are used for the selective preconcentration of chromate. Water samples are acidified, passed through membrane filters (0.45 μm) and then through ion-exchange columns. The columns are then eluted with 5.0 mol · L^{-1} HNO_3 and 1.0 mol · L^{-1} NH_4NO_3 + 0.1 mol · L^{-1} HNO_3 for chromium(Ⅵ) and chromium(Ⅲ), respectively. The concentrations of both chromium species can then be determined (Johnson C A. Rapid ion-exchange technique for separation and preconcentration of chromium(Ⅵ) and chromium(Ⅲ) in fresh waters. Analytica Chimica Acta, 1990, 238: 273).

5. Separation of Amino Acids

Ion-exchange methods also find applications in separating species of biological importance. Amino acids can be cationic or anionic by adjusting pH. A mixture of amino acids can be separated with a cation-exchange column (for example, Dowex 50) at low pH. Depending on the isoelectric point, the amino acids will be eluted with a buffer of increasing pH.

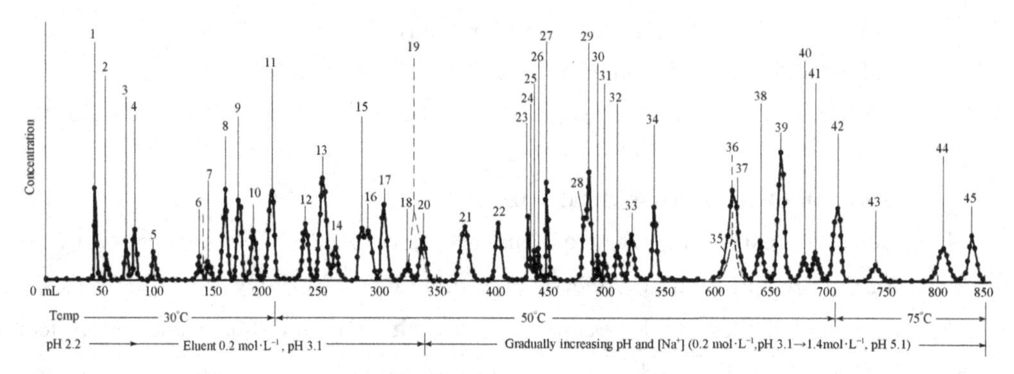

Figure 9.13 Separation of amino acids and related compounds on Dowex 50-x4 column, 0.9 by 150 cm (Moore S, Stein W H. J Biol Chem, 1951, 192:663—681; 1954,211:893—906).

1,Cysteic acid + Phosphoserine; 2, Glycerophosphoethanolamine; 3, Phosphoethanolamine; 4, Tourine; 5, Urea; 6, Hydroxyproline; 7, Methionine sulfoxides; 8, Aspartic acid; 9, Threonine; 10, Serine; 11, Glutamine asparagine sarcosine; 12, Proline; 13, Glutamic acid; 14, Lanthionine; 15, Citrulline felinine; 16, Glycine; 17, Alanine; 18, α-NH_2-adipic acid; 19, Glutathione(—S—S—); 20, α-NH_2-n-butyric acid; 21, Valine; 22, Cystine; 23, Cystathionine; 24, Methionine; 25, Diaminopimelic acid; 26, Isoleucine; 27, Leucine; 28, β-Alanine; 29, Glucosamine; 30, β-NH_2-isobutyric acid; 31, Homocystine; 32, Tyrosine; 33, Phenylalanine; 34, γ-NH_2-butyric acid; 35, Hydroxylysine; 36, Ethanolamine; 37, NH_3; 38, Ornithine; 39, Lysine; 40, Creatinine; 41, Histidine; 42, 1-Methylhistidine anserine carnosine; 43, 3-Methylhistidine; 44, Tryptophan; 45, Arginine.

9.5 SEPARATION BY CHROMATOGRAPHY

Chromatography is a technique in which the components of a mixture are separated based on differences in the rate at which they are carried through a stationary phase with a gaseous or liquid mobile phase. In 1903, Michael Tswett described the separation of plant pigments into colored bands on a calcium carbonate ($CaCO_3$) column. Tswett passed the plant pigments dissolved in petroleum ether through the calcium carbonate column and eluted the pigments with petroleum ether. He observed colored bands which were the different forms of chlorophylls separated based on the difference of absorption capability of the pigments on $CaCO_3$. In this chromatographic separation procedure, $CaCO_3$ is called the **stationary phase** which is fixed in either in a column or on a planar surface, and petroleum ether is called the **mobile phase** which carries the analyte mixture and moves over or through the stationary phase.

9.5.1 Classification

In chromatography, the mobile phase can be liquid or gas, and the stationary phase can be solid or liquid film coated over a solid surface. Chromatographic methods can be classified by the physical state of the mobile and stationary phases. The chromatographic methods are often named by listing the type of mobile phase followed by the type of stationary phase. Thus, in gas-liquid chromatography, the mobile phase is a gas and the stationary phase is a liquid. Gas chromatography means the mobile phase is a gas.

Chromatographic methods can also be classified in terms of the approaches of contact between the mobile phase and stationary phase: **column chromatography** and **planar chromatography**. In column chromatography the stationary phase is held in a narrow column and the mobile phase is forced to pass through the column by gravity. In planar chromatography, the stationary phase is coated on a flat glass or simply is a porous paper, and the mobile phase moves through the stationary phase by capillary action or under the influence of gravity.

Chromatographic methods can also be classified by the chemical or physical mechanisms responsible for separation.

- **Adsorption chromatography**: Separation is based on the ability of solutes to adsorb on a solid stationary phase.

- **Partition chromatography**: Separation is based on the difference of the partition equilibrium of the solute between the liquid stationary and mobile phases.
- **Ion-exchange chromatography**: Separation is based on the ion-exchange. The stationary phase is covalently attached with cationic or anionic functional groups as ion exchangers. Ions to be separated are attracted to the stationary phase by electrostatic forces and eluted by mobile phase that dissociates the adsorbed ions on the stationary phase.
- **Size-exclusion chromatography**: Separation is based on the size of the solutes through a porous stationary phase. Large solutes cannot penetrate through the pores and pass the column rapidly. The small solutes are entrapped in the porous stationary phase and spend more time in the column.
- **Electrophoretic Separation**: Separation is based on the mobility of the solutes in their charged forms migrating under the influence of applied potentials.

Chromatographic methods are the most widely used in modern analytical laboratories, because the separation and determination can be integrated, making the analysis convenient. There are many books and courses which particularly introduce modern chromatography. In this section, the classical column chromatography and thin layer chromatography are introduced because these two techniques are still widely used.

9.5.2 Chromatogram

In 1952, Martin and R. L. M. Synge were awarded the Nobel Prize in chemistry for their theoretical development of chromatographic separation—partition chromatography. The basis of the chromatography is that a distribution equilibrium occurs at each point on the column, leading to a separation of the components similar to that in liquid-liquid extraction. The chromatographic separation of a mixture containing A and B is illustrated in Figure 9. 14. At some specific time, the column packing can be removed and the individual component can be extracted from the bands and identified. The concentration profiles of A and B can be obtained by the concentration versus distance the component migrates through the column. The elution time profile (chromatogram) can also be obtained by monitoring the concentration of the components at the column exit. Then a plot of concentration versus the time required for a component to reach the detector at the end of the column is plotted. All modern chromatographic methods except planar chromatography employ elution time profile analysis.

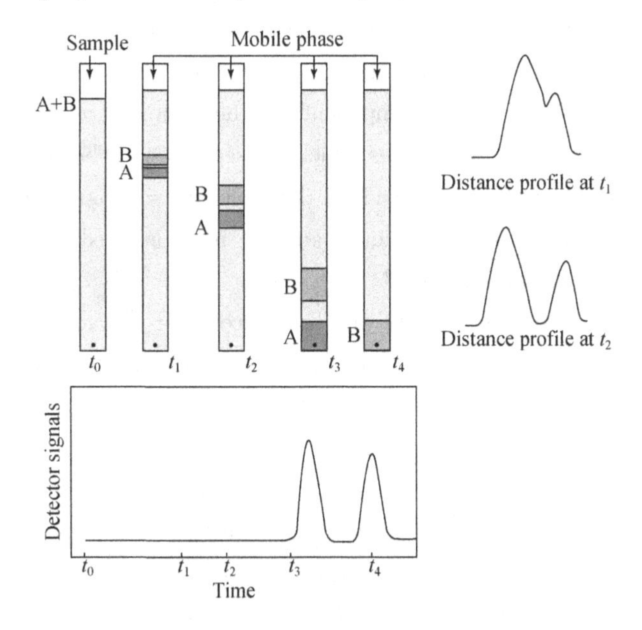

Figure 9.14 Schematic drawing of chromatographic separation of a mixture containing A and B by column elution chromatography.

9.5.3 Column Chromatography

Column chromatography is a separation technique in which the stationary bed is placed in a tube (packed column). Column chromatography is frequently used by synthetic organic chemists to separate products of their reactions, and for preparative applications of microgram to kilogram quantities. The column can be prepared by dry packing or wet packing. For the dry method, the column is first filled with dry stationary phase powder, followed by the addition of mobile phase, which is then flushed through the column until completely wet. For the wet method, a slurry is prepared of the eluent with the stationary phase powder and then carefully poured into the column.

The separation can be based on adsorption, partition, ion-exchange and size-exclusion, depending on the stationary phase used. The most common stationary phase for column chromatography are adsorbents such as silica gel, alumina, and charcoal (activated carbon).

The mixture to be analyzed by column chromatography is loaded on the top of the column. The solvent (the eluent) is passed through the column by gravity (gravity column chromatography) or by the application of air pressure (flash column chromatography). At each point, the equilibrium between the solute adsorbed on the adsorbent and the eluting solvent passing down through the column is established. Because interactions with the stationary and mobile phases of the components in the mixture are different, the solutes pass through the column at different rates to achieve separation. The effluent is collected in fractions. Fractions are typically analyzed by thin-layer chromatography to determine the extent of component separation. Fractions can also be monitored with a UV detection system and the desired fraction collected.

When selecting an eluting solvent, one has to consider the adsorption power of the stationary phase and the polarity of the solutes. The polarity of the eluting solvent affects the relative rates at which compounds move through the column. Proper choice of an eluting solvent is thus crucial to the successful application of column chromatography as a separation technique. If very strong solvents are used, all solutes pass through the column together without separation. Ether/ hydrocarbons / carbonyl solvents are of common use. The polarity of solvents in the increasing order is: petroleum ethers & hexanes < cyclohexane < carbon tetrachloride < toluene < benzene < dichloromethane < chloroform < ethyl ether < ethyl acetate < n-propanol < ethanol < methanol < water.

9.5.4 Planar Chromatography

1. Paper Chromatography

In paper chromatography, a piece of moist paper is used as the stationary phase. A suitable liquid solvent or mixture of solvents can be used for mobile phase. There are several types of paper chromatography named for the direction that the mobile phase moves through the paper: ascending, descending, horizontal, and two-way. Ascending paper chromatography is introduced here as an example (Figure 9.15).

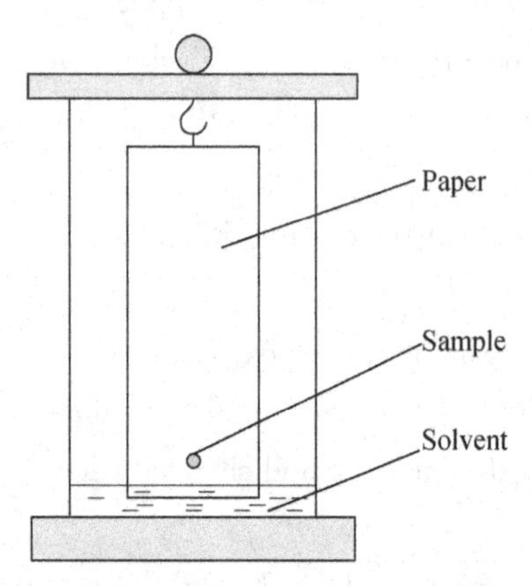

Figure 9.15 Illustration of ascending paper chromatography.

The sample is spotted 1～2 cm from the end of a chromatographic paper strip. The paper is hung vertically in a sealed chromatographic chamber with the spotted end dipped about 5 mm into the solvent thus there is solvent below the spots. The solvent rises up the paper via capillary action. When the solvent is close to the top of the paper, the paper is removed from the chamber. The solvent front is marked with a soft pencil and the paper is dried. If the eluted substances are colorless, a chemical reaction must be employed to visualize the spots (solutes) developed during separation. Generally, silver nitrate is used for reducing sugars, acid-base indicators for organic acids, and ninhydrin (triketohydrindene) for amino acids. The separated substances can also be disclosed under the UV light if these substances are fluorescent.

The ratio of fronts (R_f) is defined as the ratio of the distance travelled by the substance to the distance travelled by the solvent.

$$R_f = \frac{\text{Distance traveled by the substance}}{\text{Distance traveled by the solvent}} \tag{9-30}$$

For example, if one component of a mixture travelled 8. 4 cm from the base line while the solvent had travelled 12. 0 cm, then the R_f value for that component is $R_f = 8.4/12 = 0.70$.

The R_f value is in the range of $0 \sim 1$. $R_f = 0$ means that the solute does not move with the solvent as the solvent advanced, and $R_f = 1$ means the solute travels with the solvent front without retention on the stationary phase. R_f is valuable in judging the extent of the resolution and the approximate position of spots under investigation. One must keep in mind that R_f is not an absolute value unless all experimental conditions are controlled. It is always best to spot the knowns and the unknowns on the same sheet of paper. If two spots in the final paper chromatogram have the same R_f values, this does not necessarily true that they are the same compound. For final identification, chromatography with two or more mobile phases of different polarity is required.

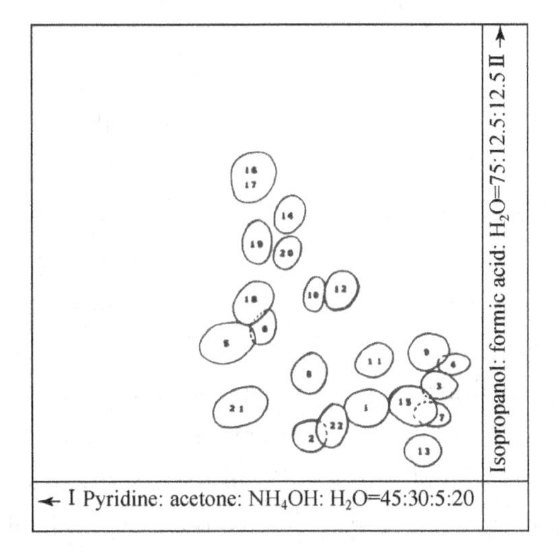

Figure 9. 16 Two-way paper chromatogram of urinary amino acids (Carmine J Spinella. Clinical Chemistry, 1969, 1011-1016).

Two-way paper chromatography can circumvent the problem of separating substances which have very similar R_f values. In two-way paper chromatography, the solvents are used sequentially to separate the components in a mixture. The sample is spotted in a corner of a square or rectangular paper. The separation is then performed as described in the ascending paper chromatography. After the first separation, the solvent front is marked and the paper is turned 90°, and a second solvent of different polarity is used for separation. For example, a high degree of separation is obtained for urinary amino acids (Figure 9.16). The spots are visualized by spraying ninhydrin reagent, air drying and heating in an oven at 110~115℃ for about 1 min.

2. Thin Layer Chromatography

Thin layer chromatography (TLC) is a liquid-solid absorption technique where the mobile phase ascends by capillary action on the thin layer of the stationary phase, usually silica gel or alumina, coated onto a glass, metal or rigid plastic plate. A small amount of the mixture to be analyzed is spotted near the bottom of this plate. The TLC plate is then placed in a shallow pool of a solvent in a closed development chamber so that the liquid is below the starting line where the sample spots have been applied. The mobile phase solvent in the chamber slowly rises up the TLC plate. As the solvent moves past a spot, an equilibrium is established at each point for each component of the mixture between the molecules of that component which are adsorbed on the stationary phase and the molecules which are in solution. Because the components of mixtures hopefully differ in solubility and in the strength of their adsorption to the adsorbent, some components will be carried farther up the plate than others. When the solvent has reached the top of the plate, the plate is removed from the developing chamber, solvent front marked, plate dried, and the separated components of the mixture are visualized. R_f values are then obtained.

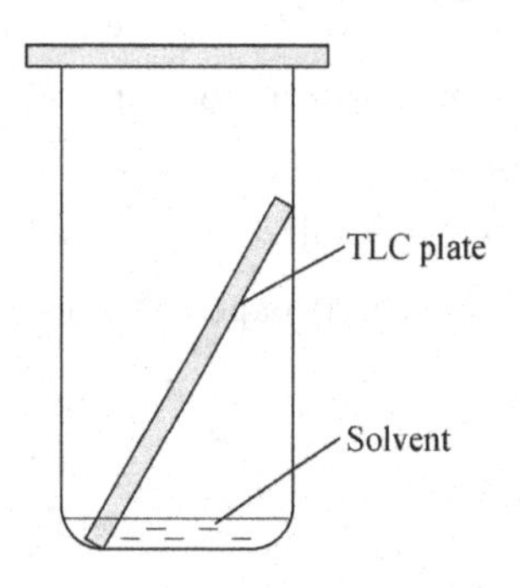

Figure 9.17 A diagram of a thin layer chromatography apparatus.

The illustration of a thin layer chromatographic apparatus is given in Figure 9.17. The TLC plates are commercially available. If you prepare your own plates, they must be placed in an oven for 3~4 hours at 120℃ to remove excess moisture. This drying step is called activation. A good solvent system is one that can move all components of a mixture to be analyzed off the baseline without carrying any component to the solvent front. The optimum separation of compounds by TLC is usually achieved when R_f values are in the range of 0.3 to 0.5.

Solvent system is equally important for TLC separation as to the column chromatography. The polarity of compounds to be separated and the eluting power of solvent, as well as the power of adsorption of the stationary phase must be considered when selecting a solvent system.

Colored components can be easily observed. If the components are colorless, the developed TLC separation can be visualized with: spray reagents, fluorescence, and charring.

Some reagents will react with the functional groups on compounds to produce a color or fluorescence, e. g. , ninhydrin forms a violet color with an amine group. After TLC separation, ninhydrin solution is then sprayed on the TLC plate to reveal the spots of amino acids in a similar way to the paper chromatography previously described.

If the components are naturally fluorescent, then the spots can be easily visualized by placing the plate under the UV irradiation (254 nm or 366 nm); however, there are many compounds that cannot be detected by this technique. An approach to detect almost any compound by fluorescence is to use fluorescent plates that have inorganic phosphors added to the TLC adsorbent. These compounds fluoresce under UV irradiation. The sample compounds that are not fluorescent cover this background fluorescence and appear as dark spots on the TLC plates. Those compounds that fluoresce naturally add their fluorescence on the fluorescent plate coating and produce colored spots and often distinctive fluorescence. For example, zinc-cadmium sulfide fluorescences with excitation at both 254 nm and 366 nm.

Charring is destructive and is the last choice for location of TLC spots. For example, concentrated sulfuric acid (H_2SO_4) is sprayed over the plate. Organic compounds are dehydrated to a charcoal black residue. The advantage of charring is that almost all the organic compounds are detectable in this process. The disadvantage is that the compounds are destroyed in the process so no further analysis is possible.

Chapter 9 Questions and Problems

9.1 What is the basic requirement for a compound to be extracted?

9.2 The iodine residue from a water purification compound can be detected qualitatively by extraction into CCl_4. Assuming that $D=4$, how many mL of CCl_4 is needed for a 50% extraction for a 10.0 mL water sample?

9.3 The weak acid HL ($K_a=2.0\times10^{-5}$) has a partition coefficient of $K_D=31$ between an aqueous phase and organic phase. A 50 mL HL aqueous solution is mixed with 5.0 mL of organic solvent. Calculate percent extraction at (1) pH=1.0; (2) pH=5.0.

9.4 To determine trace amount of chloroform ($CHCl_3$) in drinking water, 100 mL water is contacted with 1.0 mL pentane to extract $CHCl_3$. The percent extraction is found to be 53%. If 1.0 mL pentane is used to extract 10 mL the same sample, what percent extraction is expected to be?

9.5 The distribution ratio of iodine between an organic solvent and an aqueous solution is 8.0. A 100 mL such organic solvent is contacted with 50.0 mL of 0.0500 mol • L^{-1} aqueous solution. When the two phases reach equilibrium, 10.0mL organic phase is pipetted and I_2 is titrated using 0.0600 mol • L^{-1} $Na_2S_2O_3$. Calculate the volume of $Na_2S_2O_3$ needed for titration of I_2.

9.6 If 99% of a solute in 50.0 mL water is going to be removed by solvent extraction, what is the minimum distribution coefficient that permits two 25.0 mL extractions with toluene? And what about five 10.0 mL extractions?

9.7 A solution containing the following ions, Na^+, NO_3^-, Ba^{2+}, $FeCl_4^-$, Ag^+, Co^{2+} and NH_4^+. If the solution is passed through a cation exchange column. What is the order of elution?

9.8 A series of dyes is separated by thin layer chromatography. Calculate the R_f for each dye based on the data given below. Also predict the distance rhodamine B will move if the same solvent is used and moves 11.5 cm.

Substance	Solvent	Sudan Ⅳ	Bismarck brown	Rhodamine B	Fast green FEF
Distance moved/cm	6.6	0.0	1.6	3.8	5.6

CHAPTER 10

SOLVING A REAL ANALYTICAL PROBLEM

10.1 Definition of the Analytical Problem

10.2 Literature Review

10.3 Choosing a Method

10.4 Developing and Evaluating the Method

10.4.1 Selectivity

10.4.2 Accuracy

10.4.3 Sensitivity and Linear Dynamic Range

10.5 Conclusion

Up to now, we have marched through the necessary steps in the development of an analytical method, that is: choosing the analytical method; collecting a representative sample; processing the collected sample; eliminating possible interferences; calibrating the system and obtaining a standard curve; measuring the sample concentration; calculating results; and evaluating the reliability of the results. In this chapter, we will review the steps and general considerations in developing an analytical method by using one example of method development published in the literature of analytical chemistry.

10.1 DEFINITION OF THE ANALYTICAL PROBLEM

Arsenic is a ubiquitous element that ranks 20th in abundance in the earth's crust, 14th in seawater, and 12th in the human body. Arsenic is increasingly being found in drinking water in many parts of the world including Bangladesh, England, India, Thailand, and the United States. Diseases such as skin, bladder, lung, liver and kidney cancer can result from continued consumption of elevated levels of arsenic in drinking water. The World Health Organization (WHO) has reported the maximum level of arsenic in drinking water should be 10 $\mu g \cdot L^{-1}$. In some countries and regions of the world where the arsenic-contamination is widespread and alarming, an interim working limit of 50 $\mu g \cdot L^{-1}$ has been set (Keisuke Morita and Emiko Kaneko. Spectrophotometric determination of trace arsenic in water samples using a nanoparticle of ethyl violet with a molybdate-iodine tetrachloride complex as a probe for molybdoarsenate. Anal Chem, 2006, 78 (22): 7682-7688; Purnendu K Dasgupta, Huiliang Huang, Genfa Zhang, et al. Photometric measurement of trace As(Ⅲ) and As(Ⅴ) in drinking water. Talanta, 2002, 58(1): 153-164).

Erosion and leaching of arsenic from soil has led to arsenic in the aquatic compartment of countries and thence to the contamination of the drinking water. Arsenic is usually removed by processing in the water treatment plants. Simple, easy-to-handle, inexpensive analytical methods for analyzing arsenic with sufficient sensitivity are needed to provide quality control of the water treatment facilities of communities around the world.

Before selecting an analytical method, an analytical chemist needs to define the problem by answering the following questions:

- What forms of arsenic exist in the drinking water?

- What is the concentration range of arsenic to be determined?
- What analytical sensitivity and accuracy are desired?
- What other chemicals are present in the water sample?

The chemical forms of an analyte present in the sample are fundamental to the selection of the sample preparation method and analytical procedure. The concentration range of the analyte may limit the number of feasible methods. In natural water, the arsenic occurs mainly in the As(Ⅲ) (arsenite) and As(Ⅴ) (arsenate) inorganic forms. The concentration of arsenic after water processing at a water plant may be a few dozen of micrograms per liter, which is in the normal working range for spectrophotometric methods and thus spectrophotometry can be used for the determination of arsenic compounds in drinking water.

Because most analytical methods are based on the reactions and physical properties that may be shared by more than one chemical entity, it is important to know what other species may be present before choosing an analytical method for the determination of the analyte. If such information is not available, qualitative analysis is necessary to identify the chemical species present that are likely to interfere with the analytical methods being considered. With this information, proper sample preparation procedures and the most cost effective analytical methods can be selected.

10.2 LITERATURE REVIEW

There are general references as well as specific publications describing analytical methods for the desired sample (drinking water) and the analyte (arsenic compounds). Monographs, protocols, and published procedures each can be very helpful in better understanding the problem, deciding the sample preparation procedure, and selecting an analytical method.

We usually start a literature search with one or more of the review papers published in analytical journals which deal with the specific entities to be determined in the above situation needing analysis. It is always helpful to conduct a general literature search on the entity to be analyzed to get a clearer picture of the analytical problem, including all or most of the analytical steps involved in an analysis. Then the searched database can be refined and some of the publications can be obtained for more detail information on specific steps. The internet makes this search and review of literature easier and efficient. Available database for the analytical chemistry field could include: Chemical Abstracts, ISI Web of Science®, American Chemical Society, Royal Society of Chemistry, SpringerLink, Elsevier ScienceDirect and VIP Information (Chinese).

Analytical methods used in the published literature for the determination of arsenic include: (1) volumetric titrations such as complexometric titration, redox titration, and potentiometric titration; (2) instrumental methods such as atomic absorption spectrometry, spectrophotometry, electrochemical methods, and chromatographic methods; and (3) hyphenated methods such as Inductively Coupled Plasma-Mass Spectrometry (ICP-MS). In this text, spectrophotometry is selected for discussion based on the following consideration: (1) the concentration level of arsenics in drinking water is several to a few dozen micrograms per liter; (2) the volumetric titration methods deal with macroamounts of analyte, thus these methods are not sensitive enough for water arsenic determination; and (3) among the more sensitive, available instrumental analytical methods, spectrophotometry is used as an example in this chapter.

Spectrometry is a simple and sensitive method that is convenient for routine analysis. A few relevant references from literature searches are:

- The molybdenum blue colorimetric method has been extensively used for arsenic measurement with a detection limit of 15 $\mu g \cdot L^{-1}$ (Rupasinghe T, Cardwell T J, Cattrall R W, et al. Anal Chim Acta, 2004, 445: 229-238). A light-emitting, diode-based photometric method was established for the differential determination of ppb levels of As(Ⅲ) and As(Ⅴ) in the presence of ppm levels of phosphate(Dasgupta P K, Huang H, Zhang G, et al. Talanta, 2002, 58: 153-164). An in-line pre-concentration step in the arsenic analysis can improve the sensitivity of total inorganic arsenic concentration (Toda K, Ohba T, Takaki M, et al. Anal Chem, 2005, 77: 4765-4773).
- The formation of ion-association complex of the malachite green (MG^+) with arsenoantimonimolybdenum has been used for the spectrophotometric determination of arsenic in aqueous solutions (Wu Q F, Liu P F. Talanta, 1983, 30: 275-276). This work was extended to the use of ethyl violet (EV) for spectrophotometric determination of arsenic (Zhao X, Xu S, Yuan X. Huanjing Huaxue, 1987, 6: 70-75).
- Methods have been developed for the arsenic determination based on the formation of

EV^+-molybdoarsenate particles. A visual colorimetry with a detection limit of 10 $\mu g \cdot L^{-1}$ and a spectrophotometric determination with a detection limit of 4 $\mu g \cdot L^{-1}$ were developed (Morita K, Kaneko E. Anal Sci, 2006, 22: 1085-1089).

10.3 CHOOSING A METHOD

With the problem defined and the information from the literature search, a method suitable for the specific analytical purpose can then be selected. Typically, there may not be an available method in the literature designed specifically for the required type of sample and situation, so modifications of the procedure are usually necessary. If none of the methods in the literature meet the requirements of the sample and the analytical situation, a new method must be developed based on the literature review.

Returning to the case of monitoring arsenic levels after water treatment plant processing, we selected from the American Chemical Society journal—*Analytical Chemistry*, complementary methods— visual colorimetry and spectrophotometric methods for arsenic compounds. These methods are each based on the formation of nanoparticles of EV^+ with an isopolymolybdate-iodine tetrachloride complex in acidic conditions.

The chemical composition for the purple particles was $[EV\text{-}MoO_2ICl_2]$, and the molar ratio of EV^+-molybdoarsenate was shown to be the 3 : 1 ratio that was previously determined in gravimetric analysis. The nanoparticles, $[EV\text{-}MoO_2ICl_2]_{agg}$, formed stable microparticles with EV^+-molybdoarsenate, giving an apparently homogeneous purple color that was dependent on the arsenic concentration. Quantitative spectrophotometric and visual colorimetric methods for arsenic were developed according to the aggregations of $[EV\text{-}MoO_2ICl_2]$ and EV^+-molybdoarsenate as shown in the expression below

$$H_3AsO_4 + \text{molybdate} \xrightarrow{HCl} AsMo_{12}O_{40}^{3-}$$

$$EV^+ + AsMo_{12}O_{40}^{3-} \longrightarrow EV_3\text{-}AsMo_{12}O_{40}$$

$$EV^+ + [MoO_2ICl_2]^- \longrightarrow EV\text{-}MoO_2ICl_2 \longrightarrow [EV\text{-}MoO_2ICl_2]_{agg}$$

$$[EV\text{-}MoO_2ICl_2]_{agg} + EV_3\text{-}AsMo_{12}O_{40} \longrightarrow \{[EV\text{-}MoO_2ICl_2]_{agg}[EV_3\text{-}AsMo_{12}O_{40}]\}_{agg}$$

These methods were expected to be simple and highly sensitive; however, the phosphate potentially present in drinking water will interfere by undergoing an analogue reaction to that of arsenic. Iron (Ⅲ) also common to drinking water would interfere with the analysis. Plans were to separate these potentially interfering substances from the drinking water sample before arsenic compound analysis. The phosphate would be removed with a weak base anion-exchange resin. The interference of iron (Ⅲ) would be masked by adding EDTA to the water sample.

10.4 DEVELOPING AND EVALUATING THE METHOD

Once the procedure for analysis was selected, the variables must be tested to establish the final procedure that will produce the most accurate and sensitive results. Most important is to establish that the chosen procedure has the expected selectivity and sensitivity which will determine with the necessary accuracy the concentration of arsenic in drinking water.

10.4.1 Selectivity

In the case of arsenic in water, the selectivity of the method has been evaluated by investigating the effect of foreign ions on the spectrophotometric determination of arsenic. The results are summarized in Table 10. 1. The recovery of arsenic was calculated to evaluate the accuracy of the method in the presence of the foreign ions. The most potent interfering chemical species were iron (Ⅲ), phosphate, and silica. The other ions tested had no significant effect on the analyses. In the chosen analysis, iron (Ⅲ) was masked with EDTA. The phosphate interference was eliminated by passing the water sample through a column of the anion-exchange resin. In the presence of NaCl (0. 02 mol · L^{-1}) as a masking reagent, silica was tolerated up to 20 mg · L^{-1}.

Table 10. 1 Effect of Foreign Ions on the Determination of Arsenic (10 μg · L^{-1})

Ions	Added/μg · L^{-1}	Molar Ratio(Foreign Ion/As)	As Found/μg · L^{-1}	Recovery/%
Na^{+}	2300	750 000	10. 2	102
K^{+}	3900	750 000	10. 5	105
Ca^{2+}	400	75 000	10. 4	104
Mg^{2+}	200	61 000	10. 1	101
Cu^{2+}	100	12 000	9. 6	96
Mn^{2+}	100	14 000	10. 1	101
Zn^{2+}	100	11 000	10. 4	104
Al^{3+}	80	22 000	10. 5	105

(continued)

Ions	Added/μg · L^{-1}	Molar Ratio(Foreign Ion/As)	As Found/μg · L^{-1}	Recovery/%
Fe^{3+} [a]	50	6 500	10.1	101
Cl^-	6200	750 000	10.2	102
NO_3^-	6200	750 000	10.1	101
SO_4^{2-}	960	75 000	10.4	104
PO_4^{3-}	0.08	20	14.3	143
PO_4^{3-} [b]	50	12 000	9.8	98
SiO_2^{2-}	0.5	130	15.9	159
SiO_2^{2-} [c]	25	6 600	10.4	104

[a] Masked using EDTA; [b] Pretreated using an anion-exchange resin; [c] Masked using sodium chloride.

10.4.2 Accuracy

The accuracy of the selected analytical procedure can be evaluated by analyzing one or more standard samples whose analytical compositions are known to determine whether a systematic error exists. It is essential that the standards used closely resemble the samples to be analyzed both to analyte concentration and overall composition of the sample.

The systematic error of an analytical method can also be evaluated by comparing the analysis of the sample using the developed procedure with another entirely different method.

When standard reference materials and different analytical methods are not available, the standard addition method can be used to evaluate the accuracy of the selected method. In addition to the selected representative sample to be analyzed, a known amount of the analyte can be added to one portion of the selected water sample to make two samples: one containing an unknown concentration of analyte and the other containing both an unknown and known amounts of analyte. The two samples are then analyzed using the established method and the recovery of the added quantity of added analyte is obtained. The standard addition method may reveal systematic errors arising from the way the sample is prepared or from the presence of other species in the matrix.

In the example for arsenic in water, the standard addition method had been used for tap water and groundwater to evaluate the accuracy of the selected more sensitive spectrophotometric method. The analytical recovery of arsenic added to tap water and groundwater was from 96% to 104% (Table 10.2). Analyses of waters from abandoned mines and the river water samples were compared with hydride generation atomic absorption spectrometry (HG-AAS). The selected spectrophotometric method tended to give comparable but somewhat higher results than HG-AAS method (Table 10. 2).

Table 10. 2 Analytical Results of Water Samples

Samples	As Added /μg · L^{-1}	As Found /μg · L^{-1}	Recovery /%	HG -AAS[a] /μg · L^{-1}
Tap water[b]	10	9. 8	9. 8	
	5	4. 8	96	
Groundwater[b]	10	10. 2	102	
	5	5. 2	104	
Water in abandoned mine 1[c]		26. 0		21. 2
Water in abandoned mine 2[c]		3. 2		2. 6
River water 1[c]		31. 1		29. 4
River water 2[c]		16. 5		15. 1
River water 3[c]		9. 7		7. 9
River water 4[c]		4. 3		2. 3
River water 5[c]		2. 0		0. 9

[a] Hydride generation atomic absorption spectrometry; [b] Arsenic was not detected in the tap waters or groundwaters; [c] With 2-fold dilution.

Because arsenic in the form of arsenite (As(Ⅲ)) is significantly more toxic than arsenate (As(Ⅴ)), the determination of arsenic in both oxidation states is imperative. The selected spectrophotometric method was used to determine As(Ⅴ) and As(Ⅲ) in samples taken from Amemasu River (41. 59°N, 140. 46°E). Total arsenic in both ionic forms was determined after pre-oxidizing As(Ⅲ) to As(Ⅴ) with potassium iodate. As(Ⅴ) in the same original water sample was determined without pre-oxidization. , thus the amount of As(Ⅲ) was obtained by subtracting As(Ⅴ) from the total arsenic (Table 10. 3).

Table 10. 3 Analytical Results (μg · L^{-1}) of River Water Samples

Samples	Total As	As(Ⅴ)	As(Ⅲ)	HG-AAS
A	251. 1	172. 7	78. 3	266. 0
B	79. 8	61. 4	18. 4	73. 6
C	30. 4	27. 0	3. 4	26. 4
D	18. 5	16. 9	1. 6	15. 7

10.4.3 Sensitivity and Linear Dynamic Range

The sensitivity of an analytical method is often expressed in the **calibration sensitivity**, i. e. the slope of the calibration curve ($\Delta R/\Delta c$), as shown in Figure 10.1. If the calibration curve is linear, the sensitivity is constant throughout the linear range. If the calibration curve is not linear, the sensitivity changes with the concentration of the analyte in the tested concentration range.

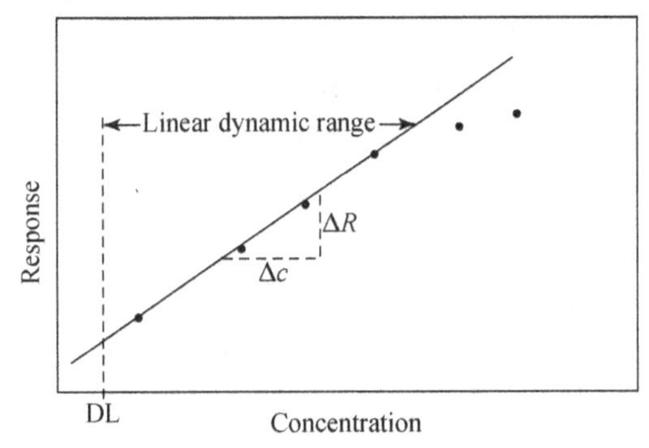

Figure 10.1 The Calibration curve of response (R) versus concentration.

The detection limit (DL) is the smallest concentration that can be determined with a certain level of confidence. Generally, the detection limit of an analytical method is defined as the analyte concentration yielding a response of 3 times greater than the standard deviation of the blank (S_b) to a confidence level of 98.3% as shown in Equation (10-1).

$$\mathrm{DL} = \frac{3S_b}{\text{Slope of calibration curve}} \tag{10-1}$$

The linear dynamic range of an analytical method often refers to the concentration range that can be determined with a linear calibration curve. The lower limit is generally considered the detection limit. The upper end of the range is often taken as the concentration at which the analytical signal deviates by 5% from the linear curve.

In this example, the linear range is found to be linear over the concentration range up to 20 $\mu g \cdot L^{-1}$. The detection limit is 0.5 $\mu g \cdot L^{-1}$. For visual determination using a 350 μL polystyrene white well, the upper limit of the linear curve is about 20 $\mu g \cdot L^{-1}$ and the detection limit is 1 $\mu g \cdot L^{-1}$.

10.5 CONCLUSION

We hope that this example describing the steps necessary in the selection of an analytical method will be a useful reminder to you when you are asked to develop an important analytical method. When you have the opportunity, consider reading the original journal articles describing the steps in the thought process of the authors that by their research linked the necessary steps to develop a cost effective analytical method to measure arsenic compounds in the drinking water anywhere in the world.

Up to now, we have learned some of the analytical methods and the necessary steps to establish a practical analytical method. We will continue to develop analytical methods and the coupling techniques to link analytical need to research activities, to shape lectures/presentations and to all facets of the educational process. Modern analytical chemistry encounters continuing challenges, such as (1) improving sensitivity, e. g. single-molecule spectroscopy thus offers valuable information that cannot be obtained by studying a bulk sample; (2) improving selectivity in analysis of chemical and biological mixtures of extreme complexity and heterogeneity; (3) obtaining spatial and time information of an analytical sample; (4) miniaturization and automation of instrumentation; and (5) remote and non-invasive determination. Challenges are where the opportunity lies. Analytical chemistry will continue to develop to solve the challenging problems with more selective, sensitive and automated analytical methods to serve the needs of other sciences and human activity.

APPENDICES

Appendix A References

1. 李克安，主编. 分析化学教程[M]. 北京：北京大学出版社，2005
   Edited by Li Ke'an. Analytical Chemistry Course. Peking University Press, 2005
2. 李克安，主编. 分析化学教程习题解析[M]. 北京：北京大学出版社，2006
   Edited by Li Ke'an. Analytical Chemistry Course-Problems and Solutions. Peking University Press, 2006
3. 彭崇慧，冯建章，张锡瑜；李克安，赵凤林，修订. 分析化学——定量化学分析简明教程[M]. 第三版. 北京：北京大学出版社，2009
   Peng Chonghui, Feng Jianzhang, Zhang Xiyu, Li Ke'an and Zhao Fenglin. Analytical Chemistry—The Concise Textbook of Quantitative Chemical Analysis (Third Edition). Peking University Press, 2009
4. 彭崇慧，冯建章，张锡瑜，李克安，赵凤林. 定量化学分析简明教程[M]. 第二版. 北京：北京大学出版社，1997
   Peng Chonghui, Feng Jianzhang, Zhang Xiyu, Li Ke'an and Zhao Fenglin. Concise Textbook of Quantitative Chemical Analysis (Second Edition). Peking University Press, 1997
5. 彭崇慧，冯建章，张锡瑜. 定量化学分析简明教程[M]. 第一版. 北京：北京大学出版社，1983
   Peng Chonghui, Feng Jianzhang, Zhang Xiyu. Concise Textbook of Quantitative Chemical Analysis (First Edition). Peking University Press, 1983
6. 李娜，刘锋，赵凤林，李克安. 分析化学 I（电子课件）[CD]. 北京：北京大学出版社，2007
   Li Na, Liu Feng, Zhao Fenglin, and Li Ke'an. Analytical Chemistry I (Lecture Slides CD ROM). Peking University Press, 2007
7. 北京大学化学系分析化学教学组. 基础分析化学实验[M]. 第二版. 北京：北京大学出版社，1997
   Peking University Analytical Chemistry Teaching Group. Fundamentals of Analytical Chemistry Laboratory (Second Edition). Peking University Press, 1997
8. Douglas A. Skoog, Donald M. West, F. James Holler, Stanley R. Crouch. Fundamentals of Analytical Chemistry (Eighth Edition). Brooks Cole, 2004
9. Gary D. Christian. Analytical Chemistry (International Edition). Wiley, 2004
10. John H. Kennedy. Analytical Chemistry Principles. Harcourt Brace Jovanovich, 1984
11. Edmund Bishop. Indicators. Pergamon Press, 1972

12. David Harvey. Modern Analytical Chemistry (International Edition). McGraw Hill Higher Education, 2000

13. R. Kellner, J.-M. Mermet, M. Otto, and H. M. Widmer. Analytical Chemistry. Wiley-VCH, 1998

14. Adon A. Gordus. Schaum's Outline of Theory and Problems of Analytical Chemistry. McGraw-Hill, 1985

15. Daniel C. Harris. Quantitative Chemical Analysis (Sixth Edition). W. H. Freeman, 2002

16. I. M. Kolthoff. Quantitative Chemical Analysis (Forth Edition). Macmillan, 1969

Appendix B Indicators

B1 Selected Indicators In Acid-Base Titrimetry

Indicator	Color			p*K* (HIn)	pT	Transition Range(pH)	Drops of 0.1% Indicator Aqueous Soln. /10 mL of the Soln. to be Titrated
	Acid	Transition	Base				
Thymol blue	Red	Orange	Yellow	1.7(pK_{a1})	2.6	1.2~2.8	1~2[a]
Dimethyl yellow	Red	Orange yellow	Yellow	3.3	3.9	2.9~4.0	1
Bromophenol blue	Yellow		Purple	4.1	4	3.0~4.4	1
Methyl orange	Red	Orange	Yellow	3.4	4	3.1~4.4	1
Bromocresol green	Yellow	Green	Blue	4.9	4.4	3.8~5.4	1
Methyl red	Red	Orange	Yellow	5.0	5.0	4.4~6.2	1
Bromocresol purple	Yellow		Purple		6	5.2~6.8	1
Bromothymol blue	Yellow	Green	Blue	7.3	7	6.0~7.6	1
Phenol red	Yellow	Orange	Red	8.0	7	6.4~8.0	1
Thymol blue	Yellow		Blue	8.9(pK_{a2})	9	8.0~9.6	1~5
Phenolphthalein	Colorless	Pink	Red	9.1		8.0~9.8	1~2[b]
Thymolphthalein	Colorless	Pale blue	Blue	10.0	10	9.4~10.6	1[b]

[a] Sodium salt; [b] ethanol solution.

B2 Selected Metal Indicators for Complexometric Titrimetry

Indicator	Dissociation Constants	Metal	Color Change	Recipe
Acid chrome blue K	$pK_{a1}=6.7$ $pK_{a2}=10.2$ $pK_{a3}=14.6$	Mg(pH 10) Ca(pH 12)	Red~blue	0.1% Ethanol solution
Calconcarboxylic acid	$pK_{a2}=3.8$ $pK_{a3}=9.4$ $pK_{a4}=13\sim14$	Ca(pH 12~13)	Wine red~blue	1 : 100 NaCl
Eriochrome black T	$pK_{a1}=3.9$ $pK_{a2}=6.4$ $pK_{a3}=11.5$	Ca(pH 10, with EDTA-Mg) Mg(pH 10) Pb(pH 10, with potassium tartrate) Zn(pH 6.8~10)	Red~blue Red~blue Red~blue Red~blue	1 : 100 NaCl

(continued)

Indicator	Dissociation Constants	Metal	Color Change	Recipe
Murexide	$pK_{a1}=1.6$ $pK_{a2}=8.7$ $pK_{a3}=10.3$ $pK_{a4}=13.5$ $pK_{a5}=14$	Ca(pH>10, 25% ethanol) Cu(pH 7～8) Ni(pH 8.5～11.5)	Red～purple Yellow～purple	1 : 100 NaCl
o-PAN	$pK_{a1}=2.9$ $pK_{a2}=11.2$	Cu(pH 6) Zn(pH 5～7)	Red～yellow Pink～yellow	0.1% Ethanol solution
Sulfosalicylic acid	$pK_{a1}=2.6$ $pK_{a2}=11.7$	Fe(Ⅲ)(pH 1.5～3)	Red-violet～yellow	1%～2% Aqueous solution
Xylenol orange	$pK_{a2}=2.6$ $pK_{a3}=3.2$ $pK_{a4}=6.4$ $pK_{a5}=10.4$ $pK_{a6}=12.3$	Bi(pH 1～2) La(pH 5～6) Pb(pH 5～6) Zn(pH 5～6)	Red～yellow	0.5% Ethanol solution

B3 Selected Redox Indicators

Indicator	$\varphi^{\ominus'}(In)/V$ $[H^+]=1\ mol \cdot L^{-1}$	Color		Recipe
		Reduced Form	Oxidized Form	
Methylene blue	+0.52	Colorless	Blue	0.05% aqueous solution
Sodium diphenylamine sulfonate	+0.85	Colorless	Red-violet	0.8 g indicator + 2 g Na_2CO_3, dissolved and brought to 100 mL by H_2O
N-phenylanthranilic acid	+0.89	Colorless	Red-violet	0.11 g indicator is dissolved in 20 mL 5% Na_2CO_3, brought to 100 mL by H_2O
Tris(1,10-phenanthroline) iron (Ⅱ) sulfate (also known as ferroin)	+1.06	Red	Pale blue	1.485 g phenanthroline+0.695 g $FeSO_4 \cdot 7H_2O$, dissolved and brought to 100 mL by H_2O

Appendix C Activity Coefficients(γ) for Ions at 25℃

Ion	γ /nm $\mathring{a}$ /nm	Activity Coefficient at Indicated Ionic Strength			
		0.005	0.01	0.05	0.1
H^+	0.9	0.934	0.914	0.854	0.826
Li^+, $C_6H_5COO^-$	0.6	0.930	0.907	0.834	0.796
Na^+, HCO_3^-, IO_3^-, $H_2PO_4^-$, Ac^-	0.4	0.927	0.902	0.817	0.770
$HCOO^-$, ClO_3^-, ClO_4^-, F^-, MnO_4^-, OH^-, SH^-	0.35	0.926	0.900	0.812	0.762
K^+, Br^-, CN^-, Cl^-, I^-, NO_3^-, NO_2^-	0.3	0.925	0.899	0.807	0.754
Ag^+, Cs^+, NH_4^+, Rb^+, Tl^+	0.25	0.925	0.897	0.802	0.745
Be^{2+}, Mg^{2+}	0.8	0.756	0.690	0.517	0.446
Ca^{2+}, Cu^{2+}, Zn^{2+}, Fe^{2+}, $C_6H_4(COO)_2^{2-}$	0.6	0.748	0.676	0.484	0.402
Ba^{2+}, Cd^{2+}, Hg^{2+}, Pb^{2+}, S^{2-}, $C_2O_4^{2-}$	0.5	0.743	0.669	0.465	0.377
Hg^{2+}, CO_3^{2-}, CrO_4^{2-}, HPO_4^{2-}, SO_3^{2-}, SO_4^{2-}	0.4	0.738	0.661	0.445	0.351
Al^{3+}, Cr^{3+}, Fe^{3+}, La^{3+}	0.9	0.540	0.443	0.242	0.179
Cit^{3-}	0.5	0.513	0.404	0.179	0.112
$Fe(CN)_6^{3-}$, PO_4^{3-}	0.4	0.505	0.394	0.162	0.095
Ce^{4+}, Th^{4+}, Zr^{4+}	1.1	0.348	0.253	0.099	0.063
$Fe(CN)_6^{4-}$	0.5	0.305	0.200	0.047	0.020

$\mathring{a}$: effective diameter of the hydrated ion in nanometers.

Appendix D Constants for Acid-base, Complexometric, Redox, and Precipitation Titrimetry

D1 Dissociation Constants of Acids and Bases at 25℃

Acid	Formula		$I=0$		$I=0.1$	
			K_a	pK_a	K_a^M	pK_a^M
Arsenic acid	H_3AsO_4	K_{a1}	6.5×10^{-3}	2.19	8×10^{-3}	2.1
		K_{a2}	1.15×10^{-7}	6.94	2×10^{-7}	6.7
		K_{a3}	3.2×10^{-12}	11.50	6×10^{-12}	11.2
Arsenious acid	H_3AsO_3	K_{a1}	6.0×10^{-10}	9.22	8×10^{-10}	9.1
Boric acid	H_3BO_3	K_{a1}	5.8×10^{-10}	9.24		
Carbonic acid	$H_2CO_3(H_2O+CO_2)$	K_{a1}	4.2×10^{-7}	6.38	5×10^{-7}	6.3
		K_{a2}	5.6×10^{-11}	10.25	8×10^{-11}	10.1
Chromic acid	H_2CrO_4	K_{a2}	3.2×10^{-7}	6.50		
Hydrocyanic acid	HCN		4.9×10^{-10}	9.31	6×10^{-10}	9.2
Hydrofluoric acid	HF		6.8×10^{-4}	3.17	8.9×10^{-4}	3.1
Hydrosulphuric acid	H_2S	K_{a1}	8.9×10^{-8}	7.05	1.3×10^{-7}	6.9
		K_{a2}	1.2×10^{-13}	12.92	3×10^{-13}	12.5
Phosphoric acid	H_3PO_4	K_{a1}	6.9×10^{-3}	2.16	1×10^{-2}	2.0
		K_{a2}	6.2×10^{-8}	7.21	1.3×10^{-7}	6.9
		K_{a3}	4.8×10^{-13}	12.32	2×10^{-12}	11.7
Silicic acid	H_2SiO_3	K_{a1}	1.7×10^{-10}	9.77	3×10^{-10}	9.5
		K_{a2}	1.6×10^{-12}	11.80	2×10^{-13}	12.7
Sulfuric acid	H_2SO_4	K_{a2}	1.2×10^{-2}	1.92	1.6×10^{-2}	1.8
Sulfurous acid	$H_2SO_3(H_2O+SO_2)$	K_{a1}	1.29×10^{-2}	1.89	1.6×10^{-2}	1.8
		K_{a2}	6.3×10^{-8}	7.20	1.6×10^{-7}	6.8
Formic acid	HCOOH		1.7×10^{-4}	3.77	2.2×10^{-4}	3.65
Acetic acid	CH_3COOH		1.75×10^{-5}	4.76	2.2×10^{-5}	4.65
Propionic acid	C_2H_5COOH		1.35×10^{-5}	4.87		
Chloroacetic acid	$ClCH_2COOH$		1.38×10^{-3}	2.86	2×10^{-3}	2.7
Dichloroacetic acid	$Cl_2CHCOOH$		5.5×10^{-2}	1.26	8×10^{-2}	1.1
Aminoacetic acid	$NH_3^+CH_2COOH$	K_{a1}	4.5×10^{-3}	2.35	3×10^{-3}	2.5
(Glycine)	$NH_3^+CH_2COO^-$	K_{a2}	1.7×10^{-10}	9.78	2×10^{-10}	9.7
Benzoic acid	C_6H_5COOH		6.2×10^{-5}	4.21	8×10^{-5}	4.1

(**continued**)

Acid	Formula		$I=0$		$I=0.1$	
			K_a	pK_a	K_a^M	pK_a^M
Oxalic acid	$H_2C_2O_4$	K_{a1}	5.6×10^{-2}	1.25	8×10^{-2}	1.1
		K_{a2}	5.1×10^{-5}	4.29	1×10^{-4}	4.0
Phenol	C_6H_5OH		1.12×10^{-10}	9.95	1.6×10^{-10}	9.8
Acetyl acetone	$CH_3COCH_2COCH_3$		1×10^{-9}	9.0	1.3×10^{-9}	8.9
α-Tartaric acid	CH(OH)COOH–CH(OH)COOH	K_{a1}	9.1×10^{-4}	3.04	1.3×10^{-3}	2.9
		K_{a2}	4.3×10^{-5}	4.37	8×10^{-5}	4.1
Succinic acid	CH_2COOH–CH_2COOH	K_{a1}	6.2×10^{-5}	4.21	1.0×10^{-4}	4.00
		K_{a2}	2.3×10^{-6}	5.64	5.2×10^{-6}	5.28
p-Phthalic acid	$C_6H_4(COOH)_2$	K_{a1}	1.12×10^{-3}	2.95	1.6×10^{-3}	2.8
		K_{a2}	3.91×10^{-6}	5.41	8×10^{-6}	5.1
Citric acid	HO–C(CH_2COOH)$_2$COOH	K_{a1}	7.4×10^{-4}	3.13	1×10^{-3}	3.0
		K_{a2}	1.7×10^{-5}	4.76	4×10^{-5}	4.4
		K_{a3}	4.0×10^{-7}	6.40	8×10^{-7}	6.1
Ethylenediamine tetraacetic acid	H_2C–N(CH_2COOH)$_2$ / H_2C–N(CH_2COOH)$_2$	K_{a1}			1.3×10^{-1}	0.9
		K_{a2}			3×10^{-2}	1.6
		K_{a3}			8.5×10^{-3}	2.07
		K_{a4}			1.8×10^{-3}	2.75
		K_{a5}	5.4×10^{-7}	6.27	5.8×10^{-7}	6.24
		K_{a6}	1.12×10^{-11}	10.95	4.6×10^{-11}	10.34
8-Hydroxy quinoline	(quinoline ring, N, OH)	K_{a1}	8×10^{-6}	5.1	1×10^{-5}	5.0
		K_{a2}	1×10^{-9}	9.0	1.3×10^{-10}	9.9
Malic acid	HO–CH(COOH)CH_2COOH	K_{a1}	4.0×10^{-4}	3.40	5.2×10^{-4}	3.28
		K_{a2}	8.9×10^{-6}	5.05	1.9×10^{-5}	4.72
Salicylic acid	C_6H_4(OH)COOH	K_{a1}	1.05×10^{-3}	2.98	1.3×10^{-3}	2.9
		K_{a2}			8×10^{-14}	13.1
Sulfosalicylic acid	HOOC(HO)C_6H_3–SO_3^-	K_{a1}			3×10^{-3}	2.6
		K_{a2}			3×10^{-12}	11.6
Maleic acid	CHCOOH=CHCOOH	K_{a1}	1.2×10^{-2}	1.92		
		K_{a2}	6.0×10^{-7}	6.22		

(continued)

Base	Formula		$I=0$		$I=0.1$	
			K_b	pK_b	K_b^M	pK_b^M
Ammonia	NH_3		1.8×10^{-5}	4.75	2.3×10^{-5}	4.63
Hydrazine	H_2N-NH_2	K_{b1}	9.8×10^{-7}	6.01	1.3×10^{-6}	5.9
		K_{b2}	1.32×10^{-15}	14.88		
Hydroxylamine	NH_2OH		9.1×10^{-9}	8.04	1.6×10^{-8}	7.8
Methylamine	CH_3NH_2		4.2×10^{-4}	3.38		
Ethylamine	$C_2H_5NH_2$		4.3×10^{-4}	3.37		
Anilin	$C_6H_5NH_2$		4.2×10^{-10}	9.38	5×10^{-10}	9.3
Ethylene diamine	$H_2NCH_2CH_2NH_2$	K_{b1}	8.5×10^{-5}	4.07		
		K_{b2}	7.1×10^{-8}	7.15		
Triethanolamine	$N(CH_2CH_2OH)_3$		5.8×10^{-7}	6.24	1.3×10^{-8}	7.9
Aminoform urotropine	$(CH_2)_6N_4$		1.35×10^{-9}	8.87	1.8×10^{-9}	8.74
Pyridine	C_5H_5N		1.8×10^{-9}	8.74	1.6×10^{-9} ($I=0.5$)	8.79
1,10-Phenanthroline monohydrate			6.9×10^{-10}	9.16	8.9×10^{-10}	9.05

D2 Stability Constants of Metal Complexes

Metal Ion	Ionic Strength	n	$\lg\beta_n$
Ammonia(NH_3)			
Ag^+	0.1	1, 2	3.40, 7.40
Cd^{2+}	0.1	1, …, 6	2.60, 4.65, 6.04, 6.92, 6.6, 4.9
Co^{2+}	0.1	1, …, 6	2.05, 3.62, 4.61, 5.31, 5.43, 4.75
Cu^{2+}	2	1, …, 4	4.13, 7.61, 10.48, 12.59
Ni^{2+}	0.1	1, …, 6	2.75, 4.95, 6.64, 7.79, 8.50, 8.49
Zn^{2+}	0.1	1, …, 4	2.27, 4.61, 7.01, 9.06
Hydroxide(OH^-)			
Ag^+	0	1, 2, 3	2.3, 3.6, 4.8
Al^{3+}	2	4	33.3
Bi^{3+}	3	1	12.4
Cd^{2+}	3	1, …, 4	4.3, 7.7, 10.3, 12.0
Cu^{2+}	0	1	6.0
Fe^{2+}	1	1	4.5

(**continued**)

Metal Ion	Ionic Strength	n	$\lg\beta_n$
Fe^{3+}	3	1, 2	11.0, 21.7
Mg^{2+}	0	1	2.6
Ni^{2+}	0.1	1	4.6
Pb^{2+}	0.3	1, …, 3	6.2, 10.3, 13.3
Zn^{2+}	0	1, …, 4	4.4, —, 14.4, 15.5
Zr^{4+}	4	1, …, 4	13.8, 27.2, 40.2, 53
Fluoride(F^-)			
Al^{3+}	0.53	1, …, 6	6.1, 11.15, 15.0, 17.7, 19.4, 19.7
Fe^{3+}	0.5	1, 2, 3	5.2, 9.2, 11.9
Th^{4+}	0.5	1, 2, 3	7.7, 13.5, 18.0
TiO^{2+}	3	1, …, 4	5.4, 9.8, 13.7, 17.4
Sn^{4+}	*	6	25
Zr^{4+}	2	1, 2, 3	8.8, 16.1, 21.9
Chloride(Cl^-)			
Ag^+	0.2	1, …, 4	2.9, 4.7, 5.0, 5.9
Hg^{2+}	0.5	1, …, 4	6.7, 13.2, 14.1, 15.1
Iodide(I^-)			
Cd^{2+}	*	1, …, 4	2.4, 3.4, 5.0, 6.15
Hg^{2+}	0.5	1, …, 4	12.9, 23.8, 27.6, 29.8
Cyanide(CN^-)			
Ag^+	0~0.3	1, …, 4	—, 21.1, 21.8, 20.7
Cd^{2+}	3	1, …, 4	5.5, 10.6, 15.3, 18.9
Cu^+	0	1, …, 4	—, 24.0, 28.6, 30.3
Fe^{2+}	0	6	35.4
Fe^{3+}	0	6	43.6
Hg^{2+}	0.1	1, …, 4	18.0, 34.7, 38.5, 41.5
Ni^{2+}	0.1	4	31.3
Zn^{2+}	0.1	4	16.7
Thiocyanate(SCN^-)			
Fe^{3+}	*	1, …, 5	2.3, 4.2, 5.6, 6.4, 6.4
Hg^{2+}	1	1, …, 4	—, 16.1, 19.0, 20.9
Thiosulfate($S_2O_3^{2-}$)			
Ag^+	0	1, 2	8.82, 13.5
Hg^{2+}	0	1, 2	29.86, 32.26

(continued)

Metal Ion	Ionic Strength	n	$\lg\beta_n$
Citric acid($C_3H_4(OH)(COOH)_3$)			
Al^{3+}	0.5	1	20.0
Cu^{2+}	0.5	1	18
Fe^{3+}	0.5	1	25
Ni^{2+}	0.5	1	14.3
Pb^{2+}	0.5	1	12.3
Zn^{2+}	0.5	1	11.4
Sulfosalicylic acid($C_6H_4(OH)(SO_3H)COOH$)			
Al^{3+}	0.1	1, 2, 3	12.9, 22.9, 29.0
Fe^{3+}	3	1, 2, 3	14.4, 25.2, 32.2
Acetylacetone($CH_3COCH_2COCH_3$)			
Al^{3+}	0.1	1, 2, 3	8.1, 15.7, 21.2
Cu^{2+}	0.1	1, 2	7.8, 14.3
Fe^{3+}	0.1	1, 2, 3	9.3, 17.9, 25.1
1,10-phenanthroline($C_{12}H_8N_2$)			
Ag^{+}	0.1	1, 2	5.02, 12.07
Cd^{2+}	0.1	1, 2, 3	6.4, 11.6, 15.8
Co^{2+}	0.1	1, 2, 3	7.0, 13.7, 20.1
Cu^{2+}	0.1	1, 2, 3	9.1, 15.8, 21.0
Fe^{2+}	0.1	1, 2, 3	5.9, 11.1, 21.3
Hg^{2+}	0.1	1, 2, 3	—, 19.65, 23.35
Ni^{2+}	0.1	1, 2, 3	8.8, 17.1, 24.8
Zn^{2+}	0.1	1, 2, 3	6.4, 12.15, 17.0
Ethylenediamine($NH_2CH_2CH_2NH_2$)			
Ag^{+}	0.1	1, 2	4.7, 7.7
Cd^{2+}	0.1	1, 2	5.47, 10.02
Cu^{2+}	0.1	1, 2	10.55, 19.60
Co^{2+}	0.1	1, 2, 3	5.89, 10.72, 13.82
Hg^{2+}	0.1	2	23.42
Ni^{2+}	0.1	1, 2, 3	7.66, 14.06, 18.59
Zn^{2+}	0.1	1, 2, 3	5.71, 10.37, 12.08

Data with "*" lack of ionic strength information.

D3 Logarithmic Stability Constants of Metal Ion Complexes with Aminocarboxylic Acids

Metal Ion	EDTA			EGTA		HEDTA	
	$\lg K^{H}$(MHL)	$\lg K$(ML)	$\lg K^{OH}$(MOHL)	$\lg K^{H}$(MHL)	$\lg K$(ML)	$\lg K$(ML)	$\lg K^{OH}$(MOHL)
Ag^{+}	6.0	7.3					
Al^{3+}	2.5	16.1	8.1				
Ba^{2+}	4.6	7.8		5.4	8.4	6.2	
Bi^{3+}		27.9					
Ca^{2+}	3.1	10.7		3.8	11.0	8.0	
Ce^{3+}		16.0					
Cd^{2+}	2.9	16.5		3.5	15.6	13.0	
Co^{2+}	3.1	16.3			12.3	14.4	
Co^{3+}	1.3	36					
Cr^{3+}	2.3	23	6.6				
Cu^{2+}	3.0	18.8	2.5	4.4	17	17.4	
Fe^{2+}	2.8	14.3				12.2	5.0
Fe^{3+}	1.4	25.1	6.5			19.8	10.1
Hg^{2+}	3.1	21.8	4.9	3.0	23.2	20.1	
La^{3+}		15.4			15.6	13.2	
Mg^{2+}	3.9	8.7			5.2	5.2	
Mn^{2+}	3.1	14.0		5.0	11.5	10.7	
Ni^{2+}	3.2	18.6		6.0	12.0	17.0	
Pb^{2+}	2.8	18.0		5.3	13.0	15.5	
Sn^{2+}		22.1					
Sr^{2+}	3.9	8.6		5.4	8.5	6.8	
Th^{4+}		23.2					8.6
Ti^{3+}		21.3					
TiO^{2+}		17.3					
Zn^{2+}	3.0	16.5		5.2	12.8	14.5	

EDTA, ethylenediamine tetraacetic acid; EGTA, ethylene glycol tris(2-aminoethylether) tetraacetic acid; HEDTA, 2-hydroxyethylethylenediamine triacetic acid.

D4 Logarithmic Values of $\alpha_{L(H)}$ Used as Complexing, Masking and Buffering Agents

pH	EDTA	HEDTA	NH_3	CN^{-}	F^{-}
0	24.0	17.9	9.4	9.2	3.05
1	18.3	15.0	8.4	8.2	2.05
2	13.8	12.0	7.4	7.2	1.1

(continued)

pH	EDTA	HEDTA	NH_3	CN^-	F^-
3	10.8	9.4	6.4	6.2	0.3
4	8.6	7.2	5.4	5.2	0.05
5	6.6	5.3	4.4	4.2	
6	4.8	3.9	3.4	3.2	
7	3.4	2.8	2.4	2.2	
8	2.3	1.8	1.4	1.2	
9	1.4	0.9	0.5	0.4	
10	0.5	0.2	0.1	0.1	
11	0.1				
12					
13					
		Used Constants			
$\lg K_1$	10.34	9.81	9.4	9.2	3.1
$\lg K_2$	6.24	5.41			
$\lg K_3$	2.75	2.72			
$\lg K_4$	2.07				
$\lg K_5$	1.6				
$\lg K_6$	0.9				

D5 Logarithmic Values of $\alpha_{M(OH)}$ for Metal Complexes with Hydroxide Ions

Metal Ion	Ionic Strength	pH													
		1	2	3	4	5	6	7	8	9	10	11	12	13	14
Al^{3+}	2					0.4	1.3	5.3	9.3	13.3	17.3	21.3	25.3	29.3	33.3
Bi^{3+}	3	0.1	0.5	1.4	2.4	3.4	4.4	5.4							
Ca^{2+}	0.1													0.3	1.0
Cd^{2+}	3									0.1	0.5	2.0	4.5	8.1	12.0
Co^{2+}	0.1								0.1	0.4	1.1	2.2	4.2	7.2	10.2
Cu^{2+}	0.1								0.2	0.8	1.7	2.7	3.7	4.7	5.7
Fe^{2+}	1									0.1	0.6	1.5	2.5	3.5	4.5
Fe^{3+}	3			0.4	1.8	3.7	5.7	7.7	9.7	11.7	13.7	15.7	17.7	19.7	21.7
Hg^{2+}	0.1			0.5	1.9	3.9	5.9	7.9	9.9	11.9	13.9	15.9	17.9	19.9	21.9
La^{3+}	3										0.3	1.0	1.9	2.9	3.9
Mg^{2+}	0.1											0.1	0.5	1.3	2.3
Mn^{2+}	0.1										0.1	0.5	1.4	2.4	3.4

(continued)

Metal Ion	Ionic Strength	pH													
		1	2	3	4	5	6	7	8	9	10	11	12	13	14
Ni^{2+}	0.1									0.1	0.7	1.6			
Pb^{2+}	0.1							0.1	0.5	1.4	2.7	4.7	7.4	10.4	13.4
Th^{4+}	1				0.2	0.8	1.7	2.7	3.7	4.7	5.7	6.7	7.7	8.7	9.7
Zn^{2+}	0.1									0.2	2.4	5.4	8.5	11.8	15.5

D6 Transition Points, pM($(pM)_t$), and Logarithmic Values of $\alpha_{In(H)}$ for Metal Indicators

D6-1 Eriochrome Black T (EBT)

pH	6.0	7.0	8.0	9.0	10.0	11.0	12.0	13.0	Stability Constants
$\lg\alpha_{In(H)}$	6.0	4.6	3.6	2.6	1.6	0.7	0.1		$\lg K^H(HIn)$ 11.5, $\lg K^H(H_2In)$ 6.4
$(pCa)_t$ (to red)			1.8	2.8	3.8	4.7	5.3	5.4	$\lg K(CaIn)$ 5.4
$(pMg)_t$ (to red)	1.0	2.4	3.4	4.4	5.4	6.3			$\lg K(MgIn)$ 7.0
$(pZn)_t$ (to red)	6.9	8.3	9.3	10.5	12.2	13.9			$\lg\beta(ZnIn)$ 12.9, $\lg\beta(ZnIn_2)$ 20.0

D6-2 Murexide

pH	6.0	7.0	8.0	9.0	10.0	11.0	12.0	Stability Constants
$\lg\alpha_{In(H)}$	7.7	5.7	3.7	1.9	0.7	0.1		$\lg K^H(HIn)$ 10.5
$\lg\alpha_{HIn(H)}$	3.2	2.2	1.2	0.4	0.2	0.6	1.5	$\lg K^H(H_2In)$ 9.2
$(pCa)_t$ (to red)		2.6	2.8	3.4	4.0	4.6	5.0	$\lg K(CaIn)$ 5.0
$(pCu)_t$ (to orange)	6.4	8.2	10.2	12.2	13.6	15.8	17.9	
$(pNi)_t$ (to yellow)	4.6	5.2	6.2	7.8	9.3	10.3	11.3	

D6-3 Xylenol Orange (XO)

pH	1.0	2.0	3.0	4.0	4.5	5.0	5.5	6.0
$(pBi)_t$ (to red)	4.0	5.4	6.8					
$(pCd)_t$ (to red)					4.0	4.5	5.0	5.5
$(pHg)_t$ (to red)						7.4	8.2	9.0
$(pLa)_t$ (to red)					4.0	4.5	5.0	5.6
$(pPb)_t$ (to red)			4.2	4.8	6.2	7.0	7.6	8.2
$(pTh)_t$ (to red)	3.6	4.9	6.3					
$(pZn)_t$ (to red)					4.1	4.8	5.7	6.5
$(pZr)_t$ (to red)	7.5							

The $(pM)_t$ values given are experimental.

D6-4 PAN

pH	4.0	5.0	6.0	7.0	8.0	9.0	10.0	11.0	Stability Constants (20% dioxane)
$\lg\alpha_{In(H)}$	8.2	7.2	6.2	5.2	4.2	3.2	2.2	1.2	$\lg K(HIn)$ 12.2, $\lg K^{H}(H_2In)$ 1.9
$(pCu)_t$ (to red)	7.8	8.8	9.8	10.8	11.8	12.8	13.8	14.8	$\lg K(CuIn)$ 16.0

D7 Standard ($\varphi^\ominus$) and Formal ($\varphi^{\ominus\prime}$) Electrode Potentials

D7-1 Standard Electrode Potentials($\varphi^\ominus$), 25℃

Half-reaction	$\varphi^\ominus$/V
$F_2 + 2e^- \rightleftharpoons 2F^-$	+2.87
$O_3 + 2H^+ + 2e^- \rightleftharpoons O_2 + H_2O$	+2.07
$S_2O_8^{2-} + 2e^- \rightleftharpoons 2SO_4^{2-}$	+2.0
$H_2O_2 + 2H^+ + 2e^- \rightleftharpoons 2H_2O$	+1.77
$Ce^{4+} + e^- \rightleftharpoons Ce^{3+}$	+1.61
$2BrO_3^- + 12H^+ + 10e^- \rightleftharpoons Br_2 + 6H_2O$	+1.5
$MnO_4^- + 8H^+ + 5e^- \rightleftharpoons Mn^{2+} + 4H_2O$	+1.51
$PbO_2(s) + 4H^+ + 2e^- \rightleftharpoons Pb^{2+} + H_2O$	+1.46
$BrO_3^- + 6H^+ + 6e^- \rightleftharpoons Br^- + 3H_2O$	+1.44
$Cl_2 + 2e^- \rightleftharpoons 2Cl^-$	+1.358
$Cr_2O_7^{2-} + 14H^+ + 6e^- \rightleftharpoons 2Cr^{3+} + 7H_2O$	+1.33
$MnO_2(s) + 4H^+ + 2e^- \rightleftharpoons Mn^{2+} + 2H_2O$	+1.23
$O_2 + 4H^+ + 4e^- \rightleftharpoons 2H_2O$	+1.229
$2IO_3^- + 12H^+ + 10e^- \rightleftharpoons I_2 + 6H_2O$	+1.19
$Br_2 + 2e^- \rightleftharpoons 2Br^-$	+1.08
$HNO_2 + H^+ + e^- \rightleftharpoons NO + H_2O$	+0.98
$VO_2^+ + 2H^+ + e^- \rightleftharpoons VO^{2+} + H_2O$	+0.999
$NO_3^- + 3H^+ + 2e^- \rightleftharpoons HNO_2 + H_2O$	+0.94
$Hg^{2+} + 2e^- \rightleftharpoons 2Hg$	+0.845
$Ag^+ + e^- \rightleftharpoons Ag$	+0.7994
$Hg_2^{2+} + 2e^- \rightleftharpoons 2Hg$	+0.792
$Fe^{3+} + e^- \rightleftharpoons Fe^{2+}$	+0.771
$O_2 + 2H^+ + 2e^- \rightleftharpoons H_2O_2$	+0.69
$2HgCl_2 + 2e^- \rightleftharpoons Hg_2Cl_2 + 2Cl^-$	+0.63
$MnO_4^- + 2H_2O + 3e^- \rightleftharpoons MnO_2 + 4OH^-$	+0.588
$MnO_4^- + e^- \rightleftharpoons MnO_4^{2-}$	+0.57

(**continued**)

Half-reaction	$\varphi^{\ominus}$/V
$H_3AsO_4 + 2H^+ + 2e^- \rightleftharpoons HAsO_2 + 2H_2O$	+0.56
$I_3^- + 2e^- \rightleftharpoons 3I^-$	+0.54
$I_2(s) + 2e^- \rightleftharpoons 2I^-$	+0.535
$Cu^+ + e^- \rightleftharpoons Cu$	+0.52
$Fe(CN)_6^{3-} + e^- \rightleftharpoons Fe(CN)_6^{4-}$	+0.355
$Cu^{2+} + 2e^- \rightleftharpoons Cu$	+0.34
$Hg_2Cl_2 + 2e^- \rightleftharpoons 2Hg + 2Cl^-$	+0.268
$SO_4^{2-} + 4H^+ + 2e^- \rightleftharpoons H_2SO_3 + H_2O$	+0.17
$Cu^{2+} + e^- \rightleftharpoons Cu^+$	+0.17
$Sn^{4+} + 2e^- \rightleftharpoons Sn^{2+}$	+0.15
$S + 2H^+ + 2e^- \rightleftharpoons H_2S$	+0.14
$S_4O_6^{2-} + 2e^- \rightleftharpoons 2S_2O_3^{2-}$	+0.09
$2H^+ + 2e^- \rightleftharpoons H_2$	0.00
$Pb^{2+} + 2e^- \rightleftharpoons Pb$	−0.126
$Sn^{2+} + 2e^- \rightleftharpoons Sn$	−0.14
$Ni^{2+} + 2e^- \rightleftharpoons Ni$	−0.25
$PbSO_4(s) + 2e^- \rightleftharpoons Pb + SO_4^{2-}$	−0.356
$Cd^{2+} + 2e^- \rightleftharpoons Cd$	−0.403
$Fe^{2+} + 2e^- \rightleftharpoons Fe$	−0.44
$S + 2e^- \rightleftharpoons S^{2-}$	−0.48
$2CO_2 + 2H^+ + 2e^- \rightleftharpoons H_2C_2O_4$	−0.49
$Zn^{2+} + 2e^- \rightleftharpoons Zn$	−0.7628
$SO_4^{2-} + H_2O + 2e^- \rightleftharpoons SO_3^{2-} + 2OH^-$	−0.93
$Al^{3+} + 3e^- \rightleftharpoons Al$	−1.66
$Mg^{2+} + 2e^- \rightleftharpoons Mg$	−2.37
$Na^+ + e^- \rightleftharpoons Na$	−2.713
$Ca^{2+} + 2e^- \rightleftharpoons Ca$	−2.87
$K^+ + e^- \rightleftharpoons K$	−2.925

D7-2 Formal(Conditional) Electrode Potentials($\varphi^{\ominus\prime}$) at 25℃

Half-reaction	$\varphi^{\ominus\prime}$/V	Medium
$Ag^{2+} + e^- \rightleftharpoons Ag^+$	2.00	4 mol·L^{-1} $HClO_4$
	1.93	3 mol·L^{-1} HNO_3
Ce(Ⅳ) + e^- $\rightleftharpoons$ Ce(Ⅲ)	1.74	1 mol·L^{-1} $HClO_4$
	1.45	0.5 mol·L^{-1} H_2SO_4

(continued)

Half-reaction	$\varphi^{\ominus\prime}$/V	Medium
	1.28	1 mol·L^{-1} HCl
	1.60	1 mol·L^{-1} HNO_3
$Co(III)+e^- \rightleftharpoons Co(II)$	1.95	4 mol·L^{-1} $HClO_4$
	1.86	1 mol·L^{-1} HNO_3
$Cr_2O_7^{2-}+14H^++6e^- \rightleftharpoons 2Cr^{3+}+7H_2O$	1.03	1 mol·L^{-1} $HClO_4$
	1.15	4 mol·L^{-1} H_2SO_4
	1.00	1 mol·L^{-1} HCl
$Fe(III)+e^- \rightleftharpoons Fe(II)$	0.75	1 mol·L^{-1} $HClO_4$
	0.70	1 mol·L^{-1} HCl
	0.68	1 mol·L^{-1} H_2SO_4
	0.51	1 mol·L^{-1} HCl-0.25 mol·L^{-1} H_3PO_4
$Fe(CN)_6^{3-}+e^- \rightleftharpoons Fe(CN)_6^{4-}$	0.56	0.1 mol·L^{-1} HCl
	0.72	1 mol·L^{-1} $HClO_4$
$I_3^-+2e^- \rightleftharpoons 3I^-$	0.545	0.5 mol·L^{-1} H_2SO_4
$Sn(IV)+2e^- \rightleftharpoons Sn(II)$	0.14	1 mol·L^{-1} HCl
$Sb(V)+2e^- \rightleftharpoons Sb(III)$	0.75	3.5 mol·L^{-1} HCl
$SbO_3^-+H_2O+2e^- \rightleftharpoons SbO_2^-+2OH^-$	−0.43	3 mol·L^{-1} KOH
$Ti(IV)+e^- \rightleftharpoons Ti(III)$	−0.01	0.2 mol·L^{-1} H_2SO_4
	0.15	5 mol·L^{-1} H_2SO_4
	0.10	3 mol·L^{-1} HCl
$V(V)+e^- \rightleftharpoons V(IV)$	0.94	1 mol·L^{-1} H_3PO_4
$U(VI)+2e^- \rightleftharpoons U(IV)$	0.35	1 mol·L^{-1} HCl

D8 Solubility Products ($K_{sp}^{\ominus}$) and Concentration Solubility Products (K_{sp}) at 25℃

Salt	$I=0$		$I=0.1$	
	$K_{sp}^{\ominus}$	$pK_{sp}^{\ominus}$	K_{sp}	pK_{sp}
AgAc	2×10^{-3}	2.7	8×10^{-3}	2.1
AgCl	1.77×10^{-10}	9.75	3.2×10^{-10}	9.50
AgBr	4.95×10^{-13}	12.31	8.7×10^{-13}	12.06
AgI	8.3×10^{-17}	16.08	1.48×10^{-16}	15.83
Ag_2CrO_4	1.12×10^{-12}	11.95	5×10^{-12}	11.3
AgSCN	1.07×10^{-12}	11.97	2×10^{-12}	11.7
Ag_2S	6×10^{-50}	49.2	6×10^{-49}	48.2
Ag_2SO_4	1.58×10^{-5}	4.80	8×10^{-5}	4.1
$Ag_2C_2O_4$	1×10^{-11}	11.0	4×10^{-11}	10.4

(**continued**)

Salt	$I=0$		$I=0.1$	
	$K_{sp}^{\ominus}$	$pK_{sp}^{\ominus}$	K_{sp}	pK_{sp}
Ag_3AsO_4	1.12×10^{-20}	19.95	1.3×10^{-19}	18.9
Ag_3PO_4	1.45×10^{-16}	15.84	2×10^{-15}	14.7
AgOH	1.9×10^{-8}	7.71	3×10^{-8}	7.5
$Al(OH)_3$ (amorphous)	4.6×10^{-33}	32.34	3×10^{-32}	31.5
$BaCrO_4$	1.17×10^{-10}	9.93	8×10^{-10}	9.1
$BaCO_3$	4.9×10^{-9}	8.31	3×10^{-8}	7.5
$BaSO_4$	1.07×10^{-10}	9.97	6×10^{-10}	9.2
BaC_2O_4	1.6×10^{-7}	6.79	1×10^{-6}	6.0
BaF_2	1.05×10^{-6}	5.98	5×10^{-6}	5.3
$Bi(OH)_2Cl$	1.8×10^{-31}	30.75		
$Ca(OH)_2$	5.5×10^{-6}	5.26	1.3×10^{-5}	4.9
$CaCO_3$	3.8×10^{-9}	8.42	3×10^{-8}	7.5
CaC_2O_4	2.3×10^{-9}	8.64	1.6×10^{-8}	7.8
CaF_2	3.4×10^{-11}	10.47	1.6×10^{-10}	9.8
$Ca_3(PO_4)_2$	1×10^{-26}	26.0	1×10^{-23}	23
$CaSO_4$	2.4×10^{-5}	4.62	1.6×10^{-4}	3.8
$CdCO_3$	3×10^{-14}	13.5	1.6×10^{-13}	12.8
CdC_2O_4	1.51×10^{-8}	7.82	1×10^{-7}	7.0
$Cd(OH)_2$ (freshly prepared)	3×10^{-14}	13.5	6×10^{-14}	13.2
CdS	8×10^{-27}	26.1	5×10^{-26}	25.3
$Ce(OH)_3$	6×10^{-21}	20.2	3×10^{-20}	19.5
$CePO_4$	2×10^{-24}	23.7		
$Co(OH)_2$ (freshly prepared)	1.6×10^{-15}	14.8	4×10^{-15}	14.4
CoS α type	4×10^{-21}	20.4	3×10^{-20}	19.5
β type	2×10^{-25}	24.7	1.3×10^{-24}	23.9
$Cr(OH)_3$	1×10^{-31}	31.0	5×10^{-31}	30.3
CuI	1.10×10^{-12}	11.96	2×10^{-12}	11.7
CuSCN			2×10^{-13}	12.7
CuS	6×10^{-36}	35.2	4×10^{-35}	34.4
$Cu(OH)_2$	2.6×10^{-19}	18.59	6×10^{-19}	18.2
$Fe(OH)_2$	8×10^{-16}	15.1	2×10^{-15}	14.7
$FeCO_3$	3.2×10^{-11}	10.50	2×10^{-10}	9.7
FeS	6×10^{-18}	17.2	4×10^{-17}	16.4
$Fe(OH)_3$	3×10^{-39}	38.5	1.3×10^{-38}	37.9
Hg_2Cl_2	1.32×10^{-18}	17.88	6×10^{-18}	17.2
HgS (black)	1.6×10^{-52}	51.8	1×10^{-51}	51
(red)	4×10^{-53}	52.4		
$Hg(OH)_2$	4×10^{-26}	25.4	1×10^{-25}	25.0
$KHC_4H_4O_6$	3×10^{-4}	3.5		
K_2PtCl_6	1.10×10^{-5}	4.96		
$La(OH)_3$ (freshly prepared)	1.6×10^{-19}	18.8	8×10^{-19}	18.1

(continued)

Salt	$I=0$		$I=0.1$	
	$K^\ominus_{sp}$	$pK^\ominus_{sp}$	K_{sp}	pK_{sp}
$LaPO_4$			4×10^{-23}	22.4*
$MgCO_3$	1×10^{-5}	5.0	6×10^{-5}	4.2
MgC_2O_4	8.5×10^{-5}	4.07	5×10^{-4}	3.3
$Mg(OH)_2$	1.8×10^{-11}	10.74	4×10^{-11}	10.4
$MgNH_4PO_4$	3×10^{-13}	12.6		
$MnCO_3$	5×10^{-10}	9.30	3×10^{-9}	8.5
$Mn(OH)_2$	1.9×10^{-13}	12.72	5×10^{-13}	12.3
MnS(amorphous)	3×10^{-10}	9.5	6×10^{-9}	8.8
(crystalline)	3×10^{-13}	12.5		
$Ni(OH)_2$ (freshly prepared)	2×10^{-15}	14.7	5×10^{-15}	14.3
NiS α type	3×10^{-19}	18.5		
β type	1×10^{-24}	24.0		
γ type	2×10^{-26}	25.7		
$PbCO_3$	8×10^{-14}	13.1	5×10^{-13}	12.3
$PbCl_2$	1.6×10^{-5}	4.79	8×10^{-5}	4.1
$PbCrO_4$	1.8×10^{-14}	13.75	1.3×10^{-13}	12.9
PbI_2	6.5×10^{-9}	8.19	3×10^{-8}	7.5
$Pb(OH)_2$	8.1×10^{-17}	16.09	2×10^{-16}	15.7
PbS	3×10^{-27}	26.6	1.6×10^{-26}	25.8
$PbSO_4$	1.7×10^{-8}	7.78	1×10^{-7}	7.0
$SrCO_3$	9.3×10^{-10}	9.03	6×10^{-9}	8.2
SrC_2O_4	5.6×10^{-8}	7.25	3×10^{-7}	6.5
$SrCrO_4$	2.2×10^{-5}	4.65		
SrF_2	2.5×10^{-9}	8.61	1×10^{-8}	8.0
$SrSO_4$	3×10^{-7}	6.5	1.6×10^{-6}	5.8
$Sn(OH)_2$	8×10^{-29}	28.1	2×10^{-28}	27.7
SnS	1×10^{-25}	25.0		
$Th(C_2O_4)_2$	1×10^{-22}	22		
$Th(OH)_4$	1.3×10^{-45}	44.9	1×10^{-44}	44.0
$TiO(OH)_2$	1×10^{-29}	29	3×10^{-29}	28.5
$ZnCO_3$	1.7×10^{-11}	10.78	1×10^{-10}	10.0
$Zn(OH)_2$ (freshly prepared)	2.1×10^{-16}	15.68	5×10^{-16}	15.3
ZnS α type	1.6×10^{-24}	23.8		
β type	5×10^{-25}	24.3		
$ZrO(OH)_2$	6×10^{-49}	48.2	1×10^{-47}	47.0

* $I=0.5$.

Appendix E Molecular Masses

Compound	M/g · mol^{-1}	Compound	M/g · mol^{-1}
Ag_3AsO_3	446.52	$C_7H_6O_6S$(sulfosalicyclic acid)	218.18
Ag_3AsO_4	462.52	$C_{12}H_8N_2$(phenanthroline)	180.21
AgBr	187.77	$C_{12}H_8N_2 \cdot H_2O$ (phenanthroline monohydrate)	198.21
AgSCN	165.95		
AgCl	143.32	$C_{14}H_{14}N_3O_3SNa$(methyl orange)	327.33
Ag_2CrO_4	331.73	$CaCO_3$	100.09
AgI	234.77	$CaC_2O_4 \cdot H_2O$	146.11
$AgNO_3$	169.87	CaF_2	78.08
$Al(C_9H_6ON)_3$ (8-hydroxy quinoline)	459.44	$CaCl_2$	110.99
		CaO	56.08
$AlK(SO_4)_2 \cdot 12H_2O$	474.38	$CaSO_4$	136.14
Al_2O_3	101.96	$CaSO_4 \cdot 2H_2O$	172.17
As_2O_3	197.84	$CdCO_3$	172.42
As_2O_5	229.84	$Cd(NO_3)_2 \cdot 4H_2O$	308.48
		CdO	128.41
$BaCl_2$	208.24	$CdSO_4$	208.47
$BaCl_2 \cdot 2H_2O$	244.27	$CoCl_2 \cdot 6H_2O$	237.93
$BaCO_3$	197.34	CuI	190.45
$BaCrO_4$	253.32	CuSCN	121.62
$BaSO_4$	233.39	$CuHg(SCN)_4$	496.45
BaS	169.39	$Cu(NO_3)_2 \cdot 3H_2O$	241.60
$Bi(NO_3)_3 \cdot 5H_2O$	485.07	CuO	79.55
Bi_2O_3	465.96	$CuSO_4 \cdot 5H_2O$	249.68
BiOCl	260.43		
		$FeCl_2 \cdot 4H_2O$	198.81
CH_2O(formaldehyde)	30.03	$FeCl_3 \cdot 6H_2O$	270.30
CH_3COOH	60.05	$Fe(NO_3)_3 \cdot 9H_2O$	404.00
$C_2H_5NO_2$(glycine)	75.07	FeO	71.85
$C_4H_8N_2O_2$(dimethylglyoxime)	116.12	Fe_2O_3	159.69
$C_6H_5NO_3$(nitrosophenol)	139.11	Fe_3O_4	231.54
$C_6H_{12}N_2O_4S_2$(L-cystine)	240.30	$FeSO_4 \cdot 7H_2O$	278.01
$(CH_2)_6N_4$	140.19		

(continued)

Compound	M/g · mol^{-1}	Compound	M/g · mol^{-1}
HCOOH	46.03	$KHC_4H_4O_6$ (potassium bitartrate)	188.18
H_2CO_3	62.03	$KHC_8H_4O_4$ (potassium hydrogen phthalate)	204.22
$H_2C_2O_4$	90.04		
$H_2C_2O_4 \cdot 2H_2O$	126.07	$KHSO_4$	136.16
$H_2C_4H_4O_4$ (succinic acid)	118.090	KI	166.00
$H_2C_4H_4O_6$ (tartaric acid)	150.088	KIO_3	214.00
$H_3C_6H_5O_7 \cdot H_2O$ (citric acid)	210.14	$KMnO_4$	158.03
HCl	36.46	KNO_2	85.10
$HClO_4$	100.46	KNO_3	101.10
HNO_3	63.01	KOH	56.11
HNO_2	47.01	K_2PtCl_6	485.99
H_2O_2	34.01	KSCN	97.18
H_3PO_4	98.00	K_2SO_4	174.25
H_2S	34.08	$K_2S_2O_7$	254.31
H_2SO_3	82.07		
H_2SO_4	98.07	$Mg(C_9H_6ON)_2$ (8-hydroxyquinoline)	312.61
$HgCl_2$	271.50	$MgNH_4PO_4 \cdot 6H_2O$	245.41
Hg_2Cl_2	472.09	MgO	40.30
HgO	216.59	$Mg_2P_2O_7$	222.55
HgS	232.65	$MgSO_4 \cdot 7H_2O$	246.47
$HgSO_4$	296.65	$MnCO_3$	114.95
		MnO_2	86.94
$KAl(SO_4)_2 \cdot 12H_2O$	474.38	$MnSO_4$	151.00
KBr	119.00		
$KBrO_3$	167.00	$NH_2OH \cdot HCl$ (hydroxylamine HCl)	69.49
KCN	65.116		
$K_3C_6H_5O_7$ (citrate)	306.40	NH_3	17.03
K_2CO_3	138.21	NH_4^+	18.04
KCl	74.55	$NH_4C_2H_3O_2$ (ammonium acetate)	77.08
$KClO_3$	122.55	NH_4SCN	76.12
$KClO_4$	138.55	$(NH_4)_2C_2O_4 \cdot H_2O$	142.11
K_2CrO_4	194.19	NH_4Cl	53.49
$K_2Cr_2O_7$	294.18	NH_4F	37.04
$K_3Fe(CN)_6$	329.25	$NH_4Fe(SO_4)_2 \cdot 12H_2O$	482.18
$K_4Fe(CN)_6$	368.35	$(NH_4)_2Fe(SO_4)_2 \cdot 6H_2O$	392.13

(**continued**)

Compound	$M/g \cdot mol^{-1}$	Compound	$M/g \cdot mol^{-1}$
NH_4HF_2	57.04		
$(NH_4)_2Hg(SCN)_4$	468.98	PbO	223.2
NH_4NO_3	80.04	PbO_2	239.2
NH_4OH	35.05	$Pb(C_2H_3O_2)_2 \cdot 3H_2O$	379.3
$(NH_4)_3PO_4 \cdot 12MoO_3$	1876.34	$PbCrO_4$	323.2
$(NH_4)_2S_2O_8$	228.19	$PbCl_2$	278.1
$Na_2B_4O_7$	201.22	$Pb(NO_3)_2$	331.2
$Na_2B_4O_7 \cdot 10H_2O$	381.37	PbS	239.3
Na_2BiO_3	279.97	$PbSO_4$	303.3
$NaC_2H_3O_2$ (sodium acetate)	82.03		
$Na_3C_6H_5O_7$ (citrate)	258.07	SO_2	64.06
Na_2CO_3	105.99	SO_3	80.06
$Na_2CO_3 \cdot 10H_2O$	286.14	SO_4	96.06
$Na_2C_2O_4$	134.00	SiF_4	104.08
$NaCl$	58.44	SiO_2	60.08
$NaClO_4$	122.44	$SnCl_2 \cdot 2H_2O$	225.63
NaF	41.99	$SnCl_4$	260.50
$NaHCO_3$	84.01	SnO	134.69
$NaH_2C_{10}H_{12}O_8N_2$ (disodium EDTA)	336.21	SnO_2	150.69
$Na_2H_2C_{10}H_{12}O_8N_2 \cdot 2H_2O$	372.24	$SrCO_3$	147.63
$NaH_2PO_4 \cdot 2H_2O$	156.01	$Sr(NO_3)_2$	211.63
$Na_2HPO_4 \cdot 2H_2O$	177.99	$SrSO_4$	183.68
$NaHSO_4$	120.06		
$NaOH$	39.997	$TiCl_3$	154.24
Na_2SO_4	142.04	TiO_2	79.88
$Na_2S_2O_3 \cdot 5H_2O$	248.17		
$NaZn(UO)_3(C_2H_3O_2)_9 \cdot 6H_2O$	1537.94	$ZnHg(SCN)_4$	498.28
$NiSO_4 \cdot 7H_2O$	280.85	$ZnNH_4PO_4$	178.39
$Ni(C_4H_7N_2O_2)_2$	288.91	ZnS	97.44
(nickel dimethylglyoxime)		$ZnSO_4$	161.44

ANSWERS

Chapter 1

1.1 0.08 g, 0.26 g **1.2** 4.2 mL **1.4** 1 : 2, 2 : 1, 1 : 1, 2 : 1 **1.5** 0.67g **1.6** 112.0%
1.7 $c(CH_2C_4H_4O_6)=0.03333\ mol \cdot L^{-1}$, $c(CHCOOH)=0.0833\ mol \cdot L^{-1}$ **1.8** 20.00 mL

Chapter 2

2.5 0.4, 0.2, 0.1, 0.07 **2.6** −1%, −0.2%, −0.1%, −0.05%

2.7 (A)

Results	A	B	C	D	E
Mean	20.54	0.825	2.9	70.53	0.490
Median	20.55	0.803	2.8	70.64	0.492
Range	0.12	0.108	0.6	0.44	0.031
Standard deviation (s)	0.05	0.051	0.2	0.22	0.014
Coefficient of variation (CV)	0.2	6.2	8	0.3	2.9
The 95% confidence interval (CI)	20.54±0.05	0.82±0.08	2.9±0.3	70.53±0.34	0.490±0.022

2.8 (9.32%, 9.80%), (9.44%, 9.68%), (9.48%, 9.64%)

2.9 No significant difference.

($F_{cal}=2.03<F_{0.05(4,4)}=6.39$ and $s_p=0.47$, $t_{cal}=2.02<t_{0.05(8)}=2.31$)

2.11 0.13 g, ±0.15%, no; 0.48 g, ±0.014%, yes

2.12 0.477%, 3 significant figures, increase the sample size

Chapter 3

3.1

pH	$[H^+]$	x_3	x_2	x_1	x_0
2.16	0.007	0.500	0.500	4.5×10^{-6}	3.18×10^{-16}
4.69	2.0×10^{-5}	0.003	0.994	3.0×10^{-3}	7.0×10^{-11}
7.21	6.2×10^{-8}	4.5×10^{-6}	0.500	0.500	3.9×10^{-6}
9.77	1.7×10^{-10}	6.7×10^{-11}	0.003	0.994	0.003
12.32	4.8×10^{-13}	2.7×10^{-16}	3.9×10^{-6}	0.500	0.500

3.2 $pK_a^M=0.38$, $pK_a^C=9.29$
3.5 1.96, 7.93, 9.06, 12.77, 1.84, 6.15, 6.06 **3.6** 2.60, 3.12, 4.27
3.7 8.60, 8.10, 7.21 **3.8** 11.12, 10.62, 9.53 **3.9** 5.12, 5.62, 6.59
3.10 2.29, 4.71, 7.21 **3.11** 0.75 g, 7.9 mL; 0.75 g, 6.2 mL
3.12 49.3 g **3.13** $V(H_3PO_4)=367$ mL, $V(NaOH)=633$ mL

Chapter 4

4.1 8.23, +0.02% **4.2** 8.13, 7.00, 5.88; −0.07%
4.3 5.00, 4.30, 3.60; 5.21, 5.00, 4.79 **4.4** 0.5%(0.43%), 0.3%(0.24%)
4.5 9.70, 8.88, 8.06; 1% **4.6** 5.1, +0.08%, −0.5%

Chapter 5

5.1 2.0, 6.9, 11.7
5.3 7.8×10^{-2} mol·L^{-1}; 1.4×10^{-9}, 1.9×10^{-7}, 5.8×10^{-6}, 4.1×10^{-5}, 5.3×10^{-5} mol·L^{-1}
5.4

pL	22.1	11.4	7.7	3.0
Predominant form	Fe^{3+}	[Fe]=[FeL]	FeL_2	FeL_3

5.5 14.4, $10^{-8.2}$ mol·L^{-1}, $10^{-8.5}$ · mol·L^{-1} **5.6** $10^{-7.6}$ mol·L^{-1}, pH≈3.8
5.7 $10^{9.2}$ **5.8** 5%, $10^{-6.05(6.1)}$ mol·L^{-1}, $10^{-3.3}$ mol·L^{-1}
5.9

c/mol·L^{-1}	pCa		
	−0.1%	SP	+0.1%
0.01	5.3	6.5	7.7
0.1	4.3	6.0	7.7

5.10 4.1, −0.8% **5.11** 2.1~3.4, −0.2% **5.12** +0.02%
5.13

mol·L^{-1}	[X′]	[X]	$\sum_{i=1\sim3}[H_iX]$	[Cd^{2+}]
SP	$10^{-5.3}$	$10^{-11.2}$	$10^{-6.6}$	$10^{-7.1}$
EP	$10^{-4.9}$	$10^{-10.8}$	$10^{-6.2}$	$10^{-7.1}$

5.14 0.67 g

Chapter 6

6.5 0.77 V **6.6** 0.13 V **6.7** 1.5×10^{-15} mol·L^{-1}

6.8 0.14 V, 0.23 V, 0.33 V, 0.52 V, 0.70 V **6.9** 0.02667 mol · L^{-1} **6.10** 1 : 1.1
6.11 56.08% **6.12** 37.64% **6.13** 2.454% **6.14** 19.4%, 36.2%
6.15 23.09%, 21.39% **6.16** 0.07892 mol · L^{-1}

Chapter 7

7.1 2×10^{-9}, 2×10^{-3}, 4×10^{-5}, 4×10^{-4}, 2.4×10^{-6} mol · L^{-1}
7.2 1.1×10^{-13} mol · L^{-1}
7.3 AgI first because a lower $[I^-]$ is required; no PbI_2 will be precipitated because $[Pb^{2+}][I^-]^2 < K_{sp}$.
7.4 34.15%, 65.85% **7.5** 40.84% **7.6** 65.84% **7.7** 8.16%, 44.71%
7.17 0.8999, 0.7643, 0.1110, 0.03782 **7.18** 1 g **7.19** 98.0% **7.20** 2.66%

Chapter 8

8.7 14%, 37% **8.8** 1.5×10^4 L · mol^{-1} · cm^{-1} **8.9** 0.022%, 61.0%
8.10 3.92×10^{-5} mol · L^{-1}, 2.10×10^4 L · mol^{-1} · cm^{-1} **8.11** 4.4%
8.12 44%, 5.5%, 3.4%, 2.7%, 3.4%, 5.7%
8.13 3.9×10^{-4} mol · L^{-1}, 6.3×10^{-4} mol · L^{-1}
8.14 8.4×10^2 L mol^{-1} · cm^{-1}, 6.63 **8.15** BC_2, 1.91×10^5 **8.16** 46.1%

Chapter 9

9.2 2.5 mL **9.3** 76%, 51% **9.4** 92% **9.5** 7.8 mL **9.6** 18, 7.6
9.7 NO_3^-, $FeCl_4^-$, Na^+, NH_4^+, Ag^+, Ba^{2+}, Co^{2+} **9.8** 6.7(6) cm

INDEX

1-(2-pyridylazo)-2-naphthol (PAN), 125

A

Absolute error, 24
Absorbance, 213
Absorption, 213
Absorptivity, 213
Accuracy, 23, 277
Acid, 53
Acid-base titration, 79
Acidic effective coefficient, 116
Adsorption, 194
Adsorption chromatography, 263
Adsorption indicator, 188
Aging, 193
Amphiprotic solvent, 104
Analytical balance, 11
Anion exchange resin, 257
Argentometry, 184
Autoprotolysis, 54

B

Back titration, 10
Base, 53
Bromometry, 173
Buffer, 71
Buffer capacity, 72
Buret, 14

C

Calconcarboxylic acid 126
Calibration sensitivity, 279
Carbonate error, 100
Cation exchange resin, 257
Cell, 221
Cerimetry, 174
Charge-balance equation, 62
Chelate, 110
Chelating agent, 113, 249
Chemical analysis, 7
Chemical equilibrium, 51
Chromatography, 263
Colloidal precipitate, 198
Column chromatography, 263
Common ion effect, 179
Complexation, 109
Complexometric titration, 109
Conditional formation constant, 116
Confidence interval, 35
Confidence level, 35
Conjugate acid, 53
Conjugate base, 53
Continuous extraction, 254
Coprecipitation, 193
Cross linkage, 259
Crystalline precipitate, 197
Cuvette, 221

D

Dark current, 222
Debye-Hückel equation, 52
Demasking, 140

Detection limit, 279
Detector, 221
Deuterium lamp, 218
Deviation, 24
Differentiating solvent, 54
Diprotic acid, 66
Direct titration, 10
Dissociation constant, 110
Distribution constant (distribution coefficient, partition coefficient), 246
Distribution ratio, 246
Dual beam spectrophotometer, 224

E

Electrochemical cell, 147
Electrode potential, 148
Electrode reaction, 149
Electrolyte, 51
Electromagnetic radiation, 207
Electromagnetic spectrum, 207
Emission, 210
Erio chrome black T (EBT), 125
Error, 24
Ethylenediamine tetraacetic acid (EDTA), 113
Exchange capacity, 259
Extraction, 245

F

F test, 40
Fajans method, 188
Filter, 221
Filtration, 199
Fluorescein, 188
Formal potential, 150
Formation constant, 109

G

Gaussian curve, 29
Grating, 219
Gravimetry, 7

H

Henderson-Hasselbalch equation, 72
Histogram, 27
Holographic grating, 220
Homogeneous precipitation, 198
Hydrogen lam, 218
Hydronium ion (hydrogen ion), 53

I

Ignition, 199
Indicator, 9
Indicator transition point, 84
Infrared radiation, 207
Instability constant, 110
Instrumental analysis, 8
Interference, 239
Interference filter, 221
Iodimetry, 170
Iodometry, 171
Ion exchange, 257
Ion exchange resin, 257
Ionic strength, 51
Ion-product, 104
Ion-product constant, 54

J

Jones reductor, 165

K

Karl Fischer method, 173
Kjeldahl method, 102

L

Ladder diagram (predominance-area diagram), 58
Lambert-Beer's law, 213

Leveling solvent, 54
Ligand, 109
Light absorption, 210
Light source, 218

M

Macro analysis, 8
Masking, 133
Mass-balance equation, 61
Mercury lamp, 218
Metallochromic indicator, 122
Methyl orange (MO), 80
Methyl red (MR), 81
Microanalysis, 8
Mixed crystal, 195
Mixed indicator, 81
Mobile phase, 263
Mohr method, 186
Molar absorptivity, 213
Monobase, 65
Monochromator, 219
Monoprotic acid, 64
Mother liquor, 196

N

Nernst equation, 149
Normal hydrogen electrode (standard hydrogen electrode), 148
Nucleation, 193

O

Occlusion, 195
Outlier, 42
Overall formation constant (Cumulative stability constant), 110
Oxidation, 147
Oxidizing agent (oxidant), 147

P

Paper chromatography, 266
Particle growth, 193
Phenolphthalein (PP), 80
Photocell, 221
Photodiode array, 225
Photomultiplier tube, 222
Phototube, 222
Planar chromatography, 263
Polyfunctional acid, 57
Polyfunctional base, 57
Population mean, 28
Population standard deviation, 28
Potassium bromate, 173
Potassium dichromate method, 168
Potassium permanganate method, 165
Power (intensity) of electromagnetic radiation, 208
Precipitating reagent, 178
Precipitation equilibrium, 178
Precipitation form, 191
Precipitation titration, 184
Precision, 23
Primary standard, 10
Prism, 219
Proton-condition equation, 63

R

Random error (indeterminate error), 26
Recovery, 240
Redox titration, 147
Reducing agent (reductant), 147
Reference level (zero level), 63
Reflection grating, 219
Relative error, 24
Relative supersaturation, 193
Replica grating, 219
Replicate, 23

S

Sample mean, 33
Sample standard deviation, 32
Sampling, 239
Secondary stand, 11
Selectivity, 276
Semimicro analysis, 8
Sensitivity, 279
Separation factor, 240
Side reaction, 115
Side reaction coefficient, 116
Significance level, 35
Single bean spectrophotometer, 224
Slit, 219
Solid phase extraction, 256
Solubility product, 178
Spectrophotometer, 224
Spectrophotometry, 218
Spread (range), 32
Stability constant, 110
Standard deviation, 32
Standard error of the mean, 33
Standard solution, 11
Standardization, 11
Stationary phase, 263
Stepwise formation constant (stepwise stability constant, consecutive stability constant), 110
Stoichiometric point (equivalence point), 9
Surface adsorption, 194
Systematic error (determination error), 25

T

t test, 39
The method of continuous variation, 232
The molar ratio method, 232
Thin layer chromatography, 268
Titrand, 9
Titrant, 9
Titration, 9
Titration break, 84
Titration error, 9
Transmittance, 213
Tungsten lamp, 218
Tungsten-halogen lamp, 218

U

Ultramicro analysis, 8
Ultraviolet (UV)/visible light, 207

V

Volhard method, 187
Volumetric analysis (volumetric titration), 7
Volumetric flask, 14
Volumetric pipet, 14

W

Weighing form, 191

X

Xylenol orange (XO), 125

PERIODIC TABLE OF THE ELEMENTS

Key: atomic # → 29; atomic symbol → Cu; +2,1 ← common oxidation states; 63.55 ← atomic mass (rounded)

Metals | Metalloids | Nonmetals

Period		1 I A	2 II A		3 III B	4 IV B	5 V B	6 VI B	7 VII B	8 VIII B	9 VIII B	10 VIII B	11 I B	12 II B		13 III A	14 IV A	15 V A	16 VI A	17 VII A	18 VIII A
1	1s	1 H ±1 1.008																			2 He 0 4.003
2	2s	3 Li +1 6.941	4 Be +2 9.012												2p	5 B +3 10.81	6 C ±4 12.01	7 N −3 14.01	8 O −2 16.00	9 F −1 19.00	10 Ne 0 20.18
3	3s	11 Na +1 22.99	12 Mg +2 24.31												3p	13 Al +3 26.98	14 Si ±4 28.09	15 P −3 30.97	16 S −2 32.07	17 Cl −1 35.45	18 Ar 0 39.95
4	4s	19 K +1 39.10	20 Ca +2 40.08	3d	21 Sc +3 44.96	22 Ti +4,3,2 47.87	23 V +5,2,3,4 50.94	24 Cr +3,2,6 52.00	25 Mn +2,3,4,6,7 54.94	26 Fe +3,2 55.85	27 Co +2,3 58.93	28 Ni +2,3 58.69	29 Cu +2,1 63.55	30 Zn +2 65.41	4p	31 Ga +3 69.72	32 Ge +4,2 72.64	33 As ±3,+5 74.92	34 Se +4,−2,+6 78.96	35 Br ±1,+5 79.90	36 Kr 0 83.80
5	5s	37 Rb +1 85.47	38 Sr +2 87.62	4d	39 Y +3 88.91	40 Zr +4 91.22	41 Nb +5,3 92.91	42 Mo +6,3,5 95.94	43 Tc +7,4,6 98	44 Ru +4,3,6,8 101.1	45 Rh +3,4,6 102.9	46 Pd +2,4 106.4	47 Ag +1 107.9	48 Cd +2 112.4	5p	49 In +3 114.8	50 Sn +4,2 118.7	51 Sb +3,5 121.8	52 Te +4,6,−2 127.6	53 I −1,+5,7 126.9	54 Xe 0 131.3
6	6s	55 Cs +1 132.9	56 Ba +2 137.3	† 5d	71 Lu +3 175.0	72 Hf +4 178.5	73 Ta +5 180.9	74 W +6,4 183.8	75 Re +7,4,6 186.2	76 Os +4,6,8 190.2	77 Ir +4,3,6 192.2	78 Pt +4,2 195.1	79 Au +3,1 197.0	80 Hg +2,1 200.6	6p	81 Tl +1,3 204.4	82 Pb +2,4 207.2	83 Bi +3,5 209.0	84 Po +4,2 209	85 At 210	86 Rn 0 222
7	7s	87 Fr +1 223	88 Ra +2 226	‡ 6d	103 Lr +3 262	104 Rf 261	105 Db 262	106 Sg 266	107 Bh 264	108 Hs 277	109 Mt 268	110 Ds 281	111 Rg 272	112 Uub 285	7p	113 Uut 284	114 Uuq 289	115 Uup 288	116 Uuh 292		118 Uuo 294

lanthanides (rare earth metals)	† 4f	57 La +3 138.9	58 Ce +3,4 140.1	59 Pr +3,4 140.9	60 Nd +3 144.2	61 Pm +3 145	62 Sm +3,2 150.4	63 Eu +3,2 152.0	64 Gd +3 157.3	65 Tb +3,4 158.9	66 Dy +3 162.5	67 Ho +3 164.9	68 Er +3 167.3	69 Tm +3,2 168.9	70 Yb +3,2 173.0
actinides	‡ 5f	89 Ac +3 227	90 Th +4 232.0	91 Pa +5,4 231.0	92 U +6,3,4,5 238.0	93 Np +5,3,4,6 237	94 Pu +4,3,5,6 239	95 Am +3,4,5,6 243	96 Cm +3 247	97 Bk +3,4 247	98 Cf +3 251	99 Es +3 252	100 Fm +3 257	101 Md +3,2 258	102 No +2,3 259